INVARIANCE PRINCIPLES AND ELEMENTARY PARTICLES

INVESTIGATIONS IN PHYSICS

Edited by EUGENE P. WIGNER and ROBERT HOFSTADTER

1. Ferroelectricity by E. T. JAYNES
2. Mathematical Foundations of Quantum Mechanics by JOHN VON NEUMANN
3. Shell Theory of the Nucleus by EUGENE FEENBERG
4. Angular Momentum in Quantum Mechanics by A. R. EDMONDS
5. The Neutrino by JAMES S. ALLEN
6. Photonuclear Reactions by L. KATZ, R. SKINNER, and A. S. PENFOLD (to be published)
7. Radiation Damage in Solids by D. BILLINGTON and J. CRAWFORD
8. Nuclear Structure by LEONARD EISENBUD and EUGENE P. WIGNER
9. The Physics of Elementary Particles by J. D. JACKSON
10. Invariance Principles and Elementary Particles by J. J. SAKURAI

Invariance Principles and Elementary Particles

BY

J. J. SAKURAI

PRINCETON UNIVERSITY PRESS
PRINCETON, NEW JERSEY
1964

L. C. Card 62-11964

Printed in the United States of America

Nature seems to take advantage of the simple mathematical representations of the symmetry laws. When one pauses to consider the elegance and the beautiful perfection of the mathematical reasoning involved and contrast it with the complex and far-reaching physical consequences, a deep sense of respect for the power of the symmetry laws never fails to develop.

FROM C. N. YANG's NOBEL LECTURE

PREFACE

The purpose of the present monograph is to discuss various elementary particle phenomena that can be understood from a few general principles based on invariance or symmetry considerations. I have tried to strike a reasonable balance among fundamental concepts, applications to well-known problems, and implications to problems in the twilight zone. A purely formal and axiomatic approach is avoided as much as possible, and emphasis is placed on physical and empirical aspects of the various invariance principles. Since elementary particle physics today is in a confused state, it is not surprising that certain parts of the monograph seem somewhat disorganized; it is written in such a way that there will still be room for new ideas.

The monograph is based on a graduate course given at the University of Chicago in the Spring quarter of 1959 (needless to say, numerous developments since that time in this rapidly growing field have been incorporated). The students in the original course were assumed to be familiar with *elementary* field theory including the quantization of free fields and covariant calculational techniques. However, the monograph is written in such a way that the reader who is unfamiliar with field theory may gain a great deal even if he skips certain difficult sections that require a firm understanding of field theory. For instance, it may be read profitably even if all but the earlier sections of Chapter 5 and the entire Chapter 6 are omitted. I should like to emphasize that many symmetry arguments can be understood on the phenomenological level; it is unfortunate that a number of experimentalists as well as theoreticians hold the erroneous view that one must know field theory to understand simple symmetry arguments.

In general, I have tried to make this work readable for people with different backgrounds—serious graduate students of elementary particle theory, low energy physicists who want to become acquainted with high energy physics, active experimentalists in particle physics who wish to gain insight into theoretical problems, etc. In my past contacts with various people, I have found that certain concepts in elementary particle physics are tacitly assumed in the current journals by the "experts," but are not adequately treated in existing textbooks. It is hoped that the present monograph will serve to fill in this gap which is becoming increasingly serious as it widens.

Sections of the monograph are based on lecture notes taken by Frank Chilton and P. Schlein. The help of S. G. Eckstein and S. F. Tuan is also appreciated. I would like to take this opportunity to emphasize that, without their untiring work and numerous suggestions, the present monograph would never have materialized.

J. J. Sakurai

TABLE OF CONTENTS

Preface vii

CHAPTER 1. Introduction 3

CHAPTER 2. Continuous Space-Time Transformations 9

2.1. General Considerations 9

2.2. Translation Operator and Linear Momentum 10

2.3. Rotation Operator and Angular Momentum 11

2.4. Angular Momentum Selection Rules and Spin Tests . . 15

2.5. Lorentz Invariance 18

2.6. The Dirac Equation and Bilinear Covariants 23

2.7. Covariant Spin Formalism; Helicity and Chirality. . . 29

CHAPTER 3. Parity 32

3.1. General Considerations 32

3.2. π^+-p Scattering and Phase Shift Analysis 37

3.3. Intrinsic Parity; $\tau\theta$ Puzzle 42

3.4. Multipole Expansion and Photoproduction 52

3.5. Hyperon Decay 58

3.6. Tests of Parity Conservation 65

3.7. Parity in Field Theory. 69

3.8. Beta Decay 74

CHAPTER 4. Time Reversal 79

4.1. Classical Physics 79

4.2. Wigner Time Reversal in Quantum Mechanics . . . 81

4.3. Scattering and Reaction Theory; Reciprocity Relation, K Matrix 86

4.4. Time Reversal in Field Theory; "Diracology," Beta Decay 100

4.5. Rules for Testing Time Reversal Invariance in Decay Processes 108

CHAPTER 5. Charge Conjugation 111

5.1. General Considerations 111

5.2. Charge Conjugation Parity and Selection Rules . . . 113

5.3. Charge Conjugation in the Dirac Theory 117

5.4. Quantization and the Spin-Statistics Connection . . . 120

5.5. Parity, Chirality, and Helicity under Charge Conjugation 127

5.6. Behavior of Bilinear Covariants under Charge Conjugation 129

5.7. The Breakdown of Charge Conjugation Invariance in Weak Interactions 131

CHAPTER 6. Strong Reflection and the *CPT* Theorem 136

6.1. Strong Reflection 136

6.2. Proof of the *CPT* Theorem (Hamiltonian Formalism) . 139

6.3. *CPT* and Microcausality 143

6.4. Consequences of *CPT* Invariance 147

6.5. *CP* Invariance and the Mach Principle 150

CHAPTER 7. γ_5 Invariance and Weak Interactions 152
7.1. Two Component Theory of the Neutrino 152
7.2. Lepton Conservation and the $\pi\mu e$ Sequence 155
7.3. The Form of the Four-Fermion Weak Coupling . . . 159
7.4. Nuclear Beta Decay 161
7.5. Universal Fermi Interaction 167
CHAPTER 8. Gauge Transformations and "Number Laws" . . 177
8.1. Gauge Invariance and Charge Conservation in Electrodynamics 177
8.2. Baryon Conservation 185
8.3. Superselection Rules; Concept of Charge 187
8.4. Lepton Conservation 190
CHAPTER 9. Isospin and Related Topics ($S = 0$) 195
9.1. History 195
9.2. Charge Independence in Pion Physics 197
9.3. Formal Theory of Isospin Rotations 204
9.4. Classical Pion Physics 212
9.5. G Conjugation Invariance; $N\bar{N}$ Selection Rule, Nucleon Structure 223
9.6. Isospin and Weak Interactions; Conserved Vector Theory 235
9.7. Quantum Numbers of ρ, ω and η 241
CHAPTER 10. Isospin and Related Topics ($S \neq 0$) 254
10.1. History 254
10.2. Strangeness Scheme 256
10.3. Charge Independence in Strange Particle Reactions . 260
10.4. Neutral K Particles 269
10.5. Decay Interactions of Strange Particles 277
CHAPTER 11. Unsolved Problems 286
APPENDICES
Appendix A. Useful Tables of Clebsch-Gordan Coefficients . . 297
Appendix B. Dirac Matrices 300
Appendix C. Physical Constants 302
References 305
Index 323

INVARIANCE PRINCIPLES AND ELEMENTARY PARTICLES

CHAPTER 1

Introduction

The fundamental interactions in nature can be classified in historical order into the following four groups:

(1) Gravitational interactions
(2) Electromagnetic interactions
(3) Weak interactions
(4) Strong interactions

The investigations of the gravitational interactions were initiated in 1666 by Newton, then a twenty-four-year-old student. Coulomb's pioneer work on the electromagnetic interactions appeared in 1776. Although there are numerous phenomena that can be understood as manifestations of the first two classes of interactions—the motions of planets and artificial satellites, the propagation of radio waves, molecular forces, atomic spectra, superconductivity, etc.—it is an empirically well established fact, though not clearly recognized until the 1930's, that there are forces in nature that are neither gravitational nor electromagnetic. Shortly after Chadwick's discovery of the neutron, Fermi (1934) wrote down a beta decay Hamiltonian, which, with slight modifications, is still believed to be the correct weak interaction Hamiltonian in the low energy limit. At about the same time, in order to explain nuclear forces, Yukawa (1935) spelled out what has since been recognized as the starting point of the strong interactions.

The various classes of fundamental interactions are characterized by coupling constants that differ in many orders of magnitude. The electromagnetic coupling is characterized by the well-known dimensionless constant 1/137. The analogous dimensionless constant that characterizes the gravitational repulsion between two protons is $G^2_{\text{grav.}}M_p^2/\hbar c = 2 \times 10^{-39}$, which shows that we can essentially ignore gravity in discussing elementary particle phenomena. The weak interaction constant in dimensionless units turns out to be of the order of 10^{-14} or 10^{-7} depending on whether we regard a weak decay process as a one-stage process or a two-stage process (we shall say more about this in Chapter 7), and the various strong interaction constants seem to be of the order of unity. One of the deepest mysteries in elementary particle physics is that there is such a wide gap in strength between the strong and electromagnetic couplings on the one hand and the weak couplings on the other.

The major reason why the strong and weak interactions had not been discovered before the twentieth century is that they are too short-ranged to manifest themselves in daily life. For instance, the range of the force between a proton and a neutron is expected to be of the order of the pion Compton wavelength, $\hbar/\mu_\pi c = 1.41 \times 10^{-13}$ cm. So, just from the de Broglie wave relation $\lambda\!\!\!^- = \hbar/p$, we realize that we need high-energy machines to probe the nature of such short-ranged forces. It is for this reason that the term "elementary particle physics" is synonymous with the term "high energy physics."

To be more specific, let us consider proton-proton scattering. Below 200 kev., the scattering cross section follows the Mott formula, which is computed under the assumption that the only force between the two protons is Coulomb's repulsive force. We start learning something about the tail end of the two nucleon potential by performing scatterings at 500 kev. to 1 Mev. In order to study the nature of the forces between two protons at distances of the order 0.5×10^{-13} cm., we need a beam of protons with laboratory kinetic energy ≈ 300 Mev. Similar considerations are applicable to other phenomena. For instance, if we want to study the "structure" of weak interactions at short distances, we need a high energy beam of neutrinos.

The fact that the strong and weak interactions are short-ranged implies that those interactions do not possess any macroscopic analogs whatsoever in the domain of classical physics. This situation differs drastically from the electromagnetic case. Coulomb's law that is valid between the electron and the proton in the hydrogen atom is essentially the same Coulomb's law that also holds between two charged macroscopic balls separated by macroscopic distances. In constructing quantum electrodynamics we have been guided by classical electrodynamics; we may require that quantum electrodynamics reduce to classical electrodynamics in the limit where the number of photons per volume $\lambda\!\!\!^-{}^3$ is much greater than unity (as in the case of radio waves), or we may write down the Lagrangian for the photon field in quantum electrodynamics in analogy with the Lagrangian for the classical Maxwell field. In contrast, in constructing theories of strong and weak interactions of elementary particles, there are no such guiding principles from macroscopic physics. There are no classical, macroscopic analogs of nuclear forces, nor of the beta decay couplings. In elementary particle physics we must start from the very beginning. This is why elementary particle physics is so fantastically difficult and, hence, so much more challenging.

It is true that elementary particle physics today is in a very unsatisfactory state. It is expected that many of our present-day concepts have to be either revised or demolished. However, there are reasons to believe that certain arguments in elementary particle physics today, namely those arguments which are firmly based on invariance or symmetry considerations, will be of more permanent value. In reviewing the history of elementary particle physics, which is less than thirty years old, we cannot underestimate the power of symmetry considerations, which have led to a number of non-trivial predictions and have provided us with remarkably orderly understandings of elementary particle phenomena. True, this is not the first time that invariance or symmetry considerations played such an important role; the whole theory of relativity rests on the idea of Lorentz invariance; in atomic physics the regularities revealed in the periodic table are a direct consequence of invariance under rotations. But in elementary particle physics the guidance from invariance considerations is particularly helpful mainly because we lack any guidance from macroscopic analogies and correspondence with classical theories. Perhaps in the future theory of elementary particles, invariance principles may play even more significant roles; in the future, as Wigner (1949) put it, we may well "derive the laws of nature and try to test their validity by means of the laws of invariance rather than to try to derive the laws of invariance from what we believe to be the laws of nature."

One of the most striking features of invariance principles is that many of the so-called invariance or symmetry *laws* are only *approximate.* At first sight, one may feel that, because of this approximate nature, invariance considerations might not be very fruitful after all. We believe that just the opposite is the case. For, when a certain invariance principle is violated, it is violated in a very definite, orderly manner and never in a chaotic manner. The very breakdown as well as the very existence of the invariance principle is likely to provide clues to the mysterious dynamics of elementary particle interactions. For instance, the law of parity conservation looks like an absolute invariance principle in the realms of strong and electromagnetic interactions, but it does not hold at all for weak interactions, and moreover, there definitely exist orderly patterns in the ways parity conservation is violated in weak processes. Similar situations hold for other invariance principles as summarized in Table 1.1. The precise meanings of "strangeness," "charge conjugation invariance," etc. will be clarified in the appropriate chapters. Invariance principles or conservation laws not mentioned in Table 1.1 such as G conjugation invariance and lepton conservation will also be discussed as we go along.

Before we proceed further, we must define what is meant by an "elementary particle." When we consider the motion of molecules in a room, for all practical purposes we can regard each molecule as "elementary," as tacitly assumed in the kinetic theory of gases. This is because the mean energy of the molecule per degree of freedom is of the order of 0.03 ev. and is too low for the non-elementary nature of the molecule to reveal itself. However, in analyzing the vibrational spectrum of diatomic molecules in the infra-red region, it is more convenient to have a picture in which the molecule is composed of two "elementary" atoms with a spring in between. In the realm of atomic physics, atoms are no longer elementary; instead, we talk

TABLE 1.1

The Validity of Invariance Principles

Symmetry operations or conserved quantities	Strong	Electromagnetic	Weak
Parity (space inversion)	yes	yes	no
Charge conjugation	yes	yes	no
Time reversal	yes	yes	yes?
Electric charge	yes	yes	yes
Baryon number	yes	yes	yes
Isospin	yes	no	no
Strangeness	yes	yes	no

about electrons and various kinds of nuclei. In atomic physics, all nuclei look elementary; by counting isotopes we note that there are more than two hundred "elementary particles." This is highly unsatisfactory. It is indeed gratifying that the series of milestone discoveries and experiments in nuclear physics—Becquerels' discovery of radioactivity in 1896, Rutherford's first laboratory transmutation of elements in 1919, Chadwick's discovery of the neutron in 1932, etc.—finally led us to the idea that what are "elementary" are not various nuclei but the proton and the neutron (Heisenberg, 1932).

We may naturally ask: Is the proton elementary? We know today that the physical proton itself is made up of the "core" and the "cloud" of virtual mesons surrounding the core. In photon-proton collisions with photon energies greater than 150 Mev., a pion in the cloud can be knocked out from the proton ($\gamma + p \rightarrow p + \pi^0$) just as an electron in the atom can be knocked out in a photoelectric process. In photon-proton collisions at even higher energies ($E_\gamma > 910$ Mev.) the proton can dissociate itself into a Λ particle and a K meson ($\gamma + p \rightarrow \Lambda^0 + K^+$), which reminds us of the photodisintegration of the deuteron ($\gamma + d \rightarrow p + n$). Is the proton a bound system of a Λ and a K? Some physicists say "yes," other physicists say "no;" still others regard such a question as meaningless. (One may argue that we should regard stable particles as elementary as much as possible, but this argument gets into difficulty since the deuteron is stable while the neutron is unstable.)

For the purpose of the present book we take, from Lee and Yang, a negative definition of elementary particles (Lee and Yang 1957b). "We believe we understand what is meant by an atom, a molecule, and a nucleus. Any small particle that is not an atom, not a molecule, not a nucleus (except the hydrogen nucleus) is called an elementary particle." With this definition, there are, without counting antiparticles, seventeen elementary particles, as shown in Table 1.2. It is not known whether or not there are more elementary particles as yet undiscovered.

TABLE 1.2
Table of Elementary Particles (Compiled by W. H. Barkas and A. H. Rosenfeld). Errors Are Not Shown

Family	Particle	Spin	Mass (Mev.)	Mean life (second)
Photon	γ	1	0	stable
	$\nu(\bar{\nu})$	$\frac{1}{2}$	$<2 \times 10^{-4}$	stable
Leptons	$e^-(e^+)$	$\frac{1}{2}$	0.511	stable
	$\mu^-(\mu^+)$	$\frac{1}{2}$	105.7	2.26×10^{-6}
	$\pi^+(\pi^-)$	0	139.6	2.6×10^{-8}
	π°	0	135.0	2×10^{-16}
Mesons	$K^+(K^-)$	0	494	1.2×10^{-8}
	$K^\circ(\bar{K}^\circ)$	0	498	K_1°: 1.0×10^{-10}
	$(\lvert m(K_1^\circ) - m(K_2^\circ)\rvert \sim 5 \times 10^{-12})$			K_2°: $\sim 7 \times 10^{-8}$
	$p(\bar{p})$	$\frac{1}{2}$	938.2	stable
	$n(\bar{n})$	$\frac{1}{2}$	939.5	1.0×10^3
	$\Lambda^\circ(\Lambda^\circ)$	$\frac{1}{2}$	1115.5	2.5×10^{-10}
	$\Sigma^+(\overline{\Sigma^+})$	$\frac{1}{2}$	1189	0.8×10^{-10}
Baryons	$\Sigma^-(\overline{\Sigma^-})$	$\frac{1}{2}$	1197	1.6×10^{-10}
	$\Sigma^\circ(\overline{\Sigma^\circ})$	$\frac{1}{2}$	1193	theory $\sim 10^{-19}$
	$\Xi^-(\overline{\Xi^-})$	?	1318	1.3×10^{-10}
	$\Xi^\circ(\overline{\Xi^\circ})$	?	~ 1312	$\sim 2 \times 10^{-10}$

The first thing one must learn about elementary particles is that they can be classified into four groups. First of all, there is the photon, which is a group by itself. The photon can interact only electromagnetically. Then there is the "lepton" family. It consists of neutrinos (ν), electrons (e), and muons (μ). (It is unfortunate that due to an historic accident, a muon is sometimes referred to as a *μ-meson.*) They are all fermions with masses considerably smaller than the proton mass, and they cannot interact strongly. In contrast, particles belonging to the "meson" family and the "baryon" family do interact strongly as well as weakly (or electromagnetically). Pions (π mesons) and K particles are bosons, and they belong to the meson family. A Λ particle, a Σ particle and a Ξ particle (cascade particle) are often referred to as "hyperons," and they are like the nucleon (proton or neutron) in the sense that they are heavy fermions capable of strong interactions. The baryon family is made up of nucleons and hyperons.

Note that some of the particle masses are almost degenerate. Take the triplet Σ^+, Σ^- and Σ^0, for instance. We are reminded here of a "hyperfine structure" splitting. It is generally believed that the three levels collapse, to a single level in the absence of the electromagnetic coupling. However we are far from being able to calculate such mass differences within the multiplet.

Of all the particles in Table 1.2, only γ, e, ν, and p are stable. Other particles in the table have fairly well-defined masses because their lifetimes are long. Note in this connection that a one per cent uncertainty in the mass value corresponds to a lifetime of the order of 10^{-22} sec. for a particle of a protonic mass. We have not included in Table 1.2 resonant states that are too short-lived to be regarded as "particles," although we believe that there is no sharp distinction between a "particle" and a resonant state with well-defined quantum numbers. We emphasize here that the longevity of the strongly interacting particles is due to the mass relation and/or the "strangeness" selection rule that forbid them to "decay strongly" (cf. Chapter 10). If the Σ mass were 60 Mev. higher, the Σ particle would appear as a resonant state of the $\pi\Lambda$ system with lifetime $\sim 10^{-23}$ to 10^{-24} sec., which is the characteristic time scale of the strong interactions. All unstable particles (with the exception of Σ° and π°) decay via weak couplings, whose characteristic time scale is $\sim 10^{14}$ times longer than the "nuclear" time scale.

We may close this introductory chapter by listing some examples of the various interactions.

(1) Strong interactions:

$$p + n \rightarrow p + n$$
$$\pi^- + p \rightarrow \pi^\circ + n$$
$$p + n \rightarrow \Lambda^\circ + n + K^+ \quad \text{(associated production)}$$
$$K^- + p \rightarrow \Sigma^+ + \pi^-$$
$$p + \bar{p} \rightarrow \Lambda^\circ + \overline{\Lambda^\circ}$$

(2) Electromagnetic interactions combined with strong interactions:

$$\gamma + p \rightarrow \pi^\circ + p \quad \text{(photopion production)}$$
$$\pi^- + p \rightarrow n + \gamma \quad \text{(radiative capture of } \pi^-\text{)}$$
$$\gamma + p \rightarrow \Lambda^\circ + K^+ \quad \text{(associated photoproduction)}$$
$$\pi^\circ \rightarrow 2\gamma$$
$$\Sigma^\circ \rightarrow \Lambda^\circ + \gamma$$

(3) Weak interactions:

$$\mu^+ \rightarrow e^+ + \nu + \bar{\nu}$$
$$n \rightarrow p + e^- + \bar{\nu} \quad \text{(beta decay)}$$
$$\mu^- + p \rightarrow n + \nu \quad \text{(muon capture)}$$
$$K^+ \rightarrow \pi^+ + \pi^\circ$$
$$\Sigma^+ \rightarrow p + \pi^\circ$$

CHAPTER 2

Continuous Space-Time Transformations

2.1. General Considerations

The connection between a conservation law and symmetry is well known in classical mechanics (see e.g. Goldstein, 1953). If L is the Lagrangian, then Lagrange's equations are

$$\frac{d}{dt}\left(\frac{\partial L(q_j, \dot{q}_j, t)}{\partial \dot{q}_j}\right) - \frac{\partial L(q_j, \dot{q}_j, t)}{\partial q_j} = 0 \tag{2.1}$$

Clearly, if L does not depend on q_j, then $\dfrac{\partial L}{\partial \dot{q}_j} = p_j$, the canonical momentum, is constant in time, i.e. is conserved.

In the Hamiltonian formalism, one has

$$\dot{p}_j = -\frac{\partial H}{\partial q_j} = [p_j, H]_{\text{classical}} \tag{2.2}$$

where the bracket is the classical or Poisson bracket. Under an infinitesimal transformation $q_j \to q_j + \delta q_j$, we have $\delta H = \delta q_j \dfrac{\partial H}{\partial q_j}$. If the Hamiltonian is invariant under such an infinitesimal transformation, i.e. if an infinitesimal displacement δq_j does not change H, the corresponding canonical momentum p_j is a constant of the motion. For example, the conservation of linear momentum is a result of the *homogeneity* of space while the conservation of angular momentum is due to the *isotropy* of space.

If invariance under translations and rotations holds, the laws of nature do not depend on "absolute space." The questions of which point we choose as the origin of the coordinate system and of which orientation axis we choose as the z-axis should play no essential role in the formulation of physical laws. Of course, on the cosmological scale, we may legitimately ask whether the laws of nature on the "fringes" of the universe are the same as the laws of nature in the "center" of the universe. For instance, we may examine whether the universal constants of nature such as the fine structure constant and the proton gyromagnetic ratio are truly universal in the sense that they are identical in various parts of the universe. Savedoff (1956) has shown that radio astronomical observations of the radio source Cygnus A, which is estimated to be $\sim 3 \times 10^8$ light years away, yield

$$\frac{(\alpha^2)_{\text{cyg. A}}}{(\alpha^2)_{\text{local}}} = 1.0036 \pm 0.0032$$

where α is the fine structure constant. A similar result has been obtained for the proton g-factor.

Even if the laws are the same everywhere, we might still expect that the results of measurements carried out in our solar system are influenced by the fact that the matter in our galaxy is not distributed isotropically with respect to the solar system. Cocconi and Salpeter (1958) have speculated on the possible dependence of the inertial mass M on the direction of its acceleration; because our solar system is not at the center of the galaxy, M may depend on whether its acceleration is in the direction towards the center of the galaxy or in a direction perpendicular to it. It can be estimated from nuclear resonance experiments that the fractional anisotropy of inertia is $\Delta M/M \lesssim 10^{-20}$ (Hughes, Robinson, and Beltram-Lopez, 1960).

Energy conservation is related to invariance under displacement of the time coordinate in the same way as momentum conservation is related to invariance under displacement of a space coordinate. We may ask whether the laws of nature in an early epoch of the universe are the same as the laws of nature today. It appears that energy conservation is compatible only with a static, non-expanding universe.

2.2. Translation Operator and Linear Momentum

In a quantum mechanical formalism one starts with state vectors which are eigenvectors of the Hamiltonian and of other commuting observables. A quantity is conserved if the generator of the corresponding infinitesimal transformation commutes with the Hamiltonian. We have

$$DH\Psi = HD\Psi \tag{2.3}$$

which is the mathematical expression of the quantum mechanical statement that the Hamiltonian H is unchanged by a transformation D.

In the coordinate representation an infinitesimal translation operator D is given by

$$D = 1 + \delta l_j \frac{\partial}{\partial x_j} \tag{2.4}$$

Equation (2.3) gives

$$[D, H] = 0 \text{ or } [\vec{\nabla}. H] = 0 \tag{2.5}$$

where the bracket is now a commutator bracket. The analogy to the classical equation $[p_j, H]_{\text{classical}} = 0$ is evident and leads one to identify $\vec{p}$ with $\vec{\nabla}$ except for a constant factor. There is the formal requirement for the infinitesimal transformation to be unitary because the norm of the state vector must be preserved. This leads one to choose the constant as

$$\vec{p} = -i\hbar\vec{\nabla} \tag{2.6}$$

where $\hbar$ must have the dimensions of action, e.g., erg-sec. Thus

$$D = 1 + \frac{i}{\hbar}\vec{p}\cdot\delta\vec{l} \tag{2.7}$$

That D is unitary (to first order in $\delta\vec{l}$) is evident since

$$D^{\dagger}D = \left(1 - \frac{i}{\hbar}\vec{p}\cdot\delta\vec{l}\right)\left(1 + \frac{i}{\hbar}\vec{p}\cdot\delta\vec{l}\right) = 1 + O((\delta\vec{l})^2)$$

where the Hermiticity of $\vec{p}$ has been assumed.

A finite transformation can be obtained by compounding successive infinitesimal transformations. (According to a mathematical theorem in the theory of Lie groups, the properties of the finite differentiable group transformations are completely determined by those of the corresponding infinitesimal transformations.) For a finite displacement $\vec{l}$, the translation operator D, is given by

$$D = \lim_{n\to\infty}\left(1 + \frac{i\vec{l}}{n\hbar}\cdot\vec{p}\right)^n = \exp\left(\frac{i\vec{p}\cdot\vec{l}}{\hbar}\right) \tag{2.8}$$

As an example, the plane wave solution to the Schrödinger equation $\exp\left[\frac{i}{\hbar}(\vec{p}\cdot\vec{x} - Et)\right]$, goes into $\exp\left(\frac{i}{\hbar}\vec{p}\cdot\vec{l}\right)\exp\left[\frac{i}{\hbar}(\vec{p}\cdot\vec{x} - Et)\right]$ when $\vec{x}$ goes into $\vec{x} + \vec{l}$.

There are various elementary particle reactions that are forbidden because of momentum and energy conservation. Trivial as they may seem, we list here some typical examples.

$e^- + e^+ \not\leftrightarrow \gamma$	in free space
$e^- \not\leftrightarrow e^- + e^+ + e^-$	” ”
$p \not\leftrightarrow p + \gamma, p + \pi^0$	” ”

2.3. Rotation Operator and Angular Momentum

In a way analogous to translations, to an infinitesimal rotation about some axis along $\delta\vec{w}$ by an amount $|\delta\vec{w}|$, we associate the rotation operator

$$R = 1 + i\,\delta\vec{\omega}\cdot\vec{J}, \tag{2.9}$$

where $\hbar = c = 1$ from now on. Just as $\vec{p}$ is called momentum, $\vec{J}$ is called the angular momentum and has analogous properties. The properties of $\vec{J}$ can be studied by examining the infinitesimal rotations which it generates. For the specific set of axes shown in Fig. 2.1, consider the following sequence of rotations and their effect upon the point (0, 1, 0). A rotation by ε about the x-axis gives $(0, 1, \varepsilon)$. A further rotation by η about the y-axis gives $(\varepsilon\eta, 1, \varepsilon)$. Now rotate by $-\varepsilon$ about the x-axis, this will give $(\varepsilon\eta, 1, 0)$.

A final rotation by $-\eta$ about the y-axis gives $(\varepsilon\eta, 1, 0)$. Thus the net rotation is $\varepsilon\eta$ about the z-axis. Symbolically

$$(1 - i\eta J_y)(1 - i\varepsilon J_x)(1 + i\eta J_y)(1 + i\varepsilon J_x) = (1 + i\varepsilon\eta J_z) \tag{2.10}$$

Hence we have the commutation relations

$$[J_x, J_y] = iJ_z \tag{2.11}$$

and the cyclic permutations thereof. From (2.11) all well-known properties of angular momentum — $J^2 = j(j + 1)$, $\langle j, m|J_x \pm iJ_y|j, m \mp 1\rangle = \sqrt{(j \pm m)(j \mp m + 1)}$, etc.—can be derived. Just as in the case of trans-

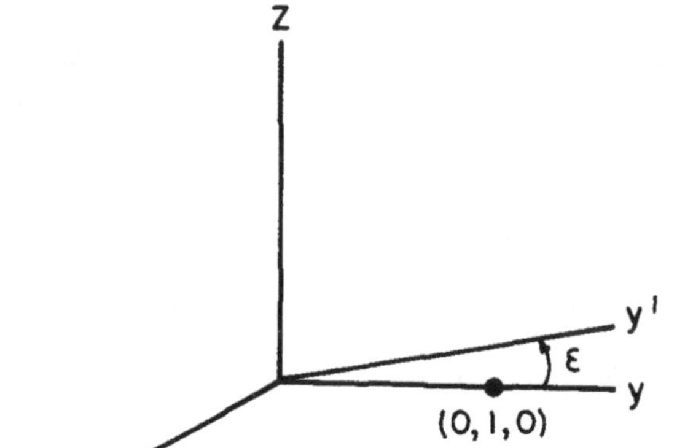

Figure 2.1. Infinitesimal rotation

lations, the infinitesimal rotations are compounded to give finite rotation operators of the form $\exp(iJ_z\omega)$ etc.

As an example, consider the plane wave expansion with the z-axis in the beam direction

$$\exp(ik|\vec{x}| \cos\theta) = \sum_{l=0}^{\infty} (2l + 1)i^l j_l(k|\vec{x}|)P_l(\cos\theta) \tag{2.12}$$

Note that the expansion contains no terms of the form $P_l^m(\cos\theta)\exp(im\varphi)$ with $m \neq 0$. The application of $1 + i\delta\omega L_z$ with $L_z = -i\left(x\frac{\partial}{\partial y} - y\frac{\partial}{\partial x}\right) = i\frac{\partial}{\partial\varphi}$ just results in the identity operation, which expresses the fact that, although the plane wave contains all possible l values, the plane wave is cylindrically symmetric in the sense that it is invariant under rotations about the z-axis; hence $l_z = m = 0$. That the component of $\vec{L}$ along the beam direction must necessarily vanish is expected classically $(\vec{x} \times \vec{p}) \cdot \vec{x} = 0$.

A scalar field $\varphi(\vec{x})$, by definition, transforms as $\varphi'(\vec{x}') = \varphi(\vec{x})$ under a rotation $\vec{x} \to \vec{x}'$. The value of the scalar field at a given point in space is the same in all coordinate frames. For the scalar field the angular momentum operator $\vec{J}$ is identical to the orbital angular momentum operator $\vec{L}$ that just acts on the coordinates. Thus the scalar field carries no intrinsic angular momentum, and describes a spinless particle.

In contrast, for a vector field $\vec{A}(\vec{x}) \to \vec{A}'(\vec{x}') \neq \vec{A}(\vec{x})$. Let us write the vector field, $\vec{A}$, at a given point as a product of a unit vector $\vec{\varepsilon}$ and a spatial function $f(\vec{x})$, and consider a rotation about the z-axis with the rotation operator $\exp[i(S_z + L_z)\omega]$. The operator L_z acts on $f(\vec{x})$. But the spin operator S_z acts on the polarization vector $\vec{\varepsilon}$. In the rotated frame

$$\begin{aligned} \varepsilon_{x'} &= \varepsilon_x \cos \omega - \varepsilon_y \sin \omega \\ \varepsilon_{y'} &= \varepsilon_x \sin \omega + \varepsilon_y \cos \omega \\ \varepsilon_{z'} &= \varepsilon_z \end{aligned} \tag{2.13}$$

A more suggestive way of writing these relations is obtained by rearrangement

$$\begin{aligned} \varepsilon_{x'} + i\varepsilon_{y'} &= \exp(i\omega)(\varepsilon_x + i\varepsilon_y) \\ \varepsilon_{x'} - i\varepsilon_{y'} &= \exp(-i\omega)(\varepsilon_x - i\varepsilon_y) \\ \varepsilon_{z'} &= \varepsilon_z \end{aligned} \tag{2.14}$$

It is evident that these combinations which correspond to states of definite circular polarization are eigenfunctions of S_z with the eigenvalues $+1$, -1, 0 respectively. The real transverse photon with momentum $\vec{k}$ requires $\vec{k} \cdot \vec{\varepsilon} = 0$, and taking the z-axis along $\vec{k}$, we see that there are only two spin states $+1$ and -1.

For two component spinors which are appropriate for describing spin $\frac{1}{2}$ particles, the rotation operator around the unit vector $\hat{n}$ can be written as $\exp\left(i \frac{\vec{\sigma}\cdot\hat{n}}{2}\omega\right)$. By expanding the exponential in a Taylor series, using the relation $(\vec{\sigma}\cdot\hat{n})^2 = \hat{n}\cdot\hat{n} + i\vec{\sigma}\cdot(\hat{n} \times \hat{n}) = 1$, and combining terms, one obtains

$$\exp\left(i \frac{\vec{\sigma}\cdot\hat{n}}{2}\omega\right) = \cos\frac{\omega}{2} + i\vec{\sigma}\cdot\hat{n}\sin\frac{\omega}{2} \tag{2.15}$$

Suppose the two component spinor u^P is given by

$$u^P = \begin{pmatrix} 1 \\ 0 \end{pmatrix} = |s = \tfrac{1}{2}, s_z = \tfrac{1}{2}\rangle$$

in the usual representation in which σ_z is diagonal.

$$\begin{aligned} \sigma_x &= \begin{pmatrix} 0 & 1 \\ 1 & 0 \end{pmatrix} \\ \sigma_y &= \begin{pmatrix} 0 & -i \\ i & 0 \end{pmatrix} \\ \sigma_z &= \begin{pmatrix} 1 & 0 \\ 0 & -1 \end{pmatrix} \end{aligned} \tag{2.16}$$

Then $\sigma_z u^P = u^P$ so the z component of the spin would be definitely $\frac{1}{2}$. In another coordinate system, say, rotated about the y-axis by ω

$$u^{P'} = \exp\left(i\frac{\sigma_y\omega}{2}\right)u^P = \begin{pmatrix} \cos\frac{\omega}{2} \\ -\sin\frac{\omega}{2} \end{pmatrix} \tag{2.17}$$

Hence the probability for $s_{z'} = \frac{1}{2}$ is $\cos^2\frac{\omega}{2}$, and for $s_{z'} = -\frac{1}{2}$ is $\sin^2\frac{\omega}{2}$. In particular for $\omega = 90°$, the observation of spin up z' (along the old x-axis) or spin down are equally probable. In other words, if we have a beam of particles with $s_z = \frac{1}{2}$, then s_x and s_y do not have sharp values. This is not surprising because s_z and s_y (or s_x) do not commute, which means that s_z and s_x cannot be simultaneously diagonalized.

Experimentally this would be found in a Stern-Gerlach type experiment. From a beam of spin $\frac{1}{2}$ atoms moving along the y-axis in an inhomogeneous field along the z-axis, select the $s_z = \frac{1}{2}$ fraction. If this part of the beam enters a region with an inhomogeneous field in the x direction, it will be split into two equal parts corresponding to $s_x = \frac{1}{2}$ and $s_x = -\frac{1}{2}$. Now select the $s_x = \frac{1}{2}$ part, and again subject it to an inhomogeneous field along the z-axis. Both $s_z = \frac{1}{2}$ and $s_z = -\frac{1}{2}$ appear, i.e. we no longer have a pure spin $s_z = \frac{1}{2}$ beam. In Chapter 10 we shall see that the behavior of neutral K particles can be discussed along similar lines.

Given *any* two component spinor, there always exists a direction $\hat{n}$ such that $\vec{\sigma}\cdot\hat{n}$ is sharp, i.e.,

$$\vec{\sigma}\cdot\hat{n}\begin{pmatrix} a \\ b \end{pmatrix} = \begin{pmatrix} \cos\theta & , & \sin\theta\exp(-i\varphi) \\ \sin\theta\exp(i\varphi), & & -\cos\theta \end{pmatrix}\begin{pmatrix} a \\ b \end{pmatrix} = \begin{pmatrix} a \\ b \end{pmatrix} \tag{2.18}$$

$$|a|^2 + |b|^2 = 1$$

where θ and φ describe the orientation of $\hat{n}$.

The reader may readily verify that within an undetermined phase factor

$$\begin{pmatrix} a \\ b \end{pmatrix} = \begin{pmatrix} \cos\frac{\theta}{2} \\ \sin\frac{\theta}{2}\exp(i\varphi) \end{pmatrix} \tag{2.19}$$

will do. This means that any linear combination of the two independent pure states

$$\begin{pmatrix} 1 \\ 0 \end{pmatrix} \quad \text{and} \quad \begin{pmatrix} 0 \\ 1 \end{pmatrix}$$

is also a pure state with a spin pointing out to *some* direction in space.

If we want to describe an *ensemble* of spin $\frac{1}{2}$ particles whose spin orientations are partially (if not completely) random, we must form an incoherent mixture of pure states. For this purpose it is convenient to use the density matrix formalism which is fully discussed in Chapter 17 of Bethe and Morrison (1956) (see also Tolhoek, 1956). Here we simply remark that a beam of spin $\frac{1}{2}$ particles is completely characterized by the 4×4 density matrix,

$$\rho = \frac{I_0}{2}(1 + \langle\vec{\sigma}\rangle\cdot\vec{\sigma}) \tag{2.20}$$

The intensity, I_0, is

$$I_0 = \mathrm{Tr}(\rho) \tag{2.21}$$

while the polarization vector $\langle\vec{\sigma}\rangle$ is given by

$$\langle\vec{\sigma}\rangle = \frac{\mathrm{Tr}(\rho\vec{\sigma})}{\mathrm{Tr}(\rho)} \tag{2.22}$$

Since $\frac{1}{2}(1 + \vec{\sigma}\cdot\hat{n})$ is the projection operator for spin along $\hat{n}$, the component of the polarization vector in the direction $\hat{n}$ is

$$\begin{aligned}\langle\vec{\sigma}\rangle\cdot\hat{n} &= \frac{\mathrm{Tr}(\rho\frac{1}{2}(1 + \vec{\sigma}\cdot\hat{n}) - \mathrm{Tr}(\rho\frac{1}{2}(1 - \vec{\sigma}\cdot\hat{n}))}{\mathrm{Tr}(\rho)} \\ &= \frac{|a_\uparrow|^2 - |a_\downarrow|^2}{|a_\uparrow|^2 + |a_\downarrow|^2}\end{aligned} \tag{2.23}$$

$|a_\uparrow|^2$ and $|a_\downarrow|^2$ refer to the probabilities of observing spin along $\hat{n}$ and opposite to $\hat{n}$. The direction of $\langle\vec{\sigma}\rangle$ is that quantization direction which makes

$$\frac{|a_\uparrow|^2 - |a_\downarrow|^2}{|a_\uparrow|^2 + |a_\downarrow|^2}$$

maximal. The magnitude of $\langle\vec{\sigma}\rangle$ is the average spin angular momentum in units of $\hbar/2$ in the polarization direction. If the beam is unpolarized, $\langle\vec{\sigma}\rangle = 0$; if the magnitude of $\langle\vec{\sigma}\rangle$ is unity, the beam is in a pure state of polarization.

2.4. Angular Momentum Selection Rules and Spin Tests

There are many applications of angular momentum considerations to elementary particle physics. Some of them are rather far-reaching.

We first show that from the two-quantum decay of the π^0, $\pi^0 \rightarrow 2\gamma$, one can conclude that its spin cannot be one (Landau, 1948; Yang, 1950). A final state wave function in momentum space for the two photons is to be constructed from the polarization vectors, $\vec{\varepsilon}_1$ and $\vec{\varepsilon}_2$ and the relative momentum vector $\vec{k}$. Also it must clearly be linear in $\vec{\varepsilon}_1$ and in $\vec{\varepsilon}_2$ and transform like a vector under rotations if the initial π^0 has spin one. Further, Bose-Einstein statistics must be obeyed because photons are bosons. There are three independent combinations of $\vec{\varepsilon}_1$, $\vec{\varepsilon}_2$, and $\vec{k}$ that are linear in $\vec{\varepsilon}_1$ and in $\vec{\varepsilon}_2$ and transform like vectors:

(a) $\vec{\varepsilon}_1 \times \vec{\varepsilon}_2$, which is not satisfactory because it is antisymmetric under interchange of the two photons

(b) $(\vec{\varepsilon}_1 \cdot \vec{\varepsilon}_2)\vec{k}$, which is not satisfactory either for the same reason ($\vec{k} \rightarrow -\vec{k}$ under interchange).

(c) $\vec{k} \times (\vec{\varepsilon}_1 \times \vec{\varepsilon}_2) = \vec{\varepsilon}_1(\vec{k} \cdot \vec{\varepsilon}_2) - \vec{\varepsilon}_2(\vec{k} \cdot \vec{\varepsilon}_1)$ which satisfies Bose-Einstein statistics but is identically zero since $\vec{k} \cdot \vec{\varepsilon} = 0$ by the transversality condition.

Thus the π^0 spin is not one if angular momentum is conserved in $\pi^0 \rightarrow 2\gamma$. This type of argument is also applicable to positronium in a $^3S_1$ state, which is forbidden to decay into two photons.

When we discuss detailed balancing in Chapter 4, we will present an argument for the pion spin $= 0$. In the following we assume that the pion is spinless.

From angular momentum conservation and Bose statistics one can also conclude that if the K_1^0 meson had odd spin, then $K_1^0 \rightarrow 2\pi^0$ would be forbidden. Since the pion is spinless, the K_1^0 spin must be equal to the orbital angular momentum L of the two pion system. But for odd L, $Y_L \rightarrow -Y_L$, for the orbital wave function of the two-pion system, when the two identical bosons are interchanged. This would violate Bose-Einstein statistics. Experimentally the decay process $K_1^0 \rightarrow 2\pi^0$ does occur; hence the K_1^0 spin must be even.

Also it can be shown that $K^+ \rightarrow \pi^+ + \gamma$ is forbidden for spin-zero K. In the center-of-mass system of the K^+ one chooses the axis of quantization along the π-γ decay line. Initially $J = 0$, $J_z = 0$. In the final state $L_z = 0$ necessarily, from the choice of the axis of quantization, and $J_z = S_z(\text{photon}) = \pm 1$. Hence this decay process is not possible if the K is spinless. Experimentally the radiative reaction in question has *not* been observed. Dalitz (1955) has estimated that if the K spin were two (or greater), the radiative reaction $K^+ \rightarrow \pi^+ + \gamma$ would compete favorably with $K \rightarrow 2\pi$, 3π. In the following we assume that the K particle is spinless. (Moreover, for the K spin $\neq 0$, decay products can be emitted anisotropically in the K rest system if the parent K particles are polarized. Experimentally there is no evidence for anisotropy.)

In the same way $\gamma + He^4 \rightarrow \pi^0 + He^4$ is forbidden for an s-wave final state. (The He^4 and π^0 spins are both zero.) Quite generally a zero-zero transition via the emission or absorption of a real photon is strictly forbidden.

Perhaps it may be appropriate to quote here an experiment of Sunyar (unpublished) from low energy nuclear physics to detect possible γ transitions between two nuclear states of spin zero. The 700 kev $0^+ \rightarrow 0^+$ transition of Ge^{22} is known to be an internal conversion electron transition.

If angular momentum conservation were violated, the transition could occur with the emission of a real photon. No evidence for γ rays has been found, and it is estimated that the angular momentum violating amplitude is $>3 \times 10^6$ weaker (Feinberg and Goldhaber, 1959).

Elementary angular momentum considerations also enable one to say something about the spins of the hyperons. For instance, the Λ particle spin can be determined as first pointed out by Adair (1955). Λ particles are produced in the reaction

$$\pi^- + p \rightarrow K^0 + \Lambda^0 \tag{2.24}$$

and the Λ subsequently decays by the reaction

$$\Lambda^0 \rightarrow \pi^- + p \tag{2.25}$$

For the initial π^--p system, $J_z = S_z^{(p)} = \pm\frac{1}{2}$ (with equal probability for unpolarized p) if the beam direction is the axis of quantization. And for Λ and K produced in the *incident beam direction* $L_z = 0$, hence $J_z = S_z^{(\Lambda)}$ is still $\pm\frac{1}{2}$. This means that for Λ spin $> \frac{1}{2}$, the Λ particles are aligned in the sense that all but $S_z^{(\Lambda)} = \pm\frac{1}{2}$ states have zero population.

Now let us consider the angular distribution of the decay products (in the Λ rest system) of those Λ's which are produced in the incident beam direction. If the Λ spin is $\frac{1}{2}$ (unpolarized Λ), the angular distribution in the decay reaction (2.25) is isotropic. This is true of either s-wave or p-wave decay, as the reader may easily verify. If both s-wave and p-wave are present, parity is not conserved in (2.25) as we shall show in Chapter 3; the distribution is also isotropic for the reason presented below. Let us consider in detail the slightly more non-trivial case, $S_\Lambda = \frac{3}{2}$. The orbital π–p state must be p or d. Consider the p state, and assume that the original proton spin had $S_z^{(p)} = \frac{1}{2}$. Using standard techniques for adding angular momenta (see Appendix A for a table of Clebsh-Gordan coefficients), we obtain for the angular momentum wave function of the decay products

$$|J = \tfrac{3}{2}, J_z = \tfrac{1}{2}\rangle = \sqrt{\tfrac{2}{3}}|L_z = 0, S_z = \tfrac{1}{2}\rangle + \sqrt{\tfrac{1}{3}}|L_z = 1, S_z = -\tfrac{1}{2}\rangle$$

But

$$|L_z = 0, S_z = \tfrac{1}{2}\rangle = -\sqrt{\frac{3}{4\pi}}\cos\theta\alpha$$

$$|L_z = 1, S_z = -\tfrac{1}{2}\rangle = \sqrt{\frac{3}{8\pi}}\sin\theta\exp(i\varphi)\beta$$

where $\alpha = |s = \frac{1}{2}, s_z = +\frac{1}{2}\rangle$ and $\beta = |s = \frac{1}{2}, s_z = -\frac{1}{2}\rangle$ represent the proton spin up and down respectively. Thus the angular distribution of the decay products is $\langle\frac{3}{2}, \frac{1}{2}|\frac{3}{2}, \frac{1}{2}\rangle \propto \frac{1}{2}(1 + 3\cos^2\theta)$. Naturally $J_z = -\frac{1}{2}$ would give the same result. It can be shown that the $d_{3/2}$-state angular distribution is also the same using essentially the same technique. If both $p_{3/2}$ and $d_{3/2}$ states are present, parity is not conserved. Recall that for given J the possible L values are $L = J \pm \frac{1}{2}$. The angular distribution would contain odd powers of $\cos\theta$ from the interference term between the two L values. But if both J_z values are present in equal numbers, which is the case experimentally for unpolarized proton targets in (2.24), the two sets of odd powers of $\cos\theta$ terms will cancel exactly, and the angular distribution will look the same as for pure $L = J + \frac{1}{2}$, or for pure $L = J - \frac{1}{2}$. To sum up, for those Λ's produced in the beam direction the decay angular distribution is a unique function of the Λ spin as shown in Table 2.1.

TABLE 2.1

Angular Distribution in Λ Decay for those Λ's produced in the Incident Beam Direction

Spin	Angular distribution
$\frac{1}{2}$	isotropic
$\frac{3}{2}$	$\frac{1}{2} + \frac{3}{2}\cos^2\theta$
$\frac{5}{2}$	$\frac{3}{4} - \frac{9}{2}\cos^2\theta + \frac{15}{4}\cos^4\theta$

Experimentally the reactions (2.24) and (2.25) were extensively studied by Eisler *et al.* (1958). They obtained decay distributions for $|\cos\theta_{\text{production}}| \gtrsim 0.6$ shown in Fig. 2.2. Also shown are the distributions expected for spins $\frac{1}{2}$, $\frac{3}{2}$, and $\frac{5}{2}$. It is evident that the spin of the Λ is most likely $\frac{1}{2}$. The same group finds that the Σ spin is also likely to be $\frac{1}{2}$ using the same method.

It is easy to see that essentially the same argument holds for the reaction

$$K^- + p \to \Sigma^+ + \pi^-$$

$$\Sigma^+ \to p + \pi^0$$

if K^- is captured at rest from an s state of the K mesic atom (Treiman, 1956). Day, Snow, and Sucher (1959) have advanced an argument (based on the rapid rates for $np \to ns$ Stark transitions) to show that in a hydrogen bubble chamber K^-'s are indeed captured from s states. The experimental distribution indicates that the decay products are emitted isotropically in the rest system of Σ^+'s, which shows that the Σ spin is $\frac{1}{2}$. $S_\Sigma = \frac{3}{2}$ seems to be ruled out by about 20 standard deviations (Leitner *et al.*, 1959).

2.5. Lorentz Invariance

In contrast to the notions of translational and rotational invariance, the notion of Lorentz invariance is not immediately obvious from our daily life. Historically, the fact that the equations of electrodynamics are invariant under what can now be recognized as relativistic transformations

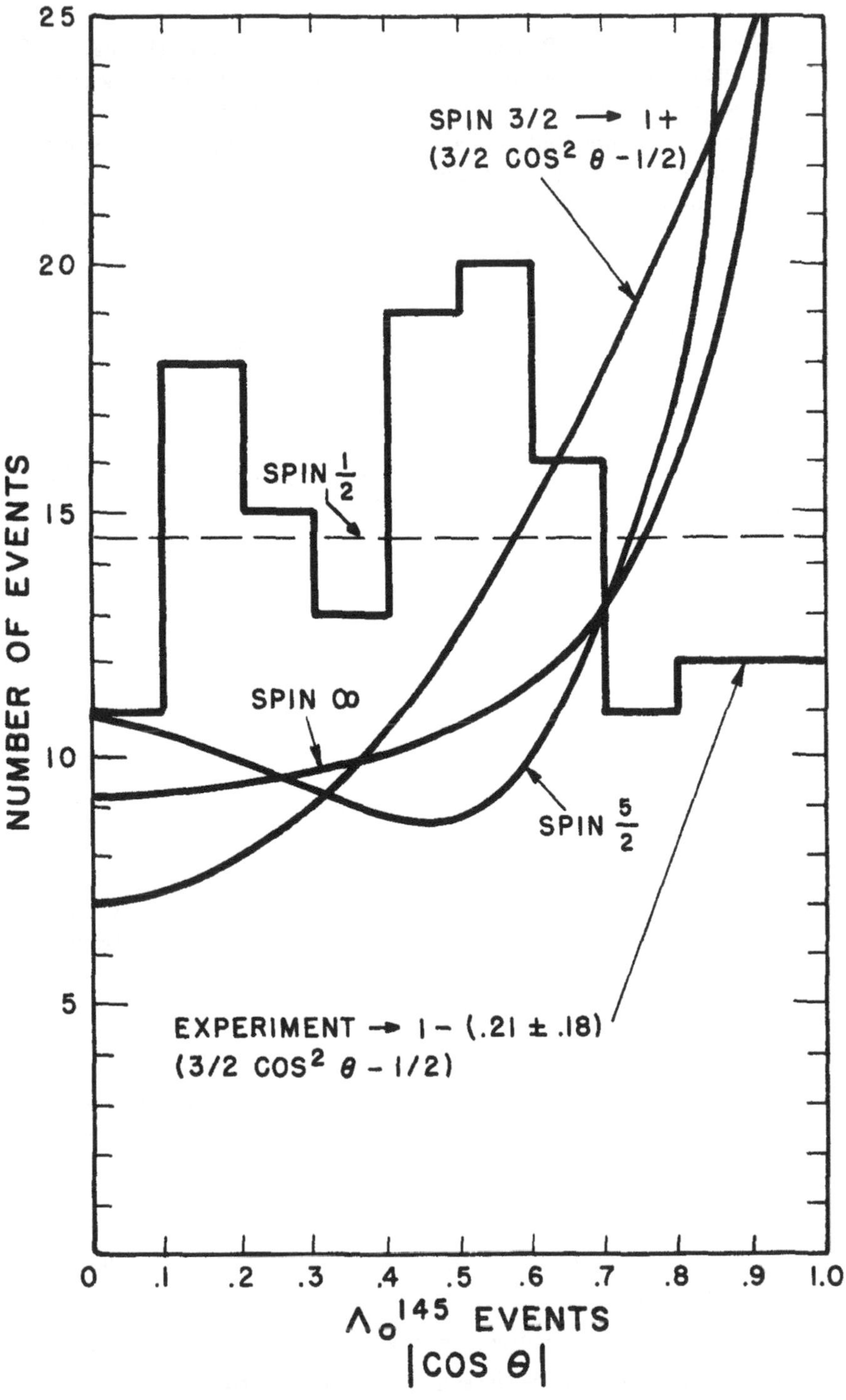

Figure 2.2. Angular distribution in Λ decay for Λ's produced in the beam direction

was first noted by H. Poincaré, who associated with it the name of Lorentz. The full significance of this new invariance, however, was not recognized until the advent of Einstein's special relativity.

The transformations that we have considered in the previous sections, translations and rotations, are Galilean, i.e., they leave $\sum_{i=1}^{3}(x_i^{(1)} - x_i^{(2)})^2$ invariant. Rotations look like

$$\begin{aligned} x_1' &= x_1 \cos\omega - x_2 \sin\omega \\ x_2' &= x_1 \sin\omega + x_2 \cos\omega \end{aligned} \tag{2.26}$$

Pure Lorentz transformations have a similar form, e.g.

$$\begin{aligned} x_1' &= \frac{1}{\sqrt{1-\beta^2}}\,x_1 - \frac{i\beta}{\sqrt{1-\beta^2}}\,(ix_0) \\ ix_0' &= \frac{i\beta}{\sqrt{1-\beta^2}}\,x_1 + \frac{1}{\sqrt{1-\beta^2}}\,(ix_0) \end{aligned} \tag{2.27}$$

or

$$\begin{aligned} x_1' &= x_1 \cosh\chi - ix_4 \sinh\chi \\ x_4' &= ix_1 \sinh\chi + x_4 \cosh\chi \end{aligned} \tag{2.27$'$}$$

with

$$\cosh\chi = \frac{1}{\sqrt{1-\beta^2}}$$

$$\sinh\chi = \frac{\beta}{\sqrt{1-\beta^2}}$$

$$x_4 = ix_0$$

The only thing different is that the angle of rotations is now purely imaginary. In general, both for pure rotations and pure Lorentz transformations

$$x_\mu' = a_{\mu\nu}x_\nu \qquad \mu = 1, 2, 3, 4$$

$$a_{\mu\nu}a_{\lambda\nu} = \delta_{\mu\lambda}$$

$a_{ij}(i, j = 1, 2, 3)$ are purely real while a_{4j} are purely imaginary. We are dealing with four-space $x_\mu = (\vec{x}, x_4) = (\vec{x}, ix_0)$, and since $x_\mu x_\mu = \vec{x}^2 + x_4^2 = \vec{x}^2 - x_0^2$ is left invariant, this transformation can be looked upon as a rotation in four-space. Using this viewpoint, we can discuss the Lorentz group just as we did for the three-dimensional group.

The generators of the infinitesimal transformations will have the commutation relations

$$[J_{\lambda\mu}, J_{\rho\sigma}] = -i(\delta_{\lambda\sigma}J_{\mu\rho} + \delta_{\mu\rho}J_{\lambda\sigma} - \delta_{\lambda\rho}J_{\mu\sigma} - \delta_{\mu\sigma}J_{\lambda\rho}) \tag{2.28}$$

Geometrically these J's have the same significance as the angular momenta J's. J_{12} is the generator for an infinitesimal rotation in the 1-2 plane and around the 3-4 plane. These commutation relations can easily be verified in the special case $L_{\lambda\mu} = -L_{\mu\lambda} = -i\left(x_\lambda \frac{\partial}{\partial x_\mu} - x_\mu \frac{\partial}{\partial x_\lambda}\right)$ but, of course, the commutation relations hold quite generally. If we define two three-vectors

$$\begin{aligned} \vec{M} &= (J_{23}, J_{31}, J_{12}) \\ \vec{N} &= -i(J_{41}, J_{42}, J_{43}) \end{aligned} \tag{2.29}$$

then the commutation relations between their components are

$$\begin{aligned} [M_i, M_j] &= iM_k \\ [N_i, M_j] &= iN_k \\ [N_i, N_j] &= -iM_k \end{aligned} \tag{2.30}$$

$(ijk) = (123)$ (meaning i, j, k, are cyclic permutations of 1, 2, 3)

The quantities $F = \frac{1}{2}(\vec{M}^2 - \vec{N}^2)$ and $G = \vec{M}\cdot\vec{N}$ are invariant under Lorentz transformations.

The particular utility of $\vec{M}$ and $\vec{N}$ lies in their relation to the two subgroups of the orthochronous proper Lorentz groups $\mathscr{L}_+^\uparrow$ where $\uparrow$ stands for $a_{44} \geq 1$, and $+$ for $\det(a_{\mu\nu}) = +1$. Representations of the infinitesimal three-dimensional rotation group (which is a subgroup of $\mathscr{L}_+^\uparrow$) have the form

$$S_{\text{rot}} = 1 + i\delta\vec{\omega}\cdot\vec{M} \tag{2.31}$$

and representations of the infinitesimal "pure" Lorentz transformation group have the form

$$S_{\text{Lor}} = 1 + i\delta\vec{\chi}\cdot\vec{N} \tag{2.32}$$

Just for completeness, let us recall that infinitesimal translations in three-space and time have the representations

$$S_{\text{trans}} = \begin{cases} 1 + i\delta\vec{l}\cdot\vec{p} \\ 1 + i\delta l_0 p_0 \end{cases} \tag{2.33}$$

From this it is clear that ten quantities, $\vec{\omega}$, $\vec{\chi}$, l_μ are needed to specify the *inhomogeneous* Lorentz transformation $x'_\mu = a_{\mu\nu}x_\nu + l_\mu$ that can be continuously generated from the identity.

As an example consider

$$J_{\mu\nu} = \tfrac{1}{2}\sigma_{\mu\nu} \tag{2.34}$$

where

$$\sigma_{\mu\nu} = \frac{1}{2i}(\gamma_\mu\gamma_\nu - \gamma_\nu\gamma_\mu)$$

The γ_μ's are 4×4 matrices satisfying

$$\{\gamma_\mu, \gamma_\nu\} \equiv \gamma_\mu\gamma_\nu + \gamma_\nu\gamma_\mu = 2\delta_{\mu\nu}, \quad \gamma_\mu^\dagger = \gamma_\mu$$

In the Dirac-Pauli representation we may write these 4×4 matrices in 2×2 block form in terms of the 2×2 identity and the Pauli matrices, σ_j,

$$\vec{\gamma} = -i\beta\vec{\alpha} = \begin{pmatrix} 0 & -i\vec{\sigma} \\ i\vec{\sigma} & 0 \end{pmatrix},$$

$$\vec{\alpha} = \begin{pmatrix} 0 & \vec{\sigma} \\ \vec{\sigma} & 0 \end{pmatrix}, \qquad \beta = \gamma_4 = \begin{pmatrix} I & 0 \\ 0 & -I \end{pmatrix} \tag{2.35}$$

In addition, there is

$$\gamma_5 = \gamma_1\gamma_2\gamma_3\gamma_4 = \frac{1}{4!}\,\epsilon_{\mu\nu\lambda\rho}\gamma_\mu\gamma_\nu\gamma_\lambda\gamma_\rho = -\begin{pmatrix} 0 & I \\ I & 0 \end{pmatrix}$$

which satisfies

$$\{\gamma_5, \gamma_\mu\} = 0, \qquad \gamma_5^2 = I, \qquad \gamma_5^\dagger = \gamma_5$$

We see that in this representation $\vec{M}$ and $\vec{N}$ are given by

$$M_k = \tfrac{1}{2}\sigma_{ij} = \frac{1}{2i}\,\gamma_i\gamma_j = \tfrac{1}{2}\begin{pmatrix} \sigma_k & 0 \\ 0 & \sigma_k \end{pmatrix}, \qquad (ijk) = (1, 2, 3) \text{ etc.}$$

$$N_k = -\frac{i}{2}\,\sigma_{4k} = \frac{i}{2}\begin{pmatrix} 0 & \sigma_k \\ \sigma_k & 0 \end{pmatrix}$$

That M_k and N_k satisfy the commutation relations (2.30) is easily verifiable. For an infinitesimal rotation around the 3-axis and an infinitesimal pure Lorentz transformation along the 1-axis, we have respectively

$$S_{\text{rot}} = 1 + \tfrac{1}{2}\gamma_1\gamma_2\,\delta\omega$$

$$S_{\text{Lor}} = 1 + \frac{i}{2}\,\gamma_1\gamma_4\,\delta\chi$$

For finite transformations we merely compound infinitesimal transformations as we did before to obtain

$$S_{\text{rot}} = \cos\frac{\omega}{2} + \gamma_1\gamma_2\sin\frac{\omega}{2} \tag{2.36}$$

$$S_{\text{Lor}} = \cosh\frac{\chi}{2} + i\gamma_1\gamma_4\sinh\frac{\chi}{2} \tag{2.37}$$

which correspond to (2.26) and (2.27). A little examination shows that $S_{\text{rot}}^\dagger = S_{\text{rot}}^{-1}$ while $S_{\text{Lor}}^\dagger = S_{\text{Lor}}$. This "difficulty" with S_{Lor} results from the four dimensional space-time being Lorentzian and not Euclidean. However, it is clear that in both cases

$$\gamma_4 S^\dagger \gamma_4 = S^{-1} \tag{2.38}$$

which is a very important relation.

PROBLEM 1. Find a 2×2 representation (using the Pauli matrices) of $\vec{M}$ and $\vec{N}$. Show that $S_\mu = (u^P \vec{\sigma} u^P,\ \pm i u^P u^P)$ transforms like a four-vector (the u^P's are Pauli spinors) under finite, proper, orthochronous Lorentz transformations (three dimensional rotations and pure Lorentz transformations).

2.6. The Dirac Equation and Bilinear Covariants

In this section we briefly discuss some of the properties of the Dirac equation, its solutions and the bilinear covariants. For a more complete discussion, the reader may refer to various standard treatises such as Pauli's *Handbuch* article (Pauli, 1933).

The Dirac equation is

$$\left(\gamma_\mu \frac{\partial}{\partial x_\mu} + m\right)\psi = 0 \tag{2.39}$$

It must be remembered that ψ is a spinor and usually represents a column of four elements. If (2.39) is multiplied by $\left(\gamma_\mu \dfrac{\partial}{\partial x_\mu} - m\right)$ from the left, then it is found that ψ satisfies

$$\left(\frac{\partial^2}{\partial x_\mu^2} - m^2\right)\psi = 0$$

which means that each of the four components satisfies the Klein-Gordon equation.

What is meant by the invariance of the Dirac equation under a Lorentz transformation? It certainly does not mean that a solution to the Dirac equation $\psi(x)$ in the original frame is identical to the corresponding solution $\psi'(x')$ in some other frame any more than the Lorentz invariance of electrodynamics implies that the electric field $\vec{E}(x)$ is the same in all frames. We assume that $\psi'(x')$ and $\psi(x)$ are related linearly as follows

$$\psi'(x') = S\psi(x), \qquad x'_\mu = a_{\mu\nu} x_\nu$$

where S is a 4×4 matrix independent of space-time. In the primed system the Dirac equation reads

$$\gamma_\mu \frac{\partial}{\partial x'_\mu} \psi' + m\psi' = 0 \tag{2.40}$$

We must show that this equation is equivalent to the Dirac equation in the original system. For this reason, we express (2.40) in terms of the original (unprimed) quantities

$$\gamma_\mu S a_{\mu\nu} \frac{\partial \psi}{\partial x_\nu} + mS\psi = 0$$

Multiply S^{-1} from the left

$$S^{-1}\gamma_\mu S a_{\mu\nu}\frac{\partial\psi}{\partial x_\nu} + m\psi = 0$$

We see that (2.40) can be made equivalent to the original Dirac equation (2.39) if there exists S with the property

$$S^{-1}\gamma_\mu S a_{\mu\nu} = \gamma_\nu$$

or

$$S^{-1}\gamma_\mu S = a_{\mu\nu}\gamma_\nu \tag{2.41}$$

We can readily show that S_{rot} and S_{Lor} discussed in the previous section are precisely the appropriate S in the sense above. For instance, for S_{rot} given in (2.36) we have

$$\left[\cos\left(-\frac{\omega}{2}\right) + \gamma_1\gamma_2\sin\left(-\frac{\omega}{2}\right)\right]\gamma_3\left[\cos\frac{\omega}{2} + \gamma_1\gamma_2\sin\frac{\omega}{2}\right] = \gamma_3, \text{ etc.}$$

and for S_{Lor} given in (2.37)

$$\left[\cosh\left(-\frac{\chi}{2}\right) + i\gamma_1\gamma_4\sinh\left(-\frac{\chi}{2}\right)\right]\gamma_4\left[\cosh\frac{\chi}{2} + i\gamma_1\gamma_4\sinh\frac{\chi}{2}\right]$$
$$= \gamma_4\cosh\chi - i\gamma_1\sinh\chi, \text{ etc.}$$

Consider now

$$j_\mu = i\bar{\psi}\gamma_\mu\psi$$

where

$$\bar{\psi} = \psi^\dagger\gamma_4$$

As $\psi \to S\psi$, we have $\bar{\psi} \to (S\psi)^\dagger\gamma_4 = \bar{\psi}S^{-1}$ because of (2.38). This means that j_μ indeed transforms like a four vector,

$$j_\mu = i\bar{\psi}\gamma_\mu\psi \to i\bar{\psi}S^{-1}\gamma_\mu S\psi = ia_{\mu\nu}\bar{\psi}\gamma_\nu\psi = a_{\mu\nu}j_\nu \tag{2.42}$$

Similarly $\bar{\psi}\sigma_{\mu\nu}\psi = \dfrac{1}{2i}\bar{\psi}[\gamma_\mu, \gamma_\nu]\psi$ can be readily seen to transform like an antisymmetric tensor of rank two. Since

$$S^{-1}_{\text{rot, Lor}}\gamma_5 S_{\text{rot, Lor}} = \gamma_5$$

(which can be readily proved using the remarkable property of γ_5, $\gamma_5\gamma_\mu = -\gamma_\mu\gamma_5$), we see that $\bar{\psi}\psi$ and $i\bar{\psi}\gamma_5\psi$ transform in the same way as a scalar, and $i\bar{\psi}\gamma_5\gamma_\mu\psi$ transform like $i\bar{\psi}\gamma_\mu\psi$ as far as orthochronous proper Lorentz transformations are concerned.

The solutions to the Dirac equation correspond to spin $\frac{1}{2}$ particles and to both positive and negative energies. The usual plane wave solutions can be written as

$$u^{(1)}\exp[i(\vec{p}\cdot\vec{x} - Ex_0)] = \begin{pmatrix} 1 \\ 0 \\ \dfrac{p_3}{E+m} \\ \dfrac{p_1 + ip_2}{E+m} \end{pmatrix}\exp[i(\vec{p}\cdot\vec{x} - Ex_0)] \tag{2.43a}$$

$$u^{(2)} \exp[i(\vec{p}\cdot\vec{x} - Ex_0)] = \begin{pmatrix} 0 \\ 1 \\ \dfrac{p_1 - ip_2}{E+m} \\ \dfrac{-p_3}{E+m} \end{pmatrix} \exp[i(\vec{p}\cdot\vec{x} - Ex_0)]$$

for positive energies $p_0 = E$ (E shall always mean $+\sqrt{|\vec{p}|^2 + m^2}$) and

$$u^{(3)} \exp[i(\vec{p}\cdot\vec{x} + Ex_0)] = \begin{pmatrix} \dfrac{-p_3}{E+m} \\ \dfrac{-p_1 - ip_2}{E+m} \\ 1 \\ 0 \end{pmatrix} \exp[i(\vec{p}\cdot\vec{x} + Ex_0)] \qquad (2.43b)$$

$$u^{(4)} \exp[i(\vec{p}\cdot\vec{x} + Ex_0)] = \begin{pmatrix} \dfrac{-p_1 + ip_2}{E+m} \\ \dfrac{p_3}{E+m} \\ 0 \\ 1 \end{pmatrix} \exp[i(\vec{p}\cdot\vec{x} + Ex_0)]$$

for negative energies $p_0 = -E$. Note that the four spinors with the plane wave dependence taken out can be written as

$$\begin{pmatrix} u^P \\ \dfrac{(\vec{\sigma}\cdot\vec{p})u^P}{E+m} \end{pmatrix} \quad \text{for } p_0 > 0, \qquad \begin{pmatrix} \dfrac{-(\vec{\sigma}\cdot\vec{p})u^P}{E+m} \\ u^P \end{pmatrix} \quad \text{for } p_0 < 0 \qquad (2.44)$$

where u^P is a two-component Pauli spinor

$$\begin{pmatrix} 1 \\ 0 \end{pmatrix} \quad \text{or} \quad \begin{pmatrix} 0 \\ 1 \end{pmatrix}$$

This means that the "small" components of the Dirac spinor can be obtained from the "large" components by applying $(\vec{\sigma}\cdot\vec{p})/(E + m)$. We could have anticipated this by writing down the Dirac equation a plane wave solution (say, $p_0 > 0$) must satisfy in the following block form

$$\begin{pmatrix} -E + m & \vec{\sigma}\cdot\vec{p} \\ -\vec{\sigma}\cdot\vec{p} & E + m \end{pmatrix} \begin{pmatrix} \psi^{(\text{large})} \\ \psi^{(\text{small})} \end{pmatrix} = 0$$

Then

$$(\vec{\sigma}\cdot\vec{p})\psi^{(\text{large})} = (E + m)\psi^{(\text{small})}$$

Using (2.44) we can easily examine the non-relativistic limits of the bilinear covariants $\bar{u}\Gamma u$ in terms of 2×2 matrices and the Pauli spinors. In the following we consider the $p_0 > 0$ case only.

If

$$\Gamma = \begin{pmatrix} I & 0 \\ 0 & I \end{pmatrix}$$

then

$$\bar{u}(\vec{p}_2)u(\vec{p}_1) = \left(u^{P\dagger}, -u^{P\dagger}\frac{(\vec{\sigma}\cdot\vec{p}_2)}{E_2+m_2}\right)\begin{pmatrix} u^P \\ \dfrac{(\vec{\sigma}\cdot\vec{p}_1)u^P}{E_1+m_1}\end{pmatrix}$$

$$= u^{P\dagger}\left(I - \frac{\vec{p}_1\cdot\vec{p}_2 + i\vec{\sigma}\cdot(\vec{p}_2\times\vec{p}_1)}{(E_1+m_1)(E_2+m_2)}\right)u^P$$

$$= u^{P\dagger}\left(I - O\left(\frac{v^2}{c^2}\right)\right)u^P$$

Similarly for $\Gamma = \gamma_4$

$$\bar{u}(\vec{p}_2)\gamma_4 u(\vec{p}_1) = u^{P\dagger}\left(I + O\left(\frac{v^2}{c^2}\right)\right)u^P$$

where the function $O\left(\frac{v^2}{c^2}\right)$ is the same in both cases. Hence for slow particles, I and γ_4 are indistinguishable. In the same sense one finds $\sigma_{ij} \sim \sigma_k$, $(ijk) = (123)$ etc. (remember σ_{ij} is 4×4, and σ_k is 2×2) and also $i\gamma_5\gamma_k \sim \sigma_k$. Meanwhile for γ_k we have

$$\gamma_k \sim -\frac{i\vec{\sigma}_k(\vec{\sigma}, \vec{p}_1)}{E_1+m_1} - \frac{i(\vec{\sigma}\cdot\vec{p}_2)\sigma_k}{E_2+m_2},$$

which is small. Other small matrices are

$$i\gamma_5 \sim -\frac{i\vec{\sigma}\cdot\vec{p}_1}{E_1+m_1} + \frac{i\vec{\sigma}\cdot\vec{p}_2}{E_2+m_2}, \text{ and also } \sigma_{i4} \text{ and } i\gamma_5\gamma_4$$

In summary, the matrices I, γ_4, σ_{ij}, $i\gamma_5\gamma_k$ are "large" and γ_k, $i\gamma_5$, σ_{i4} and $i\gamma_5\gamma_4$ are small when taken between *two positive* (or *two negative*) energy states, as summarized in Table 2.2 (in which the non-relativistic approximation $E \sim m$ has been made).

"Largeness" and "smallness" are interchanged if taken between states of opposite energy. Thus in positronium annihilation, in which a slow positive energy electron makes a transition to a negative energy state, γ_k is "large" and γ_4 is "small". Similarly γ_5 is "large" when it connects a positive energy state with a negative energy state.

We now consider different kinds of couplings that Dirac fields can have with other fields. Classically, the interaction of a current with the electromagnetic field is given by $j_\mu A_\mu$. In analogy the coupling with the Dirac field is

$$H_{\text{int}} = ieA_\mu\bar{\psi}\gamma_\mu\psi \tag{2.45}$$

The fourth-component contribution is large in the non-relativistic limit,

TABLE 2.2

Γ	$\Gamma^{(P)}$
I	I
γ_k	$\left[-\frac{i(\vec{p}+\vec{p}')}{2m} + \frac{\vec{\sigma}\cdot\{(\vec{p}'-\vec{p})\vec{x}\}}{2m}\right]_k$
γ_4	I
$i\gamma_5\gamma_k$	σ_k
$i\gamma_5\gamma_4$	$\frac{i\vec{\sigma}\cdot(\vec{p}+\vec{p}')}{2m}$
$\sigma_{ij} = \frac{1}{2i}[\gamma_i, \gamma_j]$	σ_k where $(ijk) = (123)$ etc.
$\sigma_{k4} = \frac{1}{2i}[\gamma_k, \gamma_4]$	$\left[-\frac{(\vec{p}-\vec{p}')}{2m} - \frac{i\vec{\sigma}\cdot\{(\vec{p}+\vec{p}')x\}}{2m}\right]_k$
$i\gamma_5$	$\frac{i\sigma\cdot(\vec{p}'-\vec{p})}{2m}$

The non-relativistic two-by-two reduction of the Dirac matrices for positive energy states correct to $O(v/c)$. $\Gamma^{(P)}$ is defined by

$$\bar{u}(\vec{p}')\Gamma u(\vec{p}) = u^{P\dagger}\Gamma^{(P)}u^P$$

where $u(\vec{p})$ and u^P stand for a positive energy four-component Dirac spinor and the corresponding two-component Pauli spinor respectively.

and is just $-eA_0\psi^\dagger\psi$, the scalar potential times the charge density. The space part in the non-relativistic limit, $E \sim m$, is

$$ie\vec{A}\cdot\bar{u}\vec{\gamma}u \rightarrow \frac{e}{2m}u^{P\dagger}(\vec{A}\cdot(\vec{p}_1+\vec{p}_2) + i\vec{\sigma}\cdot(\vec{p}_2-\vec{p}_1)\times\vec{A})u^P \tag{2.46}$$

The first term is that expected from the Schrödinger equation current for plane waves. But in the second term, $\vec{p}_2 - \vec{p}_1$ is just the momentum transferred to the radiation field (photon); then $\vec{p}_2 - \vec{p}_1 \rightarrow -i\vec{\nabla}$ acting on $\vec{A}$ (in coordinate space), hence this term has the form

$$\frac{e}{2m}\vec{\sigma}\cdot\left(\vec{\nabla}\times\vec{A}\right) = \frac{e}{2m}\vec{\sigma}\cdot\vec{B}$$

But since $\vec{s} = \frac{1}{2}\vec{\sigma}$, we have $2\frac{e}{2mc}\vec{s}\cdot\vec{B}$. Obtaining the required g factor of 2 in such a natural way was a great triumph for the Dirac theory.

Using Yukawa's idea, we can write down the interaction density that corresponds to the coupling of the pion field to the nucleon in analogy with electromagnetism. As we shall see in Chapter 3, the pion field φ is pseudoscalar, and one could choose a direct pseudoscalar coupling $iG\bar{\psi}\gamma_5\psi\varphi$ which, according to Table 2.2, tends to $\frac{iG}{2M_N}u^{P\dagger}\vec{\sigma}u^P\cdot\vec{q}\varphi$ where $\vec{q}$ is the pion momentum. Another possible coupling would be a derivative coupling $i \frac{F}{\mu_\pi} \bar{\psi}\gamma_5\gamma_\mu\psi \frac{\partial}{\partial x_\mu} \varphi$ which non-relativistically tends to $\frac{iF}{\mu_\pi} u^{P\dagger}\vec{\sigma}u^P \cdot \vec{q}\varphi$ where μ_π is the pion mass. Thus if the nucleons are treated non-relativistically, both of these lead to the same interaction. Actually this equivalence of pseudoscalar direct (*ps-ps*) and pseudovector-gradient (*ps-pv*) couplings holds also for first order relativistic processes since

$$\bar{u}(\vec{p}_2)i\gamma_5\gamma\mu(p_2 - p_1)\mu u(\vec{p}_1) = 2M\bar{u}(\vec{p}_2)\gamma_5 u(\vec{p}_1) \tag{2.47}$$

provided that $u(\vec{p}_2)$ and $u(\vec{p}_1)$ are free-particle spinors. The two couplings will be the same if $F = \mu_\pi G/2M_N$; so if $F^2/4\pi \approx 0\,08$, then $G^2/4\pi \approx 15$. ($F^2/4\pi$ is often denoted by f^2.)

So far we have been rather sloppy with normalization of our Dirac spinors. For covariant calculations, a convenient normalization is as follows:

$$\begin{aligned} \bar{u}^{(r)}u^{(s)} &= \delta_{rs} \qquad r, s = 1, 2 \quad (p_0 > 0) \\ \bar{u}^{(r)}u^{(s)} &= -\delta_{rs} \quad r, s = 3, 4 \quad (p_0 < 0) \end{aligned} \tag{2.48}$$

Using the equation

$$(i\gamma \cdot p + m)u = 0 \tag{2.49}$$

and its adjoint equation, one finds that

$$u^\dagger u = \frac{E}{m} \tag{2.50}$$

i.e. there is one particle per volume m/E. With this normalization, the density of states is

$$\rho = \frac{V}{(2\pi\hbar)^3} d^3p = \frac{1}{(2\pi)^3} \frac{m}{E} d^3p \tag{2.51}$$

Note $d^3p/E = d^3p/|p_0|$ is relativistically invariant since

$$\int \delta(p_0^2 - \vec{p}^2 - m^2)\, dp_0 = \frac{1}{2p_0}\bigg|_{p_0^2 = \vec{p}^2 + m^2}$$

Writing u as

$$u = N \begin{pmatrix} Iu^P \\ \dfrac{(\vec{\sigma} \cdot \vec{p})u^P}{E + m} \end{pmatrix}$$

we have

$$\bar{u}u = |N|^2\left(1 - \frac{p^2}{(E + m)^2}\right) = |N|^2 \frac{2m}{E + m}$$

Hence

$$N = \sqrt{\frac{E + m}{2m}} \tag{2.52}$$

Since $N = 1 + 0\left(\left(\frac{v}{c}\right)^2\right)$, our previous results on non-relativistic limits given in Table 2.2 are still correct.

2.7. Covariant Spin Formalism; Helicity and Chirality

In practical calculations, one is interested in $\Sigma(u_2\Gamma_A u_1)\,(\bar{u}_1\Gamma_B u_2)$, where the sum is over the two spin states of u_1, corresponding to $p_0 > 0$ (or $p_0 < 0$). We can sum over all four states if we insert the projection operator

$$\Lambda_+ = \frac{-i\gamma\cdot p + m}{2m} \tag{2.53}$$

to select the $p_0 > 0$ states. If instead of summing over the spin states, we wanted to compute the polarization, then we can still sum over the spin states but now must insert a projection operator to project out the appropriate spin state. Since $\frac{1}{2}(1 + \vec{\sigma}\cdot\hat{n})$ is the appropriate projection operator in the Pauli two-component formalism, one possibility would be to express everything in a two-component formalism and use $\frac{1}{2}(1 + \vec{\sigma}\cdot\hat{n})$, but this is troublesome.

Clearly for small momentum

$$\tfrac{1}{2}(1 + \vec{\Sigma}\cdot\hat{n})$$

where

$$\Sigma_k = \sigma_{ij} = \begin{pmatrix}\sigma_k & 0\\ 0 & \sigma_k\end{pmatrix} \qquad (i, j, k) = (1, 2, 3)$$

is acceptable. But, since

$$\left[-\frac{i\gamma\cdot p + m}{2m}, \vec{\Sigma}\cdot\hat{n}\right] \neq 0$$

in general, this would not be suitable. However, it is interesting that if $\hat{n} = \hat{p}$, they do commute; i.e. only if $p_1 = p_2 = 0$, are the solutions of the Dirac equation that we have enumerated in (2.43) also eigenfunctions of

$$\Sigma_3 = \begin{pmatrix}1 & 0 & 0 & 0\\ 0 & -1 & 0 & 0\\ 0 & 0 & 1 & 0\\ 0 & 0 & 0 & -1\end{pmatrix}$$

Hence the beam direction obtains a special significance as an axis of quantization. The helicity is defined by

$$\mathscr{H} = \langle\vec{\Sigma}\cdot\hat{p}\rangle_{p_0>0\ (\text{or } p_0<0)} \tag{2.54}$$

It is one for particles that are 100% longitudinally polarized. Since $\frac{1}{2}(1 + \vec{\Sigma}\cdot\hat{n})$ does commute with Λ_+ for $\hat{n} = \hat{p}$, this is a suitable spin projection operator for longitudinal polarization as well as for particles at rest.

Since we know how to project spin in the rest system, all we need to know in order to construct a covariant projection operator is how a unit vector in the rest system will appear in the lab. system in which the particle is moving. Let w_μ be the four-vector such that

$$w_\mu = (\hat{n}, 0)$$

in the rest system; notice that the time component is zero. Now in the rest system $w_\mu^2 = 1$ and $w_\mu p_\mu = 0$ (no matter what direction $\hat{n}$ has), but covariance requires that these quantities be invariant under Lorentz transformation. To find what w_μ looks like in the lab. frame, it is perhaps easiest to divide $\vec{w}$ into longitudinal and transverse parts and transform each separately, then recombine. It will help to recall that $\beta = \frac{p}{E}$ and $\frac{1}{\sqrt{1-\beta^2}} = \frac{E}{m}$. The answer is

$$\begin{aligned} w_\mu &= (\vec{w}, iw_0) \\ \vec{w} &= \hat{n} + \frac{\vec{p}(\vec{p}\cdot\hat{n})}{m(E+m)} \\ w_0 &= \frac{\vec{p}\cdot\hat{n}}{m} \end{aligned} \tag{2.55}$$

It can easily be verified that $w_\mu^2 = 1$ and $w_\mu p_\mu = 0$.

For the case $\hat{n} = \hat{p}$, we have

$$\begin{aligned} \vec{w} &= \hat{p}\left(1 + \frac{p^2}{m(E+m)}\right) = \hat{p}\,\frac{E}{m} \\ w_0 &= \frac{|\vec{p}|}{m} \end{aligned} \tag{2.56}$$

which is useful for computing longitudinal polarization.

In the discussion of the non-relativistic limits of the bilinear covariants, we saw that $i\gamma_5\vec{\gamma}$ and $\vec{\Sigma}$ give the same expectation value as $\frac{v}{c} \to 0$. Hence we may try $\Lambda_s = \frac{1}{2}(1 + i\gamma_5\gamma_\mu w_\mu)$. We now verify that it has the desired properties. In the rest system $\frac{1}{2}(1 + i\gamma_5\gamma_\mu w_\mu)$ and $\frac{1}{2}(1 + \vec{\Sigma}\cdot\hat{n})$ give the same result when they act on $p_0 > 0$ spinors; hence it is appropriate here. For longitudinal polarization ($\hat{n} = \hat{p}$), we find

$$i\gamma_5\gamma_\mu w_\mu \begin{pmatrix} I \\ \dfrac{\vec{\sigma}\cdot\vec{p}}{E+m} \end{pmatrix} = \begin{pmatrix} \vec{\sigma}\cdot\hat{p}\,\dfrac{E}{m} - \vec{\sigma}\cdot\hat{p}\,\dfrac{p^2}{(E+m)m} \\ \dfrac{-\vec{\sigma}\cdot\hat{p}(\vec{\sigma}\cdot\vec{p})}{(E+m)}\dfrac{E}{m} + \dfrac{|\vec{p}|}{m} \end{pmatrix} = \vec{\Sigma}\cdot\hat{p}\begin{pmatrix} I \\ \dfrac{\vec{\sigma}\cdot\hat{p}}{E+m} \end{pmatrix} \tag{2.57}$$

hence it works correctly in this case. The only thing left to check is that it commutes with Λ_+; that $[i\gamma_5\gamma_\mu w_\mu, i\gamma\cdot p] = 0$ can be shown. Hence

$\Lambda_s = \frac{1}{2}(1 + i\gamma_5\gamma_\mu w_\mu)$, where w_μ, as given by (2.55), is the appropriate spin projection operator. Thus we see that, to do practical calculations on polarized particles, we need only insert $\Lambda_s\Lambda_+$. More on this subject and related ones can be read in Tolhoek (1956), which also contains references to earlier works (e.g. Michel and Wightman, 1955).

One may note that for extreme relativistic particles ($E/m \to \infty$), w_μ blows up even though the product

$$i\gamma_5\gamma_\mu w_\mu \begin{pmatrix} I \\ \dfrac{\vec{\sigma}\cdot\vec{p}}{E + m} \end{pmatrix}$$

does not. Note, however, that

$$\gamma_5 \begin{pmatrix} I \\ \dfrac{\vec{\sigma}\cdot\vec{p}}{E + m} \end{pmatrix} = -\begin{pmatrix} 0 & I \\ I & 0 \end{pmatrix} \begin{pmatrix} I \\ \dfrac{\vec{\sigma}\cdot\vec{p}}{E + m} \end{pmatrix} = -\begin{pmatrix} \dfrac{\vec{\sigma}\cdot\vec{p}}{E + m} \\ I \end{pmatrix} \tag{2.58}$$

coincides with

$$-\vec{\Sigma}\cdot\hat{p} \begin{pmatrix} I \\ \dfrac{\vec{\sigma}\cdot\vec{p}}{E + m} \end{pmatrix} = -\begin{pmatrix} \vec{\sigma}\cdot\hat{p} & 0 \\ 0 & \vec{\sigma}\cdot\hat{p} \end{pmatrix} \begin{pmatrix} I \\ \dfrac{\vec{\sigma}\cdot\vec{p}}{E + m} \end{pmatrix} = -\begin{pmatrix} \vec{\sigma}\cdot\hat{p} \\ \dfrac{|\vec{p}|}{E + m} \end{pmatrix} \tag{2.59}$$

as $E/m \to \infty$, $|\vec{p}| \to E$. Thus $\gamma_5 \to -\vec{\Sigma}\cdot\hat{p}$ (helicity operator) as $v/c \to 1$ (for $p_0 > 0$ spinors). The wave function for right-handed particles (spin along the direction of motion) is $\frac{1}{2}(1 - \gamma_5)\psi$ if $\frac{v}{c} \sim 1$.

In the literature, γ_5 is often referred to as the chirality operator (from the Greek word χειρ meaning "hand"). The word "chirality" was coined by Lord Kelvin (in a somewhat different context). The word "helicity" was introduced by Watanabe (1957) to replace "spirality."

CHAPTER 3

Parity

3.1. General Considerations

Let us consider the operation of space inversion or parity:

$$P : \vec{x} \xrightarrow{P} \vec{x}' = -\vec{x} \tag{3.1}$$

The coordinate axes are projected back through themselves to give the inverted coordinate system as shown in Fig. 3.1. We have thereby transformed a right-handed coordinate system into a left-handed one. The

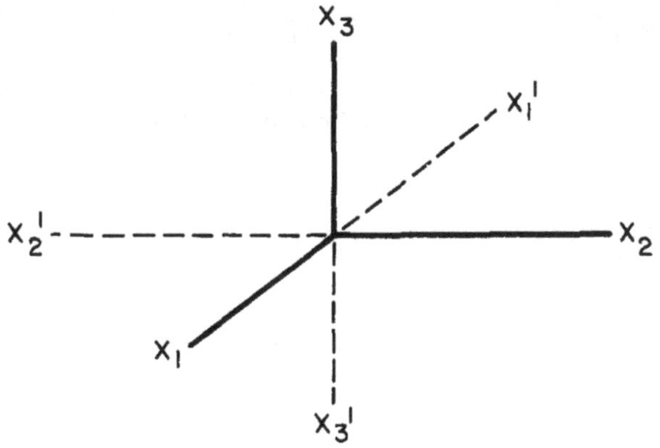

Figure 3.1. Space inversion (or parity operation) $\vec{x} \rightarrow \vec{x}' = -\vec{x}$

invariance of a physical system under space inversion implies that it is not possible to distinguish right from left. Until recent years it was believed that all physical laws were invariant under space inversion. True, we find asymmetrical objects in nature such as sugar molecules and the spiral shell of a snail; yet these asymmetries can be explained as generic accidents.

It is nevertheless interesting to note the distinction which is made between right and left in our everyday lives. For example, the Latin word for "left," *sinistrum*, has an evil connotation. Likewise, in the Far Eastern tradition, the right (右) side has been regarded as a place of honor, whereas the left (左) side has carried a connotation of being inferior. For discussions of invariance under space inversion in the history of fine arts, see Weyl (1952).

Space inversion is fundamentally different from the transformations we have discussed so far. Previously we first considered an infinitesimal transformation $q_i \to q_i + \delta q_i$, and we then saw that a finite transformation can be compounded from successive infinitesimal transformations. This is, however, not the case for space inversion.

In classical mechanics, one examines the invariance of the Hamiltonian against an infinitesimal change in the generalized coordinates. All the conservation laws stem from such considerations. One can see that, although the Hamiltonian

$$H_{\text{class}} = \frac{1}{2m}\,(|\vec{p}|^2 + m^2\omega^2|\vec{x}|^2)$$

is certainly invariant under the discrete transformation $\vec{x} \to -\vec{x}$ and $\vec{p} \to -\vec{p}$, no useful constant-of-the-motion results from such a parity consideration in classical mechanics.

In quantum mechanics, if we have $HP\Psi = PH\Psi$ (i.e. $[H, P] = 0$), then H and P are simultaneously diagonalizable. The state vector which is obtained by measuring the energy first and then inverting the axes is the same as if we invert the axes first and then measure the energy. As an example, we consider a particle moving in a spherically symmetric, central potential. Then we have

$$H = \frac{-\nabla^2}{2m} + V(|\vec{x}|)$$

In this case, the eigenstates of H are also eigenstates of parity.

Consider now the connection between mirror reflection and space inversion. If we make the reflection with respect to the $x_1 - x_2$ plane (denoted by operator R_f) then

$$x_1 \xrightarrow{R_f} x_1$$

$$x_2 \xrightarrow{R_f} x_2$$

$$x_3 \xrightarrow{R_f} -x_3$$

If we now apply to this system a continuous rotation of 180° about the x_3 axis (the operator is denoted by $R_{x_3}(\pi)$), we have:

$$x_1 \xrightarrow{R_f} x_1 \xrightarrow{R_{x_3}(\pi)} -x_1$$

$$x_2 \to x_2 \to -x_2$$

$$x_3 \to -x_3 \to -x_3$$

Thus the space inverted state can be obtained by a continuous transformation from the mirror reflected state.

We can ask what the commutation relation between rotation and space inversion is. Consider $R_{x_3}(\pi)P$

$$x_1 \xrightarrow{P} -x_1 \xrightarrow{R_{x_3}(\pi)} x_1$$
$$x_2 \rightarrow -x_2 \rightarrow x_2$$
$$x_3 \rightarrow -x_3 \rightarrow -x_3$$

Meanwhile consider $PR_{x_3}(\pi)$

$$x_1 \xrightarrow{R_{x_3}} -x_1 \xrightarrow{P} x_1$$
$$x_2 \rightarrow -x_2 \rightarrow x_2$$
$$x_3 \rightarrow +x_3 \rightarrow -x_3$$

We see that rotation and space inversion commute. In particular, the space inversion operator commutes with an infinitesimal rotation operator, and hence with the angular momentum $\vec{J}$.

$$[P, \vec{J}] = 0$$
or
$$P\vec{J}P^{-1} = \vec{J} \tag{3.2}$$

We could have guessed this from:

$$\vec{L} = \vec{x} \times \vec{p}$$
$$P(\vec{x} \times \vec{p})P^{-1} = (-\vec{x}) \times (-\vec{p}) = \vec{L}$$

This commutation relation is quite general, and applies also to spin angular momentum

$$P\vec{S}P^{-1} = \vec{S}$$

Note that the commutation rules

$$[x_i, p_j] = i\delta_{ij}$$
$$[J_i, J_j] = iJ_k;\ (ijk) = (123) \text{ etc.}$$

are thereby preserved under space inversion. P is unitary with $P^\dagger P = 1$ and $P = P^{-1}$. For a detailed discussion of this point see Wigner (1959). The angular solutions to the Schrödinger equation for a spherically symmetric potential with no spin are the usual spherical harmonics.

$$Y_{lm}(\theta, \phi) = (-1)^m \sqrt{\frac{(2l+1)}{4\pi}\frac{(l-|m|)!}{(l+|m|)!}}\, P_l^{|m|}(\cos\theta)\exp(im\phi)$$

One may ask how these functions behave under the parity operation

$$\begin{aligned} \theta &\rightarrow \pi - \theta \\ \phi &\rightarrow \pi + \phi \end{aligned} \qquad (\cos\theta \rightarrow -\cos\theta)$$

Since

$$P_l(x) = \frac{1}{2^l l!}\frac{d^l}{dx^l}[(x^2 - 1)l]$$

$$P_l^m(x) = (1 - x^2)^{m/2}\frac{d^m P_l(x)}{dx^m}$$

and $(x^2 - 1)^l$ is an even function of x,

$$P_l^m(-\cos\theta) = (-1)^{l+m}P_l^m(\cos\theta)$$

$$\exp(\mathrm{im}(p + \pi)) = (-1)^m \exp(\mathrm{im}\phi)$$

and we have

$$Y_{lm} \rightarrow (-1)^{l+m}(-1)^m Y_{lm} = (-1)^l Y_{lm} \tag{3.3}$$

Thus:

$s, d, g\ldots$ waves have even parity

$p, f, h\ldots$ waves have odd parity

Since spin is unchanged by parity, the transformation of 2-component spinors under P is $u^P \xrightarrow{P} \eta u^P$ (where the superscript ^{P} refers to Pauli spinors). The requirement is that $|\eta|^2 = 1$. If particles are neither created nor destroyed in an interaction, no complications arise. The reason is that only $u^{P\dagger}\vec{\sigma}u^P$, $u^{P\dagger}u^P$ are observable, and $|\eta|^2 = 1$. We have to be more careful when particles are created or annihilated; e.g. $\Sigma^0 \rightarrow \Lambda^0 + \gamma$; here the transformation properties of u_Λ^P and u_Σ^P are important. We will come back to this later on.

In the interaction picture which is obtained from the Schrödinger picture by the unitary transformation $S = \exp(iH_0 t)$, we have the equation

$$i\frac{\partial}{\partial t}\Psi = H_{\text{int}}(t)\Psi \text{ with } H_{\text{int}} = \exp(iH_0 t)H_I^{\text{Schröd}}\exp(-iH_0 t) \tag{3.4}$$

where $H_I^{\text{Schröd}}$ is the interaction Hamiltonian in the Schrödinger picture. If U is the operator that takes us from a state $\Psi(t_0)$ to a state $\Psi(t)$ (i.e. $U(t, t_0)\Psi(t_0) = \Psi(t)$), then we have the operator equation:

$$i\frac{\partial}{\partial t}U(t, t_0) = H_{\text{int}}(t)U(t, t_0) \tag{3.5}$$

with the boundary condition $U(t_0, t_0) = 1$. This equation can be integrated to give the integral equation:

$$U(t, t_0) - U(t_0, t_0) = -i\int_{t_0}^{t} dt' H_{\text{int}}(t')U(t', t_0)$$

$$U(t, t_0) = 1 - i\int_{t_0}^{t} dt' H_{\text{int}}(t')U(t', t_0) \tag{3.6}$$

We define the S matrix as that U for which we let $t_0 \rightarrow -\infty$ and $t \rightarrow +\infty$.

The iterated solution for the integral equation has the form:

$$S = U(\infty, -\infty) = 1 + \sum_{n=1} (-i)^n \int_{-\infty}^{\infty} dt_1 \int_{-\infty}^{t_1} dt_2 \ldots$$
$$\int_{-\infty}^{t_{n-1}} dt_n \{H_{\text{int}}(t_1) \ldots H_{\text{int}}(t_{n-1}) H_{\text{int}}(t_n)\} \quad (3.7)$$

The matrix element: $\langle f|S^{(m)}|i\rangle$ can be shown to be identical to the transition matrix of the "old-fashioned" m^{th} order time-dependent perturbation theory. For example

$$\begin{aligned}\langle f|S^{(1)}|i\rangle &= -i \int_{-\infty}^{\infty} dt \langle f|H_{\text{int}}(t)|i\rangle \\ &= -i\langle f|H_I^{\text{Schröd}}|i\rangle \int_{-\infty}^{\infty} dt \exp\,[i(E_f - E_i)t] \\ &= -2\pi i\delta(E_i - E_f)]\langle f|H_I^{\text{Schröd}}|i\rangle \end{aligned} \quad (3.8)$$

For the second-order term, we make use of the completeness relation $\sum|m\rangle\langle m| = 1$ (where $|m\rangle$ is an eigenstate of H_0), and one of the integrations yields the well-known energy denominator $E_i - E_m$. Thus we arrive at

$$\langle f|S|i\rangle = 1 - 2\pi i\delta(E_i - E_f)[\langle f|H_I^{\text{Schröd}}|i\rangle + \sum_m \frac{\langle f|H_I^S|m\rangle\langle m|H_I^S|i\rangle}{E_i - E_m} + \ldots.] \quad (3.9)$$

If H is invariant under P, then S is also invariant. We just apply

$$PH_I(t_1)P^{-1}PH_I(t_2)P^{-1}\ldots PH_I(t_n)P^{-1}$$

To sum up, if $PHP^{-1} = H$, then

$$PSP^{-1} = S \quad (3.10)$$

This statement is not as trivial as it sounds. It is a consequence of the unitarity of the P operator. The analogous statement does not hold for the time reversal operator which is antiunitary, as we shall see in the next chapter. Since

$$P\Psi_f = PS\Psi_i = PSP^{-1}P\Psi_i = SP\Psi_i$$

the S matrix connects states of the same parity (i.e. if $P\Psi_i = \pm\Psi_i$, then $P\Psi_f = \pm\Psi_f$ with the same $\pm$ sign). In other words, the invariance of the Hamiltonian under space inversion requires that the parity of the initial state be identical to the parity of the final state.

If the S matrix is invariant under parity, the distribution function W obtained from S must necessarily satisfy

$$W(\vec{p}_1, \vec{p}_2 \ldots; \vec{s}_1, \vec{s}_2 \ldots) = W(-\vec{p}_1, -\vec{p}_2 \ldots; \vec{s}_1, \vec{s}_2 \ldots) \quad (3.11)$$

This means that terms of the form $\vec{p}_1 \cdot (\vec{p}_2 \times \vec{p}_3)$, $\vec{s}_1 \cdot \vec{p}_1$ etc. cannot have non-zero expectation values. In Section 3.6 we will discuss experiments to test parity conservation.

3.2. π^+ -p Scattering and Phase Shift Analysis

Elastic scattering of a π^+ by a proton is a particularly simple case because there is no net annihilation or creation of particles.

The leading term in S (namely 1) represents no scattering (i.e., nothing happens). It is customary to define $T^{(+)}$, the transition matrix, by

$$S = 1 - 2\pi i\delta(E_i - E_f)T^{(+)} \tag{3.12}$$

(The meaning of the superscript (+) will become apparent when we discuss time reversal.) Then for no scattering, $T^{(+)} = 0$ and $S = 1$. Let us consider a particular representation of $T^{(+)}$; namely M, the 2×2 matrix that acts on the initial nucleon spinor to produce the scattered state (remember that we initially have one pion of a definite momentum far away). A precise definition of this M would then be

$$\psi_j = \exp(ipz)u_j + \frac{\exp(ipr)}{r} M_{jl}u_l \tag{3.13}$$

where ψ_j represents a two-component spinor and space function. We may ask: What is the restriction imposed on M by parity conservation? We consider the three independent vectors $\vec{\sigma}, \vec{p}_i, \vec{p}_f$ where $\vec{\sigma}$ is the nucleon spin and $\vec{p}_i, \vec{p}_f$ are the initial and final momenta respectively. From rotational invariance and from $[\sigma_i, \sigma_j] = 2i\sigma_k$, we have only three invariants. They are

$$\vec{\sigma}\cdot\vec{p}_i;\ \vec{\sigma}\cdot\vec{p}_f;\ \vec{\sigma}\cdot(\vec{p}_i \times \vec{p}_f) \propto \vec{\sigma}\cdot\hat{n}, \text{ where } \hat{n} = \frac{\vec{p}_i \times \vec{p}_f}{|\vec{p}_i \times \vec{p}_f|}$$

$\hat{n}$ is the unit vector normal to the scattering plane. If we now impose invariance under space inversion, then the pseudoscalars (quantities that change sign under space inversion) $\vec{\sigma}\cdot\vec{p}_i$ and $\vec{\sigma}\cdot\vec{p}_f$ are excluded.

Thus the most general form of M is

$$M = f(\theta) + ig(\theta)\vec{\sigma}\cdot\hat{n} \tag{3.14}$$

What is the physical meaning of the $\vec{\sigma}\cdot\hat{n}$ term? Consider the scattering to be in the x-z plane. As usual the incident beam is along the $+z$ direction. Then $\hat{n}$ is along the $+y$ direction and $ig(\theta)\vec{\sigma}\cdot\hat{n} = ig(\theta)\sigma_y$. We choose the quantization axis along the beam direction (z axis). We shall see that $g(\theta)$ is the "spin-flip" amplitude.

Since

$$ig(\theta)\sigma_y = \begin{pmatrix} 0 & g(\theta) \\ -g(\theta) & 0 \end{pmatrix}$$

under the scattering we have

$$\alpha \rightarrow M\alpha = f\alpha - g\beta$$
$$\beta \rightarrow M\beta = g\alpha + f\beta$$

where α and β are two-component spinors corresponding to nucleon spin along the $+z$ and $-z$ directions respectively.

$$\alpha = \begin{pmatrix} 1 \\ 0 \end{pmatrix}, \qquad \beta = \begin{pmatrix} 0 \\ 1 \end{pmatrix}$$

But note that if we choose the representation of $\vec{\sigma}$ such that the quantization is along the y axis, there is no "spin flip," and we have

$$\sigma_x = \begin{pmatrix} 0 & -i \\ i & 0 \end{pmatrix}, \sigma_y = \begin{pmatrix} 1 & 0 \\ 0 & -1 \end{pmatrix}, \sigma_z = \begin{pmatrix} 0 & 1 \\ 1 & 0 \end{pmatrix},$$

$$\alpha \to (f + ig)\alpha$$

$$\beta \to (f - ig)\beta$$

where we have used $\vec{\sigma}\cdot\hat{n} = \sigma_y$. Thus the scattering is different for the two initial spin states, which explicitly exhibits that scattering is spin-dependent if $g \neq 0$. If the initial proton is unpolarized, the differential cross section has the form

$$\frac{d\sigma}{d\Omega} = \tfrac{1}{2}(|f + ig|^2 + |f - ig|^2) = |f|^2 + |g|^2 \tag{3.15}$$

But if we ask what is the final state polarization along $\hat{n}$, we see that

$$\langle\vec{\sigma}\rangle\cdot\hat{n} = \frac{|f + ig|^2 - |f - ig|^2}{|f + ig|^2 + |f - ig|^2}$$

$$= \frac{2\text{Im}(fg^*)}{|f|^2 + |g|^2} \tag{3.16}$$

and we have interference between f and g. A quick way to obtain the same results is to use the density matrix formalism which is discussed in detail in Bethe and Morrison (1956) (see also Tolhoek, 1956). Using the density matrix formalism, we have

$$\frac{d\sigma}{d\Omega} = \tfrac{1}{2}\text{Tr}(MM^\dagger) = |f|^2 + |g|^2 \tag{3.15'}$$

or to calculate the polarization

$$\langle\vec{\sigma}\rangle\cdot\hat{n} = \frac{\text{Tr}(MM^+\vec{\sigma})\cdot\hat{n}}{\text{Tr}(MM^+)}$$

$$= \frac{\text{Tr}[(|f|^2 + |g|^2 + 2\text{Im}(fg^*)\vec{\sigma}\cdot\hat{n})\vec{\sigma}]\cdot\hat{n}}{2(|f|^2 + |g|^2)}$$

$$= \frac{2\text{Im}(fg^*)}{|f|^2 + |g|^2} \tag{3.16'}$$

(since $\text{Tr}(\sigma_i\sigma_j) = 2\delta_{ij}$)

Now, what is the connection between this formalism and the ordinary phase shift formalism? In particular, what are f and g in terms of the spherical harmonics Y_{lm} and the phase shifts $\delta_{l,J=l\pm 1/2}$?

Usually we classify states by their eigenvalues J, M. If parity is a goód quantum number, then we also have $w = \pm 1$ (w is here the eigenvalue of P). For elastic scattering of a spin 0 particle by a spin $\frac{1}{2}$ particle, we can talk about l instead of parity. This situation arises because for a given J, the only two l values possible are $l = J + \frac{1}{2}$ and $J - \frac{1}{2}$ (states of opposite orbital parity). Thus if P is conserved in the interaction, l is also a good quantum number. (This, of course, is no longer true for np scattering where the transition ${}^3S_1 \to {}^3D_1$ is possible through the tensor force.)

To derive the desired relation between $M = f + i\vec{\sigma}\cdot\hat{n}g$ and the phase shifts, we first note that the operators

$$P_+ = \frac{l + 1 + \vec{L}\cdot\vec{\sigma}}{2l + 1}$$
$$P_- = \frac{l - \vec{L}\cdot\vec{\sigma}}{2l + 1} \tag{3.17}$$

are respectively the projection operators for $J = l + \frac{1}{2}$ and $J = l - \frac{1}{2}$ because of the relation

$$\langle\vec{L}\cdot\vec{\sigma}\rangle = J(J + 1) - l(l + 1) - \tfrac{3}{4} = \begin{cases} l & \text{for } J = l + \frac{1}{2} \\ -l - 1 & \text{for } J = l - \frac{1}{2} \end{cases}$$

The incident plane wave can be written as

$$\exp(ikz)u_m = \sum_{l=0}^{\infty} (2l + 1)i^l j_l(kr)(P_+ + P_-)P_l(\cos\theta)u_m$$

As a result of scattering, the scattered wave has the asymptotic form

$$\frac{\exp(ikr)}{kr} \sum_{l=0} (2l + 1)[a_{l,l+1/2}P_+ + a_{l,l-1/2}P_-]P_l(\cos\theta)u_m \tag{3.18}$$

where

$$a_{l,J} = \frac{\exp(2i\delta_{lJ}) - 1}{2ik} = \frac{\exp(i\delta_{lJ})\sin\delta_{lJ}}{k} \tag{3.19}$$

On the other hand

$$\vec{\sigma}\cdot\vec{L}P_l(\cos\theta)u_m = -i\vec{\sigma}\cdot(\vec{x} \times \hat{n}_\theta)\frac{\sin\theta}{r}\frac{dP_l(\cos\theta)}{d(\cos\theta)}u_m$$
$$= i\vec{\sigma}\cdot\hat{n}\sin\theta\frac{dP_l(\cos\theta)}{d(\cos\theta)}u_m$$

where $\hat{n}_\theta$ is a unit vector along increasing θ. So by comparing (3.18) with (3.13) and (3.14) we obtain

$$f(\theta) = \frac{1}{k}\sum_{l=0} [(l + 1)a_{l,l+1/2} + la_{l,l-1/2}]\, P_l(\cos\theta)$$
$$g(\theta) = \frac{1}{k}\sum_{l=1} (a_{l,l+1/2} - a_{l,l-1/2})\sin\theta\frac{dP_l(\cos\theta)}{d(\cos\theta)} \tag{3.21}$$

In particular, if we consider s and p wave scattering only, we have

$$f(\theta) = \frac{1}{k}[a_s + (a_{p1/2} + 2a_{p3/2}) \cos \theta]$$
$$g(\theta) = \frac{1}{k}(a_{p3/2} - a_{p1/2}) \sin \theta \tag{3.22}$$

If $a_{p_{1/2}} = a_{p_{3/2}}$, we have no spin-flip scattering and we get the usual expression (i.e. $f(\theta) = (1/k) \sum_l (2l + 1)a_l P_l$) For various different $a_{lJ} \neq 0$, the differential cross section takes on the forms shown in Table 3.1. Note that odd powers of $\cos \theta$ appear in the angular distribution only if states of opposite parity interfere.

TABLE 3.1.

Angular Distribution for πp Scattering

	$\frac{d\sigma}{d\Omega}$
Pure $s_{1/2}$	isotropic
Pure $p_{1/2}$	isotropic
$s_{1/2}$ and $p_{1/2}$	isotropic + $\cos \theta$ term
Pure $p_{3/2}$	$1 + 3 \cos^2 \theta$

From (3.22) and the relations, $\vec{q}_f \cdot \vec{q}_i = q^2 \cos \theta$, $\vec{q}_i \times \vec{q}_f = q^2 \hat{n} \sin \theta$, the projection operator that will act on the initial Pauli spin function and select a scattered state of definite J and l is seen to be

$$p_{1/2} : \{\vec{q}_f \cdot \vec{q}_i - i\vec{\sigma} \cdot (\vec{q}_i \times \vec{q}_f)\}/3q^2$$
$$p_{3/2} : \{2\vec{q}_f \cdot \vec{q}_i + i\vec{\sigma}(\vec{q}_i \times \vec{q}_f)\}/3q^2 \tag{3.23}$$

For a given angular distribution from the πp scattering experiments, several ambiguities in the assignment of the phase shifts have been shown to exist.

(a) The trivial ambiguity: the signs of all the phase shifts at a given energy can be reversed (i.e. $f \to -f$, $g \to -g$). This ambiguity has been resolved (Orear, 1954) by looking at the interference between π^+p. nuclear scattering and Coulomb scattering (which we know to be repulsive). This is best done at small angles where destructive interference is found thereby indicating that the dominant part of the nuclear force is attractive.

(b) Minami ambiguity: Given a set of phase shifts $\delta_{s_{1/2}}$, $\delta_{p_{1/2}}$, etc., there exists another set obtained from the former by:

$$\begin{aligned} \delta_{s_{1/2}} &\rightleftarrows \delta_{p_{1/2}} \\ \delta_{p_{3/2}} &\rightleftarrows \delta_{d_{3/2}} \end{aligned} \tag{3.24}$$

such that the same $\frac{d\sigma}{d\Omega}$ results (Minami, 1954). This is the case because the angular distribution depends on the total J but not on the parity (or equivalently, not on l—cf. the Adair argument for the Λ^0 spin determination, Chapter 2). At low energies this ambiguity is not serious. For example, the energy dependence of $\delta_{p_{3/2}}$ at low energies is expected to vary as $q^{2l+1} = q^3$. If $\delta_{d_{3/2}}$ were the principal phase shift, one would expect the dependence to be as $q^{2l+1} = q^5$.

PROBLEM 2. Using the properties of the spherical harmonics, show explicitly that $\vec{\sigma}\cdot\hat{p}|J, w, M = \frac{1}{2}\rangle = |J, -w, M = \frac{1}{2}\rangle$ where the unit vector $\hat{p}$ specifies the angular variables of the spherical harmonics. Discuss the relevance of this relation to the Minami ambiguity. (This was first discussed by Dyson and Nambu. The existence of the ambiguity can also be readily demonstrated by using a representation in which helicity and J are diagonal (Jacob and Wick, 1959).)

(c) Fermi-Yang ambiguity: If we have a set of δ_{lj} that satisfy the observed angular distribution, then it can be shown that there exists a second set δ'_{lj} related to the first by:

$$\begin{gathered} \delta'_{s_{1/2}} = \delta_{s_{1/2}} \\ \delta'_{p_{3/2}} - \delta'_{p_{1/2}} = -(\delta_{p_{3/2}} - \delta_{p_{1/2}}) \\ \exp(2i\delta'_{p_{1/2}}) + 2\exp(2i\delta'_{p_{3/2}}) = \exp(2i\delta_{p_{1/2}}) + 2\exp(2i\delta_{p_{3/2}}) \end{gathered} \tag{3.25}$$

such that $|f'|^2 = |f|^2$ and $|g'|^2 = |g|^2$. In particular, if $\delta_{p_{3/2}}$ is the only large phase shift in the Fermi solution, it can be shown that both $\delta'_{p_{1/2}}$ and $\delta'_{p_{3/2}}$ must be large in the Yang solution (see p. 73 of Bethe and de Hoffman (1955)). Although the two sets of phase shifts give different final-state proton polarization, it has not as yet proved feasible to discriminate conclusively between the two solutions by recourse to the experimental results. However, on theoretical grounds "the spin-flip" dispersion relations show that only the Fermi solution (with large $p_{3/2}$ amplitude) is acceptable (Davidon and Goldberger, 1956).

Consider now the expression for the resulting proton polarization which we previously obtained:

$$P(\theta) = \langle\vec{\sigma}\rangle\cdot\hat{n} = \frac{2\,\mathrm{Im}(fg^*)}{|f|^2 + |g|^2} \tag{3.16}$$

We see that this polarization vanishes under any of the following conditions:

(1) All the phase shifts are small, in which case $\frac{\exp(2i\delta) - 1}{2i} = \exp(i\delta)\sin\delta \approx \delta$ and the numerator vanishes since both f and g are real.

(2) One state dominant. Then fg^* is real.

(3) There exists no l for which $\delta_{l,l+1/2} \neq \delta_{l,l-1/2}$ (i.e. there is no "spin-orbit coupling"); g vanishes.

(4) Scattering in the forward and backward direction. Then $\sin\theta = 0$ and $g = 0$. Also note that $P\left(\frac{\pi}{2}\right) \neq 0$ implies that states of opposite parity are interfering.

Let us look at some of the results of the π^+p scattering experiments. The dependences of σ_{total} on π kinetic energy and of $\frac{d\sigma}{d\Omega}$ on θ are shown in Fig. 3.2 and Fig. 3.3 It is useful here to remember that the maximum total cross section depends only on J. For $J = \frac{3}{2}$ of the πp system, $\sigma_{\max} = 8\pi\lambda\!\!\!^{-2}$. In general

$$\sigma_{\max} = 4\pi\lambda\!\!\!^{-2}\,\frac{(2J+1)}{(2s_1+1)(2s_2+1)} \tag{3.26}$$

The distributions are indicative of a $p_{3/2}$ resonance dominating the interaction at 190 Mev. The justification for this interpretation is as follows:

(a) σ_{total} reaches the maximum allowable σ for $J = \frac{3}{2}$ at ≈ 190 Mev.

(b) The angular distribution $\left(\frac{d\sigma}{d\Omega}\right) \propto 1 + 3\cos^2\theta$ at this peak. This is expected for a $p_{3/2}$ resonance.

(c) σ_{total} fits the simple resonance formula:

$$= \frac{2\pi\lambda\!\!\!^{-2}\Gamma^2}{(E-E_0)^2 + \frac{\Gamma^2}{4}}, \qquad \Gamma = \left[\frac{2(a/\lambda\!\!\!^{-})^3}{1 + \left(\frac{a}{\lambda\!\!\!^{-}}\right)^2}\right] b \tag{3.27}$$

(for $E_0 = 154$ Mev. center-of-mass energy, $a = 1.4\frac{\hbar}{\mu c}$, $b = 75$ Mev.) (see e.g. Gell-Mann and Watson, 1954).

(d) The $p_{3/2}$ resonance is suggested by the Chew-Low theory and by dispersion relations.

We will discuss low energy πp scattering in a more systematic manner in Chapter 9.

3.3. Intrinsic Parity; τθ Puzzle

(a) Photon. The charge current $j_\mu = (\vec{j}, i\rho)$ in the electromagnetic interaction $j_\mu A_\mu$ transforms under the parity operation as:

$$(\vec{j}, i\rho) \xrightarrow{P} (-\vec{j}, i\rho)$$

since $\vec{j}$ is essentially $\rho\vec{v}$.

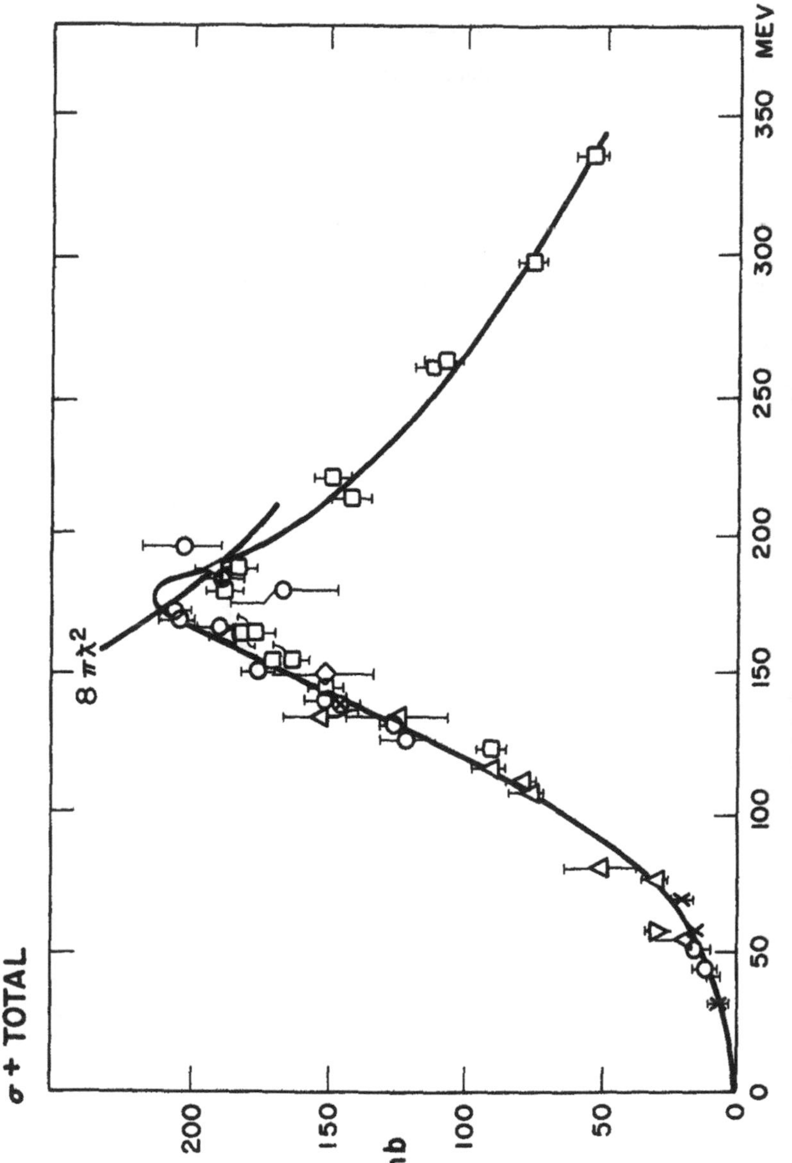

Figure 3.2. The total cross section for $\pi^+ + p \rightarrow \pi^+ + p$

If the interaction is to be invariant under P, we must have

$$\vec{A}(\vec{x}) \xrightarrow{P} -\vec{A}(-\vec{x})$$
$$A_4(\vec{x}) = iA_0(\vec{x}) \rightarrow A_4(-\vec{x})$$

Thus, the electric and magnetic field vectors transform as

$$\vec{E}(\vec{x}) = -\vec{\nabla}A_0 + \frac{\partial\vec{A}}{\partial t} \xrightarrow{P} -\vec{E}(-\vec{x})$$
$$\vec{B}(\vec{x}) = \vec{\nabla} \times \vec{A} \xrightarrow{P} +\vec{B}(-\vec{x}) \tag{3.28}$$

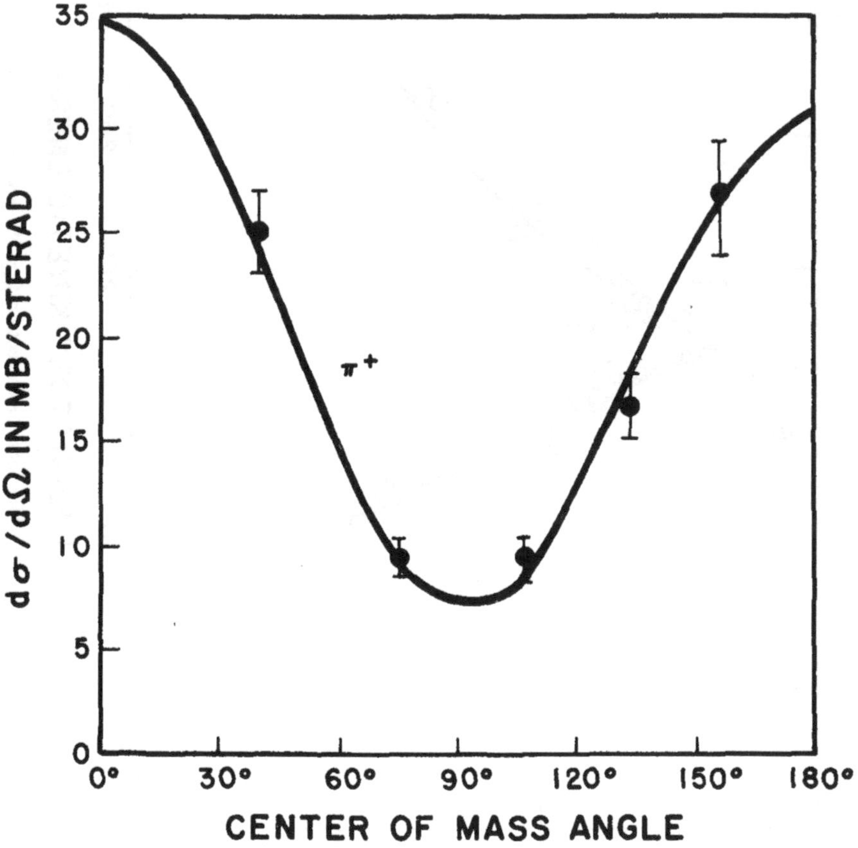

Figure 3.3. The angular distribution for $\pi^+ + p \rightarrow \pi^+ + p$ at 190 Mev

If we consider an $\vec{A}$ field of the form $\vec{A}(\vec{x}) \propto \vec{\varepsilon} f(\vec{x})$ where $f(x)$ is scalar, then we see that P causes $\vec{\varepsilon} \rightarrow -\vec{\varepsilon}$. This behavior of the polarization vector characterizes what we call the intrinsic parity of the field. The way in which one combines the polarization vector with a scalar function of space to obtain a photon eigenfunction of total (spin + orbital) angular momentum will be briefly treated in a later section when we discuss multipole expansions.

(*b*) Neutral pion. The π^0 undergoes a fast decay (lifetime $\tau \approx 2 \times 10^{-16}$ sec.) into two photons. The interaction is generally considered to be of the form $\pi^0 \rightarrow p + \bar{p} \rightarrow 2\gamma$ where $p + \bar{p}$ could, in fact, be any baryon-antibaryon pair. The following argument shows how we can in principle determine the scalar or pseudoscalar nature of the spin zero pion (Yang, 1950).

We consider the decay of a spin zero pion at rest into two photons. The final-state wave function must transform like a $J = 0$ system, and it must be linear in each of the two photon polarization vectors. We have two possibilities:

$$\vec{\varepsilon}_1 \cdot \vec{\varepsilon}_2 \qquad \text{even parity}$$

$$(\vec{\varepsilon}_1 \times \vec{\varepsilon}_2) \cdot \hat{k} \quad \text{odd parity} \tag{3.29}$$

($\hat{k}$ is the unit relative momentum vector of the two-photon system.) The first expression is seen to be invariant under P while the second changes sign. In general the final state is expected to be a linear combination of

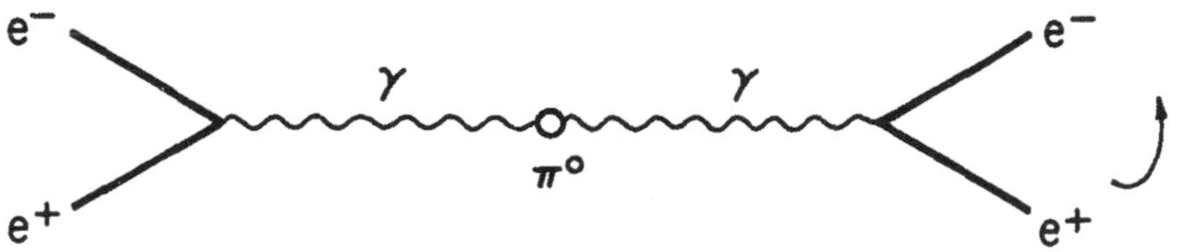

Figure 3.4. Experimental arrangement to determine the π° parity

these two forms. If parity is conserved in the decay, however, only one of the two forms is allowed, and we could determine, by measuring the relative orientations of $\vec{\varepsilon}_1$ and $\vec{\varepsilon}_2$, whether the π^0 is described by a scalar (even parity) or pseudoscalar (odd parity) field. For instance, if the π^0 parity is odd, $\vec{\varepsilon}_2$ cannot have any component in the direction of $\vec{\varepsilon}_1$. Since the pair-production cross section is known to be dependent on the photon polarization vector $\vec{\varepsilon}_1$, we may study the relative orientation of the two pair-production planes, as shown in Fig. 3.4. First measure coincidence rates. Then rotate the right half of the apparatus by 90° and measure coincidence rates again. One expects a 12% effect. This experiment is an exceedingly difficult one and has not yet been done. However, Kroll and Wada (1955) have shown that the above correlation effect persists for *internally* converted pairs in the process $\pi^0 \rightarrow e^- + e^+ + e^- + e^+$, i.e. the electron-positron pairs "remember" the polarization of the *virtual* intermediate photons. More quantitatively, let φ be the angle between the two planes of e^+e^- pairs. The theoretically expected distribution is $1 + \alpha \cos 2\varphi$ with $\alpha_{\text{th.}} = +0.47$ for a scalar π^0 and $\alpha_{\text{th.}} = -0.47$ for a pseudoscalar π^0. The observed angular distribution based on sixty events can be fitted with $\alpha_{\text{exp.}} \sim -0.7$, excluding an even π^0 parity by about 3 standard deviations (Plano *et al.* 1959).

We can also discuss the two-photon wave functions using the circular (rather than linear) polarization language. Consider the two-photon wave function with even parity. If we take $\hat{k}$ along the z axis, then we can write $\vec{\varepsilon}_1 \cdot \vec{\varepsilon}_2 = \frac{1}{2}[(\varepsilon_{1x} + i\varepsilon_{1y})(\varepsilon_{2x} - i\varepsilon_{2y}) + (\varepsilon_{1x} - i\varepsilon_{1y})(\varepsilon_{2x} + i\varepsilon_{2y})]$. The functions in the brackets refer to photon polarization states. The form is $\psi_{RR} + \psi_{LL}$. (Note $\vec{k}_1$ is along the z axis, $\vec{k}_2$ is along the $-z$ axis.) The photon wave function for the odd parity pion can be shown in a similar fashion to have the form $(\vec{\varepsilon}_1 \times \vec{\varepsilon}_2)\cdot\hat{k} = \psi_{RR} - \psi_{LL}$. Note that, since under space inversion, $R \rightleftarrows L$, $\psi_{RR} + \psi_{LL}$ is even and $\psi_{RR} - \psi_{LL}$ is odd, as expected. In either parity case, if one of the photons is observed to be right- (left-) handed, the other photon must also be right- (left-) handed, as is evident from angular momentum conservation.

(c) Charged pion. The following argument for the determination of the parity of the π^- was originally due to Ferretti (1946). The original experiments were done by Panofsky, Aamodt, and Hadley (1951,) and Chinowsky and Steinberger (1954).

The π^- captured at rest in hydrogen exhibits the following two capture modes:

$$\pi^- + p \to n + \pi^0$$

$$\pi^- + p \to n + \gamma$$

The branching ratio $(n\pi^0)/(n\gamma)$ is termed the Panofsky ratio and currently has the experimental value 1.5–1.6 (see e.g. Samios, 1960). (This number has been a function of time. Panofsky's original value was 0.9.) When π^- are stopped in deuterium, we have an additional nucleon present and the possible final states are

$$\begin{aligned} \pi^- + d &\to n + n \\ &\to n + n + \gamma \\ &\to n + n + \pi^0 \end{aligned}$$

The third mode has not been observed, but it should be comparable to the first two. Panofsky *et al.* studied the branching ratios of the various modes. Although their observation of the $n + n$ mode was indirect, Steinberger and Chinowsky later observed it directly and confirmed the original work. The occurrence of this $n + n$ mode is used in the following argument as proof of the odd π^--nucleon relative parity. (We assume in the conventional manner that the proton and the neutron have the same parity. We will discuss the arbitrariness of this in Chapter 8.)

Let us assume that the π^- is captured from an atomic s state of deuterium. This assumption has been shown to be justified by studies of mesonic X-rays and also by an extrapolation to low energy of the observed cross section of the charge symmetric reaction $\pi^+ + d \to p + p$ (Brueckner, Serber, and Watson, 1951), and more recently by the theoretical argument of Day, Snow, and Sucher (1959) and by the direct experimental determination of the π^- cascade time (Fields *et al.* 1960; Doede *et al.* 1963). Thus we have initially a spin zero particle in an s state of a spin 1 nucleus (the deuteron is in a $^3S_1$ state and has even parity). Angular momentum conservation in the interaction tells us that the only possible final states are $^3S_1$, $^3P_1$, $^1P_1$, $^3D_1$. However, since we have two identical Fermions in the final state, the necessary anti-symmetrization of the wave function implies that a triplet (singlet) spin state must combine with an odd (even) orbital state, and eliminates all but the $^3P_1$ state, which has odd parity. So parity conservation requires

$$(\pi^- \text{ parity}) \times (\text{even orbital parity}) \times (\text{even } d \text{ parity}) = \text{odd parity} \quad (3.30)$$

Thus the π^- must have odd intrinsic parity.

As for the π^+ parity, field theory requires that, if the π^- is odd, the π^+ must necessarily be odd, as we shall see later.

(d) $\tau - \theta$ puzzle. The following two heavy meson decays are observed within the indicated experimental uncertainties; the masses and lifetimes of the decaying particles are found to be identical for the two decays. (Birge *et al.* 1956; Fitch and Motley 1956; Alvarez *et al.* 1956; Orear, Harris, and Taylor 1956b).

TABLE 3.2

Masses and Lifetimes of θ and τ

Decay	Mass of decaying particle	Lifetime of decaying particle
$\theta^+ \to \pi^+ + \pi^0$	$(966.7 \pm 2.0)m_e$	$(1.21 \pm .02) \times 10^{-8}$ sec.
$\tau^+ \to 2\pi^+ + \pi^-$	$(966.3 \pm 2.0)m_e$	$(1.19 \pm .05) \times 10^{-8}$ sec.

The θ^+ decays appear to take place for about 20% of the charge +1 heavy meson decays, while the τ^+ occurs about 5% of the time. The abundance ratios for θ vs. τ are found not to depend on the production origin of the particles. Thus the reasonable conclusion is that the θ^+ snd τ^+ are merely two different decay modes of the same parent K particle. A difficulty exists, however. If parity is assumed to be a good quantum number in the decay, the following argument due to Dalitz (1953, 1954) shows that the θ and τ must have opposite intrinsic parities.

Consider the decay $\theta^+ \to \pi^+ + \pi^0$. If the θ^+ has spin S, then the orbital angular momentum of the final state must be S and the final state parity is $(-1)^S(-1)^2 = (-1)^S$. Thus the possible spin-parity assignments are $0^+, 1^-, 2^+, 3^-, \ldots$.

Now consider the decay $\tau^+ \to 2\pi^+ + \pi^0$. We call the relative orbital angular momentum of the $2\pi^+$ system, l, and of the orbital angular momentum of the π^- relative to the center of mass of the $2\pi^+$ system, L (see Fig. 3.5). For spin zero τ, $l = L$ and the final state parity is $(-1)^{l+L}$

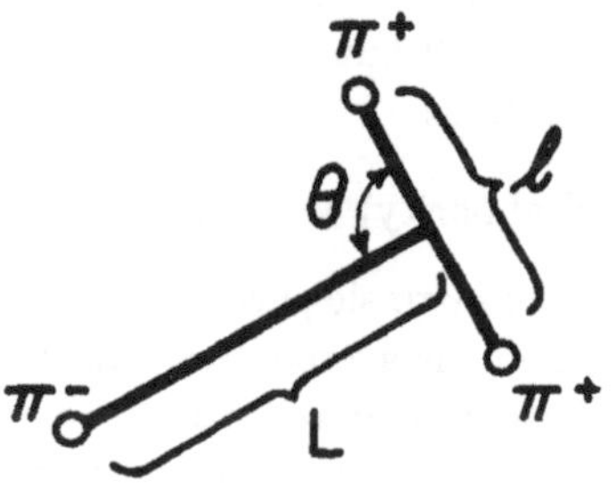

Figure 3.5 Configuration in the τ^+ decay

$(-1)^3 = -1$. So if parity is conserved, and if θ and τ have spin zero, θ and τ cannot be the same particle.

The arguments against the τ-θ identity for the K spin $\neq 0$ proceed along the following lines. Dalitz, in his analysis, showed that, if we assume the spin and parity of the τ^+ to be the same as for the θ^+, the following three decay patterns for the τ are forbidden:

(a) π^- has the maximum possible energy. Then the $2\pi^+$'s are relatively at rest, hence $l = 0$ and $S = L$. $w_\tau = -(-1)^S \neq w_\theta$,

(b) π^- has a zero kinetic energy. Then $L = 0$, and $l = S$. $w_\tau = -(-1)^S \neq w_\theta$. Note here the interesting point that since the $2\pi^+$ system is a system of two identical bosons, its wave function must be symmetric with respect to their interchange. Thus the K spin must be even if this pattern occurs (this argument is independent of parity).

(c) The three pions are collinear. Since the system orientation is completely specified by pointing out one direction in space, the wave function must transform like a spherical harmonic Y_{jm} with $j = S$. Thus $w_\tau = -(-1)^S \neq w_\theta$.

The experimental π^- energy and angular distributions obtained by a number of groups (e.g. Baldo-Ceolin *et al.*, 1957; Biswas *et al.*, 1956; Orear, Harris, and Taylor (1956a) are shown in Fig. 3.6 and 3.7. We note that both "slow" π^-'s and fast π^-'s are fully allowed; in addition, configurations in which the three pions are collinear ($\cos\theta = 1$) are also fully allowed. Of course, we must ask more quantitative questions: How slow is

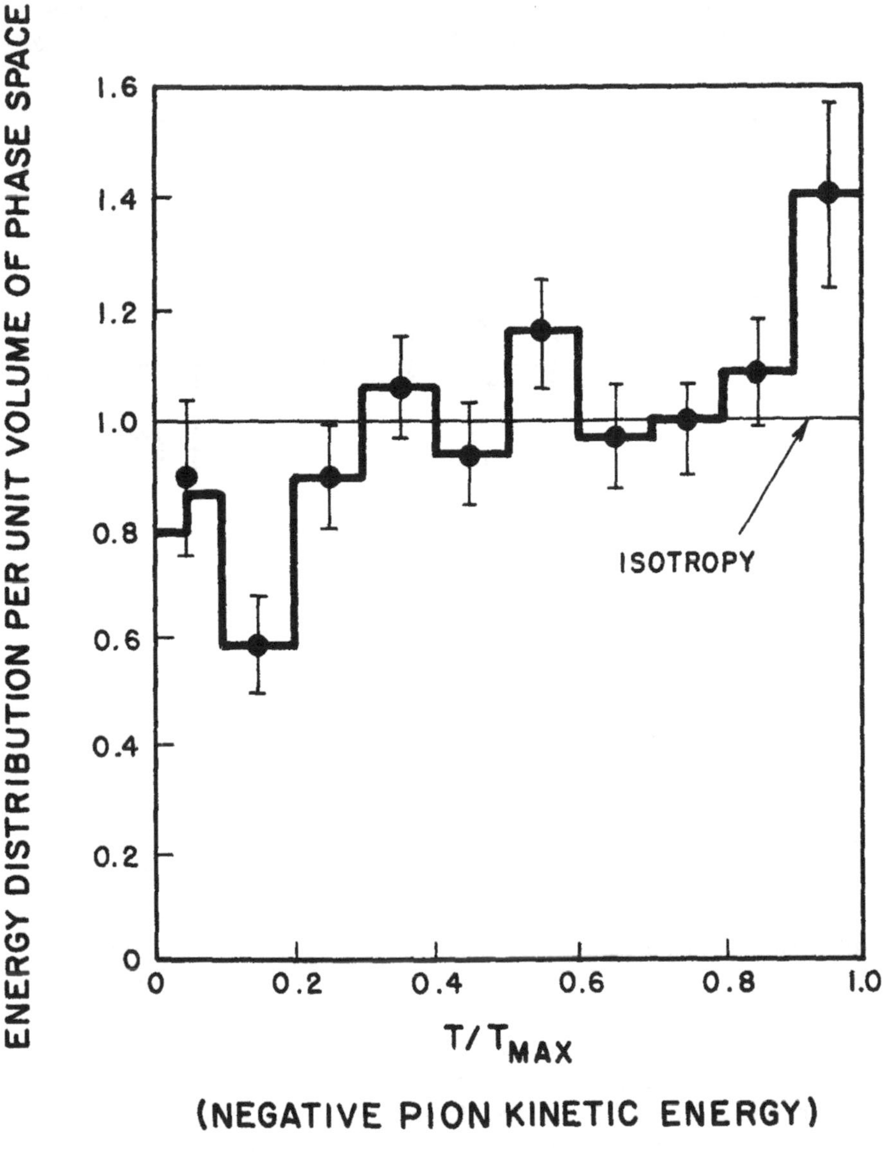

Figure 3.6 The energy distribution (per unit volume of phase space) of π^- in the decay of τ mesons based on 892 events compiled by Dalitz (1957). The distribution must vanish at $T/T_{max} = 0$ and 1 if θ and τ are the same particle and if parity is conserved.

"slow?" How collinear is "collinear?" Under "reasonable" assumptions (which can be justified since the de Broglie wavelength of the decay pion is relatively large compared to what we believe to be the range of the τ decay interaction), we can derive more-or-less model-independent expressions for the energy and angular distributions for each spin-parity combination. Experimental data indicate that the most probable spin-parity

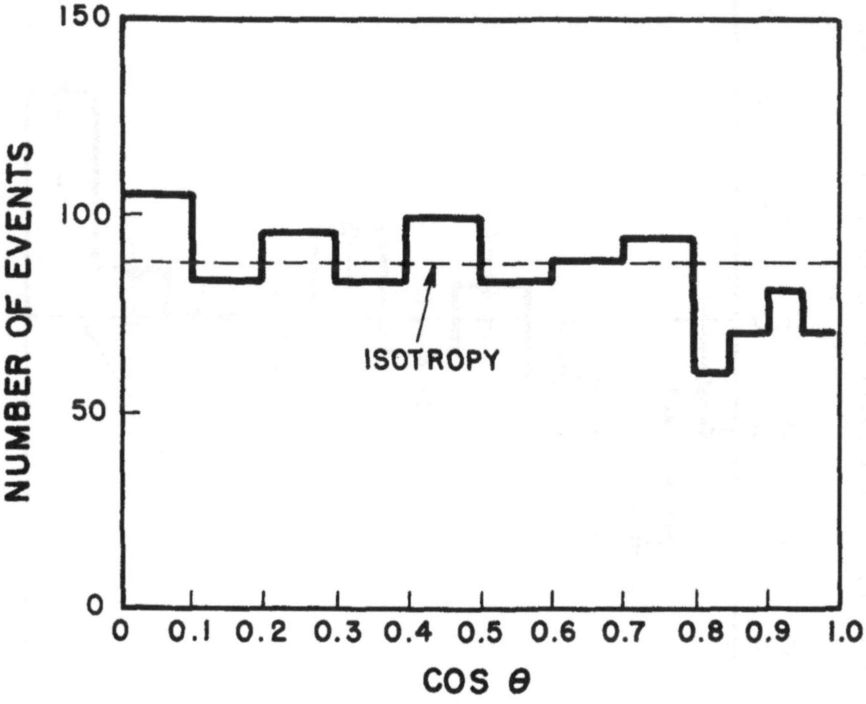

Figure 3.7 The angular distribution in decay of τ mesons. The distribution must vanish at $\cos\theta = 1$ if θ and τ are the same particle and if parity is conserved.

assignment for the τ is 0^- or possibly 2^-. The probability that the actual distribution should have been found for a 1^- particle turns out to be $< 10^{-30}$, and similar conclusions can be obtained for higher spin values with parity $(-1)^S$.

Thus the most likely conclusion from the τ decay *was* that the τ and θ had to be different particles despite the fact that they had identical masses and lifetimes. This led to the well-known "$\tau\theta$ puzzle" which was widely discussed in 1955-1956. After a few unsuccessful attempts it was suggested that parity might not be conserved in the decay interactions of K mesons. If parity is not conserved, the argument presented previously falls down, and the same parent particle can decay both into two pions and three pions.

Yet it seemed unwise to give up the time-honored, aesthetically appealing "law" of parity conservation on account of just one puzzling aspect of strange particle interactions. For this reason, Lee and Yang (1956b) undertook a systematic investigation of the validity of parity conservation. They found that parity is conserved to a high degree of accuracy in electromagnetic and strong interactions, but that the principle of parity conservation was "only an extrapolated hypothesis unsupported by experimental evidence" for those weak interactions responsible for β decay as well as for the decay of unstable particles (with the exception of π^0 and Σ^0). In addition, they suggested a number of experiments that are really sensitive to the question of whether or not parity is conserved. As is well known, the subsequent experiments have brilliantly confirmed the idea that weak interactions *in general* do not respect the principle of parity conservation. If there is any "puzzle" left at all, it centers around the following question of Pauli: "Why does the Lord still appear to be right-left symmetric when He expresses himself strongly?"

(e) K-Λ parity. Obviously we can learn nothing about the intrinsic parity of the K meson from its decays. If, however, parity is conserved in the production process

$$\pi^- + p \rightarrow K^0 + \Lambda^0$$

we can meaningfully inquire as to the relative K-Λ parity. One usually assumes a parity *convention* for the Λ so that $w_\Lambda = w_p$. The justification of this arises from the fact that (as we shall discuss later on) parity is not conserved in the Λ decay either. Thus the *relative* K-Λ parity is the *only* meaningful quantity.

If the K meson is pseudoscalar, i.e. if the relative K-Λ parity is odd, the transitions are similar to those in π^+-p scattering

$$s_{1/2} \rightarrow s_{1/2}$$
$$p_{1/2} \rightarrow p_{1/2} \qquad \text{etc.}$$

If the K is scalar, then we have

$$s_{1/2} \rightarrow p_{1/2}$$
$$p_{3/2} \rightarrow d_{3/2}$$

In this case, we see that l must change.

For the two cases the transition matrix has the forms (in the Pauli spin formalism)

$$\text{pseudoscalar } K\text{:} \quad M = f + ig(\vec{\sigma}\cdot\hat{n}) \qquad \hat{n} = \frac{\vec{p}_i \times \vec{p}_f}{|\vec{p}_i \times \vec{p}_f|}$$

$$\text{scalar } K\text{:} \quad M = \vec{\sigma}\cdot(f'\hat{p}_i + g'\hat{p}_f) \tag{3.31}$$

For the scalar K, M looks like a pseudoscalar here. What does not appear, however, are the appropriate wave functions for π and K, which behave oppositely under parity. The odd intrinsic parity of the π makes up for the apparent pseudoscalarity of M.

In both cases the resultant polarization is still along $\hat{n}$. If parity is not conserved, however, M will be a linear combination of 1, $\vec{\sigma}\cdot\hat{n}$, $\vec{\sigma}\cdot\hat{p}_i$, $\vec{\sigma}\cdot\hat{p}_f$, and one expects a polarization component *in* the production plane, i.e. the plane in which both $\vec{p}_i$ and $\vec{p}_f$ lie. Experimentally (Crawford *et al.*, 1958a), it has been found that $\langle\vec{\sigma}\rangle_{\text{plane}} \approx 0$ (within 15%).

The reaction $K^- + He^4 \rightarrow {}_\Lambda He^4 + \pi^-$ (Dalitz, 1957) provides a convenient means for determining the K-Λ relative parity. The experimental (Ammar *et al.*, 1959a) and theoretical work (Dalitz and Liu, 1959) on the ${}_\Lambda H^4$ hyperfragment which is charge-symmetric to ${}_\Lambda He^4$ indicate that the spin is probably 0 (implying that the Λ-nucleon force is stronger for the antiparallel spin alignment). Now, if all particles in the reaction are spinless, then the initial and final orbital angular momenta are the same. Since the Λ (which is even by convention) is bound to He^3 in an s state, ${}_\Lambda He^4$ has even parity. We see, therefore, that the reaction can go only if the K is odd. Experimentally the reaction is observed to take place (Block *et al.*, 1959). A weakness in this argument in favor of the pseudoscalarity of the K is that one must be certain that the ${}_\Lambda He^4$ cannot be formed in an excited state, thereby contradicting one of the initial assumptions of the argument. To the present it has not been possible to experimentally determine the existence of bound excited states of this hyperfragment.

(*Note added in proof*: We shall say more about the spin of ${}_\Lambda H^4$, ${}_\Lambda He^4$ in Section 6.)

3.4. Multipole Expansion and Photoproduction

A photon wave function with the quantization axis chosen along the beam direction can be written as

$$\vec{A}(\vec{x}, t) = \sum_{l=0}^{\infty} f_l(|\vec{x}|, t) \sum_{J=l-1}^{l+1} C_{l1}(J, \pm 1; 0, \pm 1)\vec{Y}^{\pm 1}_{J,l,1} \tag{3.32}$$

The vector spherical harmonics

$$\vec{Y}^{M}_{J,l,1} = \sum_{m=-l}^{l} C_{l1}(J, M; m, m')Y_{lm}(\theta, \phi)\chi_{m'} \tag{3.33}$$

are the eigenfunctions of the total angular momentum operator for the photon, $\vec{J} = \vec{L} + \vec{S}$, where $\vec{L}$ is the orbital angular momentum and $\vec{S}$ is the photon spin. The spin eigenfunctions $\chi_1 = -\frac{1}{\sqrt{2}}(\vec{e}_x + i\vec{e}_y)$ and $\chi_{-1} = \frac{1}{\sqrt{2}}(\vec{e}_x - i\vec{e}_y)$ are expressed in terms of the usual spherical basis. Note that a plane wave, by virtue of the mixing of the various terms in an expansion like (3.32) above, is neither an eigenstate of parity nor of J^2. Furthermore, for each eigenvalue J, there exist three $\vec{Y}^M_{J,l,1}$ with l values $l = J - 1, J, J + 1$.

One usually classifies electromagnetic waves according to their various J and l components as in Table 3.3

TABLE 3.3

Parity and Angular Momenta of Multipoles

J		l	Parity	Name
0	Impossible			
1	Dipole	$J \pm 1 = 0, 2$	$-$	$E\,1$
		$J = 1$	$+$	$M\,1$
2	Quadrupole	$J \pm 1 = 1, 3$	$+$	$E\,2$
		$J = 2$	$-$	$M\,2$

The l value, of course, determines the parity of the term. Thus the parity of the electric $2J$ pole is $-(-1)^{J\pm 1} = +(-1)^J$ and the parity of the magnetic $2J$ pole is $-(-1)^J$ where the extra minus signs have been supplied because of the odd intrinsic parity of the photon. The origin of the terminology is understood if we expand $\frac{\partial \vec{A}}{\partial t}$ and $\vec{\nabla} \times \vec{A}$ in the classical multipole fashion (see Appendix B of Blatt and Weisskopf, 1952).

If the wavelength of the emitted radiation is much larger than the dimensions of the emitting system, then only electric dipole emission ($l = 0$) is important. Since $E1$ has odd parity, then, if P is a good quantum number in the interaction, the most "favored" transition takes place between states of opposite parity. Historically, Laporte (1924) discovered the empirical rule that the energy levels of Fe atoms can be classified into "gestrichene" and "ungestrichene" types such that transitions take place only between levels of different types. Then Wigner (1927) showed that this rule was a consequence of the invariance of the electromagnetic interactions under P. The very fact that the parity selection rule works implies that parity is conserved in atomic transitions to an accuracy of $(R/\lambda)^2$ in the transition probability where R is the characteristic radius of the system. This means that the amplitude of the parity violating contribution is smaller than the parity conserving amplitude by a factor of R/λ. By considering transitions with photons of long wavelengths, we find $(R/\lambda)^2 \approx 10^{-8}$.

Consider the meson photoproduction processes

$$\gamma + N \rightarrow \pi + N$$
$$\gamma + N \rightarrow K + \Lambda \text{ (or } \Sigma)$$

where N represents either neutron or proton. If the mesons are pseudoscalar, we see that the relationships shown in Table 3.4 hold between the initial and final angular momenta and the angular distributions.

TABLE 3.4

Angular Distributions of Photoproduction of Pseudoscalar Mesons

Initial State	J_{total} (Initial)	w parity	Final State	Angular distribution (for unpolarized γ's)
$E1$	$\frac{1}{2}$	$-$	$s_{1/2}$	Constant
	$\frac{3}{2}$	$-$	$d_{3/2}$	$2 + 3\sin^2\theta$
$M1$	$\frac{1}{2}$	$+$	$p_{1/2}$	Constant
	$\frac{3}{2}$	$+$	$p_{3/2}$	$2 + 3\sin^2\theta$

We see here an obvious ambiguity that will arise in the analysis of photoproduction data. Both the $E1 \rightarrow d_{3/2}$ and the $M1 \rightarrow p_{3/2}$ transitions lead to the same angular distributions. Similarly the $E1 \rightarrow s_{1/2}$ and $M1 \rightarrow p_{1/2}$ are ambiguous. This ambiguity is the analogue of the Minami ambiguity in π-p elastic scattering (i.e. the distribution depends on J_γ, J_{total} but not on the parity). Such ambiguities are particularly serious at high energies where new resonances are being discovered (e.g. there is a peak in photoproduction at $E_\gamma^{(lab.)} \sim 800$ Mev.). We shall come back to this later.

The various angular distributions shown in Table 3.4 can be derived using the vector spherical harmonics (Feld, 1953). This is sometimes rather messy, however. A neater way (Brueckner and Watson, 1952) is to construct the matrix elements in terms of the quantities: $\vec{\sigma}$, $\vec{\varepsilon}$ (polarization vector for the photon), $\vec{k}$ (photon momentum), $\vec{q}$ (meson momentum). For example, consider the $E1 \rightarrow s_{1/2}$ transition for a pseudoscalar pion. We can write $M \propto \vec{\sigma}\cdot\vec{\varepsilon}$. This expression looks like a pseudoscalar but note that it is compensated by the odd intrinsic parity of the pion. It contains no $\vec{q}$, which is reasonable for an s-wave pion. Its effect when acting on an initial state wave function can be made more transparent if we write it in the following form

$$\vec{\sigma}\cdot\vec{\varepsilon} = \tfrac{1}{2}[(\sigma_x - i\sigma_y)(\varepsilon_x + i\varepsilon_y) + (\sigma_x + i\sigma_y)(\varepsilon_x - i\varepsilon_y)]$$

where now the brackets $(\sigma_x \pm i\sigma_y)$ have the form of the familiar "raising" (i.e., $\beta \rightarrow \alpha$) and "lowering" (i.e., $\alpha \rightarrow \beta$) operators. Note that when the nucleon spin is transformed from $\beta \rightarrow \alpha$, a photon with $s_z = 1$ is concurrently destroyed. Thus the photon gives up a projected component of angular momentum $m = 1$ to the nucleon. This is quite reasonable. Unlike the π-N scattering case, the s state is here associated with a spin flip.

The matrix for the $M1 \to p_{1/2}, p_{3/2}$ transitions can be obtained from the corresponding matrix (projection operators) for p wave πp scattering discussed earlier in Section 2 by substituting $\vec{k} \times \vec{\varepsilon}$ for $\vec{p}_i$. Thus

$$\vec{q}_f \cdot \vec{q}_i + i\vec{\sigma} \cdot (\vec{q}_f \times \vec{q}_i) \to \vec{q} \cdot (\vec{k} \times \vec{\varepsilon}) + i(\vec{\sigma} \cdot \vec{k}\vec{q} \cdot \vec{\varepsilon} - \vec{\sigma} \cdot \vec{\varepsilon}\vec{q} \cdot \vec{k}) \text{ for } J = \tfrac{1}{2}$$

$$2\vec{q}_f \cdot \vec{q}_i - i\vec{\sigma} \cdot (\vec{q}_f \times \vec{q}_i) \to 2\vec{q} \cdot (\vec{k} \times \vec{\varepsilon}) - i(\vec{\sigma} \cdot \vec{k}\vec{q} \cdot \vec{\varepsilon} - \vec{\sigma} \cdot \vec{\varepsilon}\vec{q} \cdot \vec{k}) \text{ for } J = \tfrac{3}{2} \tag{3.34}$$

The reader may verify that the eigenvalues of the operator

$$\left[\frac{\vec{\sigma}}{2} + \vec{L}_f\right]^2 \text{ where } L_f = -i(\vec{q}_f \times \vec{\nabla}_{q_f})$$

applied to the scattering matrix or to the photoproduction matrix are $J(J + 1)$ as they should be. $\vec{k} \times \vec{\varepsilon}$ appears because it is the leading "magnetic" term corresponding to $M1$ in the multipole expansion.

Let us calculate the angular distribution for the $M1 \to p_{3/2}$ transition.

$$\frac{d\sigma}{d\Omega} = \tfrac{1}{2}\mathrm{Tr}(MM^{\dagger}) \propto 4(\vec{q} \cdot (\vec{k} \times \vec{\varepsilon}))^2 + (\vec{k}\vec{q} \cdot \vec{\varepsilon} - \vec{\varepsilon}\vec{q} \cdot \vec{k})^2$$
$$= k^2q^2(1 + 3 \sin^2 \theta \sin^2 \varphi) \tag{3.35}$$

where we have chosen our spherical coordinates in such a way that the polarization vector $\vec{\varepsilon}$ lies along the x axis; $\vec{k}$ is along the $+z$ direction as usual; and θ and φ describe the direction of $\vec{q}$. If the γ ray is unpolarized, we must average over the initial polarization direction. Since $\overline{\sin^2 \varphi} = \frac{1}{2}$, we have the angular distribution $2 + 3 \sin^2 \theta$ as given in Table 3.4. If we calculate the angular distribution for the $E1 \to d_{3/2}$ transition, we obtain $1 + 3 \sin^2 \theta \cos^2 \varphi$ before we take the φ average. So the ambiguity mentioned earlier could be readily resolved if we used linearly polarized γ rays. In practice, however, an ambiguity of this kind may be more readily resolved by looking at possible $p_{3/2} - d_{3/2}$ interference effects in the recoil proton polarization rather than by using polarized γ rays (Sakurai, 1958b; Stein, 1959).

If the pion were scalar, we would have to interchange electric and magnetic multipoles (e.g., $E1 \rightleftarrows M1$ etc.). This implies that in the production matrix we must interchange $\vec{\varepsilon}$ and $\vec{k} \times \vec{\varepsilon}$.

What can we say about photopion production without using any detailed theory? Intuitively we expect that the electric dipole associated with the virtual process $p \rightleftarrows n + \pi^+$ is $\sqrt{2}M_N/\mu_\pi \sim 9$ ($\sqrt{2}$ comes from considerations of isospin) times larger than the electric dipole associated with

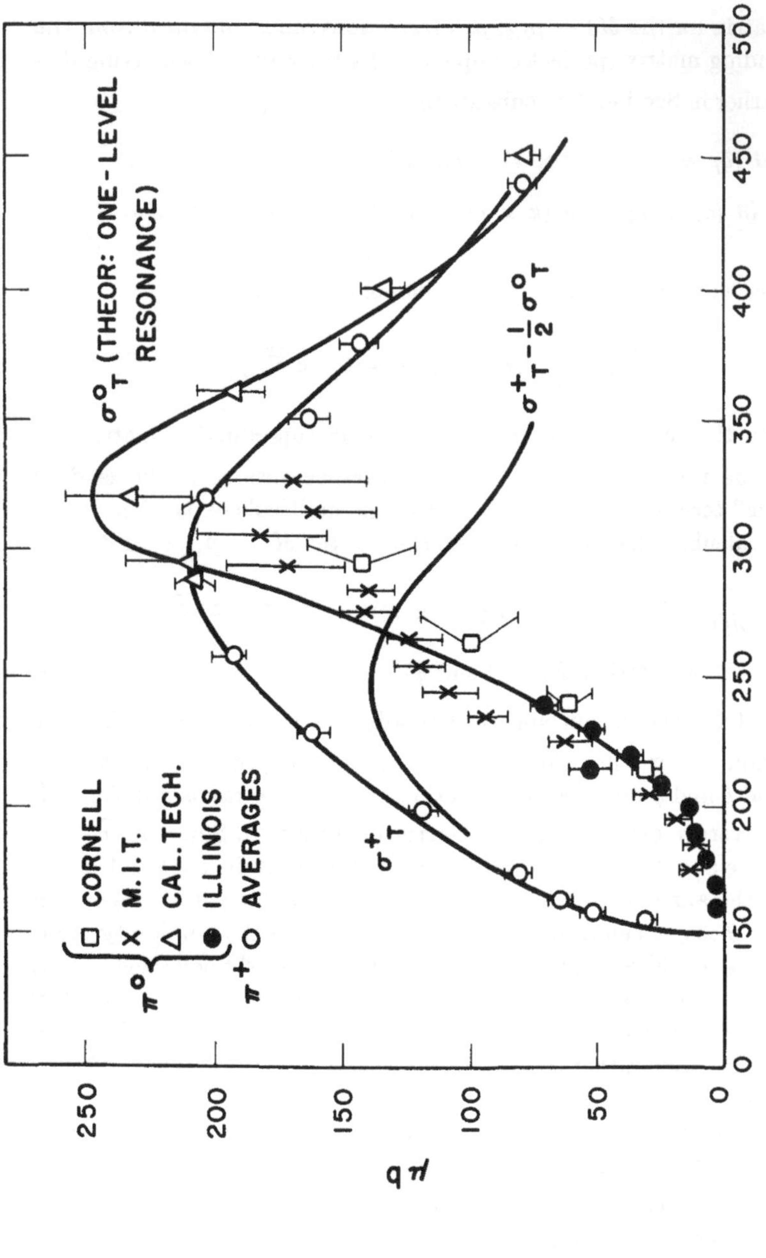

Figure 3.8. Total cross sections for photopion production

$p \rightleftarrows p + \pi^0$. So we may guess on the basis of the pseudoscalarity of the pion that a large s state interaction is present for the π^+ production but not for the π^0 production. Indeed for the reaction $\gamma + p \rightarrow \pi^+ + n$ one observes an isotropic angular distribution with energy dependence given by the phase space term $\left(q^2 \dfrac{dq}{dE} \propto q\right)$. This, of course, is characteristic of $E1 \rightarrow s_{1/2}$. For $\gamma + p \rightarrow \pi^0 + p$, on the other hand, there is no analogous s state production, and the threshold dependence seems more characteristic of p state production as shown in Fig. 3.8.

Both reactions exhibit resonant behavior at $E_\gamma \approx 330$ Mev whichroughly correspond to $E_\pi \approx 190$ Mev in the π-p scattering case. The angular distribution at the resonance peak is $2 + 3\sin^2\theta$ for the π^0 case, which we have seen is what one expects for $M1 \rightarrow p_{3/2}$. As we shall discuss later in Chapter 9, the same $T = \frac{3}{2}$, $J = \frac{3}{2}$ state (the so-called 3-3 state) of the π-N system is responsible for both the resonance in π^+-p scattering and the resonance in photopion production. Historically speaking, the existence of this resonance of the π-N system was first conjectured from the photoproduction experiments by Fujimoto and Miyazawa (1950) and by Brueckner and Case (1951). Later scattering experiments at Chicago confirmed the correctness of the idea (Anderson *et al.*, 1953)

We now turn our attention to Σ^0 decay, which is also a manifestation of both the electromagnetic and strong interactions. The decay process $\Sigma^0 \rightarrow \Lambda^0 + \gamma$ is expected to be a parity conserving, fast ($\sim 10^{-20}$ sec.) reaction. In order to construct the transition matrix for the process, we must know the relative Λ-Σ parity.

(a) If Λ and Σ transform in the same way

$$u_\Lambda \xrightarrow{P} \eta u_\Lambda$$

$$u_\Sigma \xrightarrow{P} \eta u_\Sigma$$

(i.e., $\Lambda\,\Sigma$ parity even), then the matrix element for the decay has the form

$$M = \mu_{\Lambda\Sigma}^{(\text{mag})} u_\Lambda^\dagger \vec{\sigma} \cdot (\vec{k} \times \vec{\varepsilon}) u_\Sigma \tag{3.36}$$

We have a magnetic dipole transition. $\mu_{\Lambda\Sigma}^{(\text{mag})}$ is called the "transition magnetic moment." It essentially plays the role of a coupling constant here.

(b) If Λ and Σ transform oppositely (i.e., Λ-Σ parity odd),

$$u_\Lambda \xrightarrow{P} \eta u_\Lambda$$

$$u_\Sigma \xrightarrow{P} -\eta u_\Sigma$$

then we have an electric dipole transition and the matrix element has the form

$$M = \mu_{\Lambda\Sigma}^{(\text{el.})} u_\Lambda^\dagger \vec{\sigma} \cdot \vec{\varepsilon} u_\Sigma \tag{3.37}$$

Note that the static dipole moment of a particle (a diagonal element of this matrix) must vanish if parity is conserved. In $\mu^{(\text{el.})}u^{\dagger}\vec{\sigma}\cdot\vec{\varepsilon}u$, the two u's transform in the same way under P, while $\vec{\sigma}\cdot\vec{\varepsilon}$ changes sign.

Coming back to Σ^0 decay, we may naturally ask whether we can distinguish the even $\Lambda\Sigma$ parity case (3.36) from the odd $\Lambda\Sigma$ parity case (3.37). Let us first recall that the product $\vec{\sigma}\cdot\hat{n}$ can be regarded as a rotation operator of the hyperon spin about the direction $\hat{n}$ by 180° [cf. Eq. (2.15)]. We then see that as a result of the γ emission the hyperon spin rotates itself by 180° about the direction $\vec{k}\times\vec{\varepsilon}$ in the even parity case; in contrast, for odd $\Lambda\Sigma$ parity, the hyperon spin rotates itself by 180° about the direction $\vec{\varepsilon}$. If we could measure the spin direction of the initial (Σ) and the final (Λ) hyperon, and if we knew the polarization (as well as the momentum) direction of the emitted γ ray, we could determine the $\Lambda\Sigma$ parity. Now the Σ^0 spin must be in the direction normal to the plane of the reaction responsible for the production of the Σ^0 hyperon (e.g., $\pi^- + p \rightarrow \Sigma^0 + K^0$). The Λ spin direction can be determined from the subsequent Λ decay, as will be shown in the next section. The direction of $\vec{\varepsilon}$ is very difficult to determine. Fortunately it can be shown that in the process

$$\Sigma^0 \rightarrow \Lambda + e^+ + e^-$$

the polarization vector of the intermediate *virtual* γ ray is strongly correlated with the plane of the e^+e^- pair just as in the π^0 decay case ($\pi^0 \rightarrow e^+ + e^- + e^+ + e^-$). Thus, by studying the correlation between the plane of the e^+e^- pair and the polarization direction of the Λ hyperon, it is, in principle, possible to determine the parity (Byers and Burkhardt, 1961; Snow and Sucher, 1960; Michel and Rouhaninejad, 1961; Valuev and Geshkenbein, 1960). It is also possible to determine the $\Lambda\Sigma$ parity by studying the frequency and the invariant mass distribution of the e^+e^- pairs (Feinberg, 1958a; Feldman and Fulton, 1958; Dalitz, 1962). The fraction of e^+e^- pairs with $[(E_+ + E_-)^2 - (\vec{p}_+ + \vec{p}_-)^2]^{1/2} > 10\ m_e$ (comprising about 40% of all pairs) can be shown to be about 20% greater for odd $\Lambda\Sigma$ parity ($E1$ transition) than for even $\Lambda\Sigma$ parity ($M1$ transition). Qualitatively speaking, this follows from the requirement that the matrix element must vanish as the momentum of the virtual photon $\vec{k} = \vec{p}_+ + \vec{p}_-$ goes to zero for the even parity case but not for the odd parity case. Note that this second method does not require a source of polarized Σ^0's.

3.5. Hyperon Decay

Hyperon decay is a particularly simple phenomenon that can be used to illustrate parity nonconservation. Here we see a *naturally occurring* process that serves to define the sense of right and left in an unambiguous manner. No coils, no magnetized iron, no electric current, etc. are needed.

Λ hyperons are produced in the "strong" reaction

$$\pi^- + p \rightarrow \Lambda^0 + K^0 \tag{3.38}$$

The Λ^0 exhibits the following two decay modes

$$\Lambda^0 \to \begin{cases} p + \pi^- \\ n + \pi^0 \end{cases} \tag{3.39}$$

with a lifetime of about 3×10^{-10} sec. This long lifetime cannot be accounted for by phase space limitations, and we conclude that this is an example of a "weak" process (in contrast to the decays of Σ^0 and π^0 discussed earlier).

Let us consider the production process first. The kinematics of each event can be completely characterized by two vectors, the initial pion momentum, $\vec{p}_{\pi\text{in}}$, and the Λ momentum, $\vec{p}_\Lambda$. If the Λ particle is produced in a nonforward direction, given the two vectors, $\vec{p}_{\pi\text{in}}$ and $\vec{p}_\Lambda$, we can define an axial vector $\vec{p}_{\pi\text{in}} \times \vec{p}_\Lambda$ just as the x axis and the y axis can define the z axis in the right-handed coordinate system.

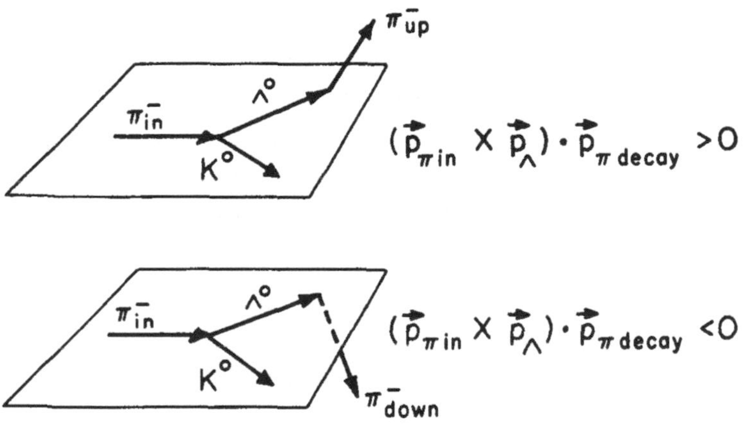

Figure 3.9. The production and decay of a Λ particle

In the subsequent decay of the Λ particle, let $\vec{p}_{\pi\text{decay}}$ be the momentum of the decay pion. To test parity conservation we examine whether there are as many "up" decays characterized by $(\vec{p}_{\pi\text{in}} \times \vec{p}_\Lambda)\cdot\vec{p}_{\pi\text{decay}} > 0$ as "down" decays characterized by $(\vec{p}_{\pi\text{in}} \times \vec{p}_\Lambda)\cdot\vec{p}_{\pi\text{decay}} < 0$. In other words, we ask whether or not the average value of the pseudoscalar quantity, $(\vec{p}_{\pi\text{in}} \times \vec{p}_\Lambda)\cdot\vec{p}_{\pi\text{decay}}$ is zero.

It is easy to see what all this means physically. Let us suppose that there are more "up" decays than "down" decays. Then in the mirror image of the world there are more "down" decays as illustrated in Fig. 3.9. So the mirror image of our world is different from our world, which is another way of saying that parity is not conserved.

The production and decay of Λ particles have been extensively studied at Brookhaven and Berkeley in propane and hydrogen bubble chambers exposed to π^- mesons of various energies, 0.88 Bev.–1.4 Bev. (Crawford *et al.*, 1957; Eisler *et al.*, 1957b; Leipuner and Adair, 1958). It was found that more decay pions are emitted upward, i.e. in the direction of $\vec{p}_{\pi \text{in}} \times \vec{p}_\Lambda$.

up:	325
down:	215

Thus a statistically significant up-down asymmetry has been established. Now, at last, it is possible to communicate to intelligent beings in outer space that the initial π^- direction, the Λ direction , and the preferential direction of the decay pion, taken in that order, form the three axes of what we mean by the right-handed coordinate system.

Let us consider this phenomenon a little more quantitatively (Morpurgo, 1956; Lee *et al.*, 1957). We first treat the case where the Λ spin is $\frac{1}{2}$. In the rest system of the decaying Λ particle the most general angular momentum conserving transition matrix is of the form

$$M = a_s + a_p \vec{\sigma} \cdot \hat{q} = a_s + a_p \begin{pmatrix} \cos\theta & \sin\theta \exp(-i\varphi) \\ \sin\theta \exp(i\varphi) & -\cos\theta \end{pmatrix} \tag{3.40}$$

where $\hat{q}$ is a unit vector in the direction of the momentum of the decay pion. Respectively, a_s and a_p are the transition amplitudes into s and p-wave final orbital angular momentum states. Note that, since $S_\Lambda = \frac{1}{2}$, the only allowed final states are $s_{1/2}$ and $p_{1/2}$. If parity were conserved (which we know now not to be the case), and the relative Λ-p parity were even, then we could have only $p_{1/2}$. For the Λ-p parity odd, we would have only $s_{1/2}$. If parity is not conserved, the relative Λ-p parity is a matter of definition (it is conveniently taken as even), and both a_s and a_p are non-vanishing.

Suppose the initial Λ spin is along the z axis. Then the final state wave function is given by

$$M\alpha = (a_s + a_p \cos\theta)\alpha + a_p \sin\theta \exp(i\varphi)\beta$$

where α and β are the two-component spinors corresponding to $m_s = \frac{1}{2}$ and $-\frac{1}{2}$ respectively. The angular distribution is given by

$$|a_s + a_p \cos\theta|^2 + |a_p|^2 \sin^2\theta = |a_s|^2 + |a_p|^2 + 2\,\text{Re}(a_s a_p^*) \cos\theta \tag{3.41}$$

In a more general case we may use the density matrix formalism to calculate the angular distribution. The density matrix for the initial state is $\rho = \frac{1}{2}(1 + \langle\vec{\sigma}\rangle \cdot \vec{\sigma})$ where $\langle\vec{\sigma}\rangle$ is the polarization of the decaying Λ defined as the average value of the Λ spin in units of $\hbar/2$. The distribution is given by

$$\tfrac{1}{2}\text{Tr}(M\rho_i M^\dagger) = |a_s|^2 + |a_p|^2 + 2\text{Re}(a_s a_p)\langle\vec{\sigma}\rangle \cdot \hat{q} \tag{3.42}$$

The expression is of the form $1 + \alpha P \cos\theta$ where

$$\alpha = \frac{2\mathrm{Re}(a_s a_p^*)}{|a_s|^2 + |a_p|^2} \tag{3.43}$$

is the asymmetry parameter inherent in the decay process, and P is the Λ^0 polarization at decay (the magnitude of $\langle\vec{\sigma}\rangle$).

In reality we must consider the laboratory system in which the target proton is at rest, and the Λ particle is moving. If the strong production process is parity conserving, the Λ particles are polarized in the direction perpendicular to the production plane. Now the component of the decay pion momentum, denoted by ξ, in the direction normal to the production plane measured in units of its maximum value is the same in the laboratory system in which the Λ is moving as in the system in which the Λ is at rest. So the numbers of "up" and "down" decays are given by

$$N_{\mathrm{up}} = \tfrac{1}{2}N \int_0^1 (1 + \bar{P}\alpha\xi)d\xi$$

$$N_{\mathrm{down}} = \tfrac{1}{2}N \int_{-1}^0 (1 + \bar{P}\alpha\xi)d\xi \tag{3.44}$$

where $\bar{P}$ stands for the average polarization of the Λ particle. From the experimental results quoted earlier, $\bar{P}\alpha \approx 0.5$ has been obtained. Since the average polarization $\bar{P}$ must be less than unity, 0.5 is a lower limit for $|\alpha|$. When the angular distribution of the production process is analyzed, an upper limit on $\bar{P}$ can be established. In this manner, it is concluded that

$$|\alpha| \geq 0.77 \pm 0.16$$

Note that since we do not know the sign of the Λ polarization, we have at this point no knowledge about the sign of α.

PROBLEM 3. Compute the polarization of the recoil proton in the decay of polarized Λ particles. Show that such polarization measurements would determine

(a) $\gamma = \dfrac{|a_s|^2 - |a_p|^2}{|a_s|^2 + |a_p|^2}$ (hence the s-to-p ratio)

(b) $\beta = \dfrac{2\mathrm{Im}(a_s a_p^*)}{|a_s|^2 + |a_p|^2}$

as well as

(c) $\alpha = \dfrac{2\mathrm{Re}(a_s a_p^*)}{|a_s|^2 + |a_p|^2}$ (both the sign and the magnitude)

For answers see Lee and Yang (1957c) and Leitner (1958).

Note added in proof. Recently precise measurements of the $\Lambda(\rightarrow p + \pi^-)$ decay parameters have been made by Cronin and Overseth (1962):

$$\alpha = -0.62 \pm 0.07$$
$$\beta = 0.19 \pm 0.19$$
$$\gamma = 0.79 \pm 0.6$$

The negative sign of α indicates that the pion is emitted preferentially in the direction opposite to the Λ spin and that the protons from unpolarized Λ's are predominantly right-handed (see also Birge and Fowler, 1960; and Leitner *et al.*, 1961). In addition the parameter α for $\Sigma^+ \rightarrow p + \pi^0$ has been measured to be

$$\alpha(\Sigma^+ \rightarrow p + \pi^0) = +0.78 \pm 0.18$$

by Beal *et al.* (1961). (Note the positive sign.) Attempts have also been made to measure α for $\Xi \rightarrow \Lambda + \pi^-$ where the subsequent Λ decay serves as a convenient analyzer for the Λ polarization; it appears that a_Σ and α_Λ have opposite signs (Fowler *et al.*, 1961; Bertanza *et al.*, 1962; Alvarez *et al.*, 1962; Pjerrou *et al.*, 1962).

Thus far we have assumed the Λ spin is $\frac{1}{2}$. If the Λ spin were $\frac{3}{2}$, then the final state would consist of $p_{3/2}$ and $d_{3/2}$ states. In the rest frame of the Λ^0, the angular distribution can be written as:

$$I(\xi) = \tfrac{1}{4} \sum_{m=-3/2}^{+3/2} \rho_{mm}(a^*\langle p, m| + b^*\langle d, m|)(a|p, m\rangle + b|d, n\rangle) \quad (3.45)$$

where the density matrix ρ_{mm} contains the information about the relative populations of the initial Λ spin states. For spin $= \frac{3}{2}$, ρ is a 4×4 matrix corresponding to the states $m = -\frac{3}{2}, -\frac{1}{2}, +\frac{1}{2}, +\frac{3}{2}$. It is permissible to disregard the cross terms in m in the above expression if we experimentally average over the azimuthal distribution (this is indeed the case). The terms of the form $\int_0^{2\pi} \exp[i(m - m')\varphi]d\varphi$ in the above distribution are zero unless $m = m'$.

In the Adair argument for the determination of the Λ spin, we have previously considered terms like ($\xi = \cos\theta$),

$$\langle p, m = \tfrac{1}{2}|p, m = \tfrac{1}{2}\rangle = \langle d, m = \tfrac{1}{2}|d, m = \tfrac{1}{2}\rangle = \tfrac{1}{2}(1 + 3\xi^2)$$

Here, however, we must consider the parity nonconserving p, d cross terms. We have terms of the form

$$\langle d, m = \tfrac{3}{2}|p, m = \tfrac{3}{2}\rangle = -\langle d, m = -\tfrac{3}{2}|p, m = -\tfrac{3}{2}\rangle = \tfrac{3}{2}(-\xi + \xi^3)$$

As expected, the odd power terms show up here. Similarly the $m = \frac{1}{2}$, p, d cross terms give $\frac{1}{2}(5\xi - 9\xi^3)$.

For any ρ_{mm} (i.e. no matter what Λ sample we consider) and for any combination (a, b), we have two additional relationships

$$\begin{aligned} &(1)\ \sum \rho_{mm} = 1 \\ &(2)\ |a|^2 + |b|^2 = 1 \end{aligned} \tag{3.46}$$

We may now maximize $|\langle\xi\rangle| = \left|\int_{-1}^{1} \xi I(\xi)d\xi\right|$ with the constraints (1) and (2) of (3.46). We get $|\langle\xi\rangle| \le \frac{1}{5}$ (Lee and Yang, 1958). In general, for any S_Λ it can be shown that

$$\langle\xi\rangle_{S_\Lambda} \le \frac{1}{2S_\Lambda + 2} \tag{3.47}$$

The experimental ξ distribution (Crawford *et al.*, 1959b) can be approximated by $1 + 0.57\xi$ (see Fig. 3.10). The observed large asymmetry implies that the Λ spin is $\frac{1}{2}$. The test functions defined by Lee and Yang for the purpose of comparing these experimental results with their predictions for the various possible Λ spins give the following results:

$S \neq \frac{3}{2}, \neq \frac{5}{2}$ by three and five standard deviations, respectively.

Consider the decay of the $_\Lambda H^4$ hypernucleus, $_\Lambda H^4 \to He^4 + \pi^-$. Since the Λ has spin $\frac{1}{2}$ and the spin of H^3 is $\frac{1}{2}$, the spin of $_\Lambda H^4$ can be either 0 or 1. The relative occurrences of this decay mode and the other possible π^- ones (e.g., $_\Lambda H^4 \to H^3 + p + \pi^-$, $2d + \pi^-$ or $He^3 + n + \pi^-$) can be used to determine the spin of $_\Lambda H^4$ (Dalitz, 1957). Qualitatively, the argument goes as follows. Since both the final state particles have spin 0, the final state orbital angular momentum must equal the spin of $_\Lambda H^4$ (i.e., s or p wave). If we consider the case, $_\Lambda H^4$ spin $= 1$, and if we take the quantization axis along the $He^4 - \pi^-$ relative momentum, then we know that the $m = \pm 1$ states must have zero population. (The plane wave cannot have a component of angular momentum in its propagation direction.) Thus, if the p wave orbital forms a significant part of the Λ final state, this two-body mode should be suppressed by a factor of three if $_\Lambda H^4$ spin $= 1$. Actually the theoretical predictions (Dalitz and Liu, 1959) of the (two-body decays)/(all decays) ratio are seen to depend strongly on the p/s wave ratio in the *free* Λ decay as well as on the spin of $_\Lambda H^4$. The experimental results (Ammar *et al.*, 1959b) are that the ratio is $\geq 0.6 \pm 0.1$. The discrimination between $_\Lambda H^4$ spin 0 or 1 is not as good as had been hoped, but these results indicate that the spin is probably 0, thereby implying a preferential antiparallel spin alignment for the Λ-nucleon system. This conclusion is of crucial importance in our earlier discussions of the reaction $K^- + He^4 \to {}_\Lambda H^4 + \pi^-$ to determine the relative $K\Lambda$ parity.

Note added in proof. Beal *et al.* (1961) and Cronin and Overseth (1962) have demonstrated that the p/s ratio in *free* Λ decay is indeed small by measuring the parameter γ (cf. Problem 3). Moreover, Block, Lendinara,

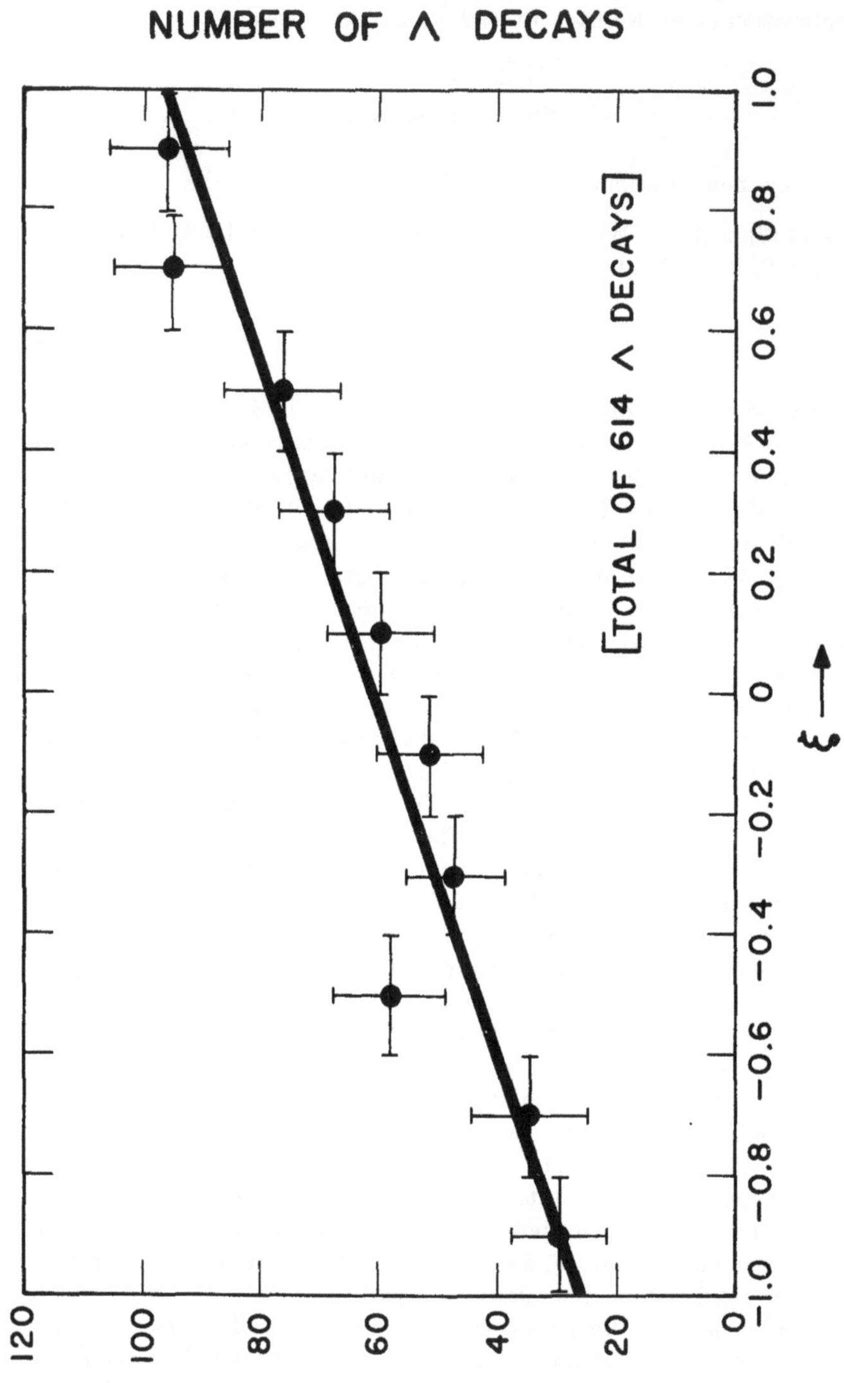

Figure 3.10. Angular distribution in Λ decay

and Monari (1962) have shown that the ${}_{\Lambda}H^4({}_{\Lambda}He^4)$ spin is zero by directly observing an isotropic angular distribution of the decay products (relative to the momentum of the hypernucleus) in the reaction ${}_{\Lambda}H^4 \rightarrow He^4 + \pi^-$ following $K^- + He \rightarrow {}_{\Lambda}H^4 + \pi^0$ (capture at rest). (We expect a pure $\cos^2\theta$ distribution for $J({}_{\Lambda}H^4) = 1$ and an isotropic distribution for $J({}_{\Lambda}H^4) = 0$ for s state capture of K^-.) It therefore appears that the $K\Lambda$ parity is most probably odd. This conclusion can be avoided only through one very remote possibility: Block *et al.* are observing $J = 1$ excited states of ${}_{\Lambda}H^4$.

A few words about $\Sigma^{\pm}$ decay, which can be treated in the same manner. We present here the experimental results obtained. There is no experimental evidence for a large up-down asymmetry in $\Sigma^- \rightarrow n + \pi^-$ for Σ^- hyperons produced in $\pi^- p$ collisions in bubble chambers. This may be due to either (a) the strong interactions responsible for production do not produce polarized Σ^-'s, or (b) the decay asymmetry inherent in the weak interactions is nearly zero or (c) both. Counter experiments have been performed by Cool *et al.* (1959) and Cork *et al.* (1960) to detect up-down asymmetries in $\Sigma^+ \rightarrow \pi^+ + n$ and in $\Sigma^+ \rightarrow \pi^0 + p$ following the reaction $\pi^+ + p \rightarrow \Sigma^+ + K^+$. A large up-down asymmetry has been reported for the $\pi^0 p$ mode but not for the $\pi^+ n$ mode. Since the Σ^+'s are produced under the same condition, we can conclude $|\alpha(\pi^0 p)| \gg |\alpha(\pi^+ p)| \approx 0$, and that the production reaction is an excellent polarizer for Σ^+'s. If we assume that the experimental angular distribution follows a linear law $1 + \alpha\bar{P}\cos\theta$, their experimental data indicate

$$\alpha\bar{P} = 0.75 \pm 0.17 \quad \text{for } \Sigma^+ \rightarrow p + \pi^0$$

$$\alpha\bar{P} = 0.03 \pm 0.08 \quad \text{for } \Sigma^+ \rightarrow n + \pi^-$$

Using the Lee-Yang argument we see that the observed large asymmetry for the $\pi^0 p$ mode strongly favors the Σ spin $= \frac{1}{2}$.

3.6. Tests of Parity Conservation

If parity is not conserved, energy eigenstates ψ are not expected to be eigenstates of parity. We may write

$$\psi = \psi_{\text{regular}} + \mathscr{F}\psi_{\text{irregular}} \tag{3.48}$$

where ψ_{regular} and $\psi_{\text{irregular}}$ are of opposite parity. $\mathscr{F}$ is called the parity-mixing amplitude. There have been several experiments to set limits on $\mathscr{F}$ in nuclear interactions.

The very fact that it has been possible to attribute definite parities to nuclear states in consistent ways already indicates that parity conservation is satisfied to a fair degree of accuracy. If parity conservation broke down, the so-called parity selection rule would be only approximate. So one of the methods of testing parity conservation is to look for transitions which are "forbidden." The first attempt to deliberately test parity conservation along this line was made by Tanner (1957), who studied the reaction

$$p + F^{19} \rightarrow O^{16} + \alpha$$

in the neighborhood of the 13.19 Mev 1^+ state of Ne^{20}. Since O^{16} and α are both 0^+, the allowed final states are 0^+, 1^-, 2^+, etc. Tanner found no evidence for α particles that fit into the interpretation

$$p + F^{19} \rightarrow Ne^{20*}_{(1+)} \rightarrow O^{16} + \alpha$$

His estimate shows $|\mathscr{F}|^2 \lesssim 4 \times 10^{-8}$. Similar experiments have been carried out by Wilkinson (1958).

We have already remarked that the static electric dipole moment of a particle must vanish if parity is conserved. Several years before the discovery of parity nonconservation in weak interactions, Purcell and Ramsey (1955) proposed a method that would be suitable for setting up an upper limit on the electric dipole moment of the neutron. In a subsequent experiment carried out by Smith, Purcell, and Ramsey (1957), a polarized neutron beam passed through an externally applied electric field whose polarity was switched back and forth. The spin directions were unaffected by this, which showed that the electric dipole moment must be vanishingly small. They concluded that

$$\frac{\mu^{(\text{el. dip.})}}{e} < (-0.1 \pm 2.4) \times 10^{-20}\ \text{cm}.$$

This is a very stringent limit in view of the fact that the natural "size" of the neutron is of the order of 10^{-14}–10^{-13} cm. As we shall show in Chapter 4, however, the vanishing dipole moment follows also from time reversal invariance.

In order to test parity conservation in more direct and unambiguous ways, one may perform experiments to see whether elementary particle interactions distinguish between the right and the left. The law of parity conservation implies that for every phenomenon there exists an exact mirror image which is physically realizable with the same probability. If pseudoscalar quantities such as $(\vec{p}_1 \times \vec{p}_2)\cdot\vec{p}_3$, $\vec{J}\cdot\vec{p}$, etc. have nonzero expectation values, then the law of parity conservation is violated.

As an example, consider a double scattering of a proton by unpolarized nuclei. In the first scattering the proton with momentum $\vec{p}_1$ is scattered into momentum $\vec{p}_2$ and in the second scattering into momentum $\vec{p}_3$. If parity is conserved, the proton distribution in the second scattering should be independent of $(\vec{p}_1 \times \vec{p}_2)\cdot\vec{p}_3$; in other words, there should not be any "up-down" asymmetry in the second scattering with respect to the plane of the first scattering. Experiments show that this is indeed the case to an accuracy of $\sim 1\%$ (Chamberlain *et al.*, 1954), which implies that $|\mathscr{F}|^2$ is less than 10^{-4}. This experiment should be contrasted with the Λ

up-down experiment discussed in the previous section in which the quantity $(\vec{p}_{\pi in} \times \vec{p}_{\Lambda})\cdot\vec{p}_{\pi decay}$ has been shown to have a nonzero average value.

In $p + \bar{p} \to n\pi$ with $n \geq 3$ ($n \geq 4$ if the annihilation takes place at rest) there are enough independent momenta so that quantities such as $(\vec{p}_1 \times \vec{p}_2)\cdot\vec{p}_3$ may be formed. A practical method for testing parity conservation in such a reaction has been suggested by Pais (1959).

There are essentially two methods for detecting terms of the form $J\cdot\vec{p}$.

(A) Start with polarized parent particles so that the screw sense is defined initially. Then ask whether the distribution of final-state particles depends on odd powers of cosine of the angle between the initial spin direction and the final momentum direction.

(B) Start with unpolarized systems, and look at the helicity of the final-state particle. $\langle\vec{\sigma}_{\text{final}}\ \vec{p}_{\text{final}}\rangle$ will have a nonvanishing expectation value only if P is not conserved.

As an example of Method (A) from low energy nuclear physics, we consider an experiment proposed and carried out by Haas, Leipuner, and Adair (1959). They obtained polarized Cd^{114*} by bombarding Cd^{113} with polarized neutrons, and studied the angular distribution of γ's which are subsequently emitted:

$$n + Cd^{113} \to Cd^{114*} \to Cd^{114} + \gamma$$

Since Cd^{114*} and Cd^{114} are 1^+ and 0^+ respectively, the γ transition must be $M1$ as long as parity is conserved. If parity were not conserved, then the transition would be a mixture of $M1$ and $E1$ with an asymmetry term ($\cos\theta$ term) proportional to $EM/(|E|^2 + |M|^2)$. This is a very sensitive test for parity conservation since E would be intrinsically large compared to M if there were no parity selection rule. They found the relative amplitude $E/M = (4 \pm 8) \times 10^{-4}$. This corresponds to $|\mathscr{F}|^2 < 10^{-16}$ for the *fundamental* nucleon-nucleon interaction. This experiment provides by far the strongest evidence for parity conservation in strong and electromagnetic interactions.

A classical example of Method (A) from the weak interactions is the β decay of Co^{60}. (See Fig. 3.11.) Wu *et al.* (1957) polarized the Co^{60} at low temperatures, and, much to their surprise, observed a $1 + aP\cos\theta$ angular distribution for the electrons. The Co^{60} polarization P was estimated from the angular distribution (polar vs. equitorial anisotropy) of the subsequent γ emission. The asymmetry parameter was unexpectedly large: $a \approx -\left(\frac{v}{c}\right)_{\text{el.}}$. This experiment of Wu *et al.*, carried out in December 1956 at the National Bureau of Standards, provided the first conclusive evidence that parity is not conserved in weak decays.

As an example of Method (B) one may ask whether various hyperons produced in "strong" reactions are longitudinally polarized. Recall that the subsequent hyperon decay ($\Lambda^0 \to p + \pi^-, \Sigma^+ \to p + \pi^0$) is an excellent analyzer of the possible polarization. Although the possible violation of parity conservation in the strong interactions of *strange* particles was speculated on at one time by both experimentalists and theoreticians, with better statistics there does not seem to be evidence for parity violation in such reactions. For instance, $a\bar{P}_{\text{longitudinal}} = 0.03 \pm 0.09$ for hyperons produced in K^-He interactions (Block *et al.*, 1960).

The decay products from weak processes such as $\pi^\pm$ decay and β decay are in general longitudinally polarized. Since the beginning of 1957, several groups have measured the polarization of electrons (positrons) in beta decays of unpolarized nuclei by numerous methods (see for example a

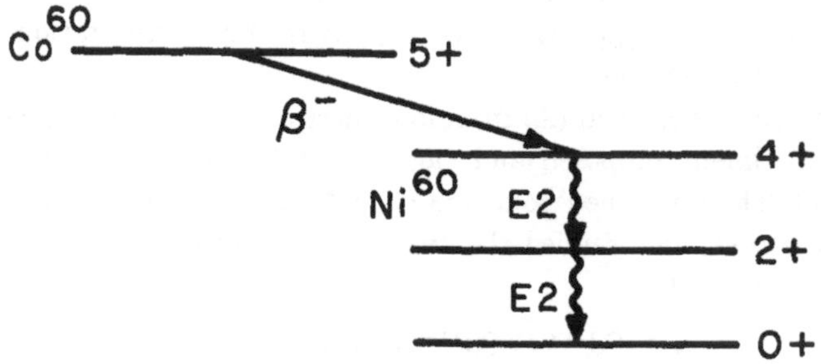

Figure 3.11. The decay scheme of Co^{60}

review article by Grodzins (1959).) It has been established that in all allowed (and most forbidden) beta decays relativistic electrons are left-handed while relativistic positrons are right-handed, or, more quantitatively, $\mathscr{H}(e^\mp) = \mp \frac{v}{c}$ to accuracies of about 10%.

Method (A) and method (B) have been combined in some experiments. Consider the $\pi\mu e$ sequence:

$$\pi^+ \rightarrow \mu^+ + \nu, \qquad \mu^+ \rightarrow e^+ + \nu + \bar{\nu}$$

If parity is not conserved in the π^+ decay, the μ^+ is generally polarized in the direction of motion, $\langle\vec{\sigma}_{\mu^+}\rangle\cdot\vec{p}_{\mu^+} \neq 0$. If parity is not conserved in the subsequent μ^+ decay either, the positron distribution depends on the cosine of the angle between the muon spin and the positron momentum. In practice, the muon from the pion decay is slowed down (e.g. in graphite or emulsions) and stops, but the muon spin still "remembers" the original muon direction, so that a forward-backward asymmetry of e^+ with respect to the muon direction is expected. Garwin, Lederman, and Weinrich (1957) and Friedman and Telegdi (1957) were first to observe such an asymmetry effect. We shall discuss the $\pi\mu e$ sequence in full detail in Chapter 7.

The reader is urged to read the classical paper by Lee and Yang (1956b) in which various methods to test parity conservation are discussed. No exposition of this subject does full justice to that paper which played an historic role in the development of elementary particle physics.

3.7. Parity in Field Theory

The free-field Klein-Gordon equation reads

$$(\Box^2 - \mu^2)\varphi = 0 \tag{3.49}$$

As we switch to the space-inverted system, $\vec{x}' = -\vec{x}$, $x_4 = x_4$,

$$(\Box^2 - \mu^2) \to \Box'^2 - \mu^2 = \Box^2 - \mu^2$$

If $\varphi(x, x_0)$ is a solution to (3.49), then the corresponding solution in the space-inverted system is $\varphi'(\vec{x}', x_0) = \pm\varphi(+\vec{x}, x_0)$. In other words, under the parity operation, $\varphi(\vec{x}, x_0) \to \pm\varphi'(\vec{x}, x_0) = \pm\varphi(-\vec{x}, x_0)$. The $+$ $(-)$ sign reflects even (odd) intrinsic parity.

A field theory in which the symbol φ stands for a relativistic wave function is called the *c*-number theory while a field theory in which φ stands for a quantized field operator is called the *q*-number theory (or second-quantized theory).

We consider the *c*-number theory first. A typical solution to (3.49) is the usual plane-wave solution

$$\varphi = \exp(i\vec{p}\cdot\vec{x} - ip_0x_0), \qquad \vec{p}^2 + p_0^2 = \mu^2.$$

The wave function in the inverted system is

$$\varphi'(\vec{x}', x_0') = \pm\varphi(+\vec{x}, x_0) = \pm\exp(-i\vec{p}\cdot\vec{x}' - ip_0x_0')$$

which indeed describes a particle with opposite momentum.

In the *q*-number theory the field operator can be expanded in creation operators $a^\dagger(\vec{p})$ and annihilation operators $a(\vec{p})$.

$$\varphi(\vec{x}, x_0) = \frac{1}{(2\pi)^{3/2}} \int \frac{d^3p}{\sqrt{2p_0}} \times \{a(\vec{p}) \exp[i(\vec{p}\cdot\vec{x} - p_0x_0)] + a^\dagger(\vec{p}) \exp[-i(\vec{p}\cdot\vec{x} - p_0x_0)]\} \tag{3.50}$$

The field operator in the space-inverted system is given by

$$P\varphi(\vec{x}, x_0)P^{-1} = \pm\varphi(-\vec{x}, x_0) \tag{3.51}$$

hence

$$\int \frac{d^3p}{\sqrt{2p_0}} \{Pa(\vec{p})P^{-1} \exp[i(\vec{p}\cdot\vec{x} - p_0x_0)] + Pa^\dagger(\vec{p})P^{-1} \exp[-i(\vec{p}\cdot\mathbf{a} - p_0x_0)]\}$$

$$= \pm \int \frac{d^3p}{\sqrt{2p_0}} \{a(\vec{p}) \exp(-i\vec{p}\cdot\vec{x} - ip_0x_0) + a^\dagger(\vec{p}) \exp(i\vec{p}\cdot\mathbf{a} + ip_0x_0)\}$$

$$= \pm \int \frac{d^3p}{\sqrt{2p_0}} \{a(-\vec{p}) \exp(i\vec{p}\cdot\vec{x} - ip_0x_0) + a^\dagger(-\vec{p}) \exp(-i\vec{p}\cdot\vec{x} + ip_0x_0)\}$$

This means that the individual annihilation and creation operators must transform as

$$Pa(\vec{p})P^{-1} = \pm a(-\vec{p}), \qquad Pa^\dagger(\vec{p})P^{-1} = \pm a^\dagger(-\vec{p}) \tag{3.52}$$

which is reasonable. In particular, the state of one particle at rest $a^\dagger_{(\vec{p}=0)}\Phi_0$ where Φ_0 stands for the vacuum state is even or odd under P depending on whether the intrinsic parity is even or odd.

The formalism we have described is adequate for a particle which is identical to its antiparticle. The field operators corresponding to π^+, K^+, K^0 must be represented by non-Hermitian fields $\varphi \neq \varphi^\dagger$. The field operator φ annihilates the particle (say π^+) and creates its antiparticle (π^-). $\varphi^\dagger$ annihilates the antiparticle and creates the particle. If φ transforms as

$$P\varphi P^{-1} = \eta\varphi(-\vec{x}\, x_0), \qquad |\eta|^2 = 1$$

then

$$P\varphi^\dagger P^{-1} = \eta^*\varphi^\dagger(-\vec{x}, x_0)$$

We will comment on the phase factor η in later parts of the book. At the present we simply note that if the charged boson, say π^+, has odd intrinsic parity corresponding to $\eta = -1$, then the π^- intrinsic parity must necessarily be odd. The charge current density constructed out of

$$j_\mu = ie\left[\frac{\partial\varphi^\dagger}{\partial x_\mu}\varphi - \varphi^\dagger\left(\frac{\partial\varphi}{\partial x_\mu}\right)\right] \tag{3.53}$$

has the desired property

$$\begin{aligned} P\vec{j}(\vec{x}, x_0)P^{-1} &= -\vec{j}(-\vec{x}, x^0) \\ Pj_0(\vec{x}, x_0)P^{-1} &= j_0(-\vec{x}, x_0) \end{aligned} \tag{3.54}$$

Unlike the Klein-Gordon equation, or the Schrödinger equation, the Dirac equation is first order in the space coordinates. For this reason, we do not expect $\psi'(\vec{x}') = \pm\psi(\vec{x})$. Just as in the Lorentz invariance case considered in Chapter 2, the invariance of the Dirac equation under parity implies that

$$\left(\gamma_\mu \frac{\partial}{\partial x'_\mu} + m\right)\psi' = 0 \qquad \vec{x}' = -\vec{x}, \quad x_4 = x_4 \tag{3.55}$$

must follow from

$$\left(\gamma_\mu \frac{\partial}{\partial x_\mu} + m\right)\psi = 0$$

We assume that ψ and ψ' are related linearly as $\psi' = S\psi$ where S is a four-by-four matrix whose form is to be determined. Writing the Dirac equation in the original system in terms of the primed variables

$$\left(-\gamma_k \frac{\partial}{\partial x'_k} + \gamma_4 \frac{\partial}{\partial x'_4} + m\right)S^{-1}\psi = 0$$

Multiply S from the left, and compare with (3.55). We note that the two equations are equivalent if

$$S\gamma_k S^{-1} = -\gamma_k$$
$$S\gamma_4 S^{-1} = \gamma_4 \tag{3.56}$$

We see that $S = \eta\gamma_4$ will do. $|\eta|^2$ must be unity if the charge density $\bar{\psi}\gamma_4\psi = \psi^\dagger\psi$ is to be the same in both coordinate systems.

An alternative approach which will determine S for us is the following. We first note that a three-dimensional rotation must commute with the parity operation and that a "pure" Lorentz transformation must anticommute with the parity operation. The four-dimensional rotation operators in the Dirac spinor representation are

$$S_{\text{Lor}}(\delta\vec{\chi}) \quad \text{contains} \quad \gamma_4\gamma_k\delta\chi_k$$
$$S_{\text{rot}}(\delta\vec{\omega}) \quad \text{contains} \quad \gamma_i\gamma_j\delta\omega_k$$

Since we must have (S_P is the four-by-four matrix associated with parity operator)

$$S_{\text{Lor}}(\delta\vec{\chi})S_P = S_P S_{\text{Lor}}(-\delta\vec{\chi})$$
$$S_{\text{rot}}(\delta\vec{\omega})S_P = S_P S_{\text{rot}}(\delta\vec{\omega}) \tag{3.57}$$

S_P must anticommute with $\gamma_4\gamma_k$ and commute with $\gamma_i\gamma_j$. We see that $S_P = \eta\gamma_4$ satisfies these conditions.

In the c-number language, the wavefunctions transform as: $\psi(\vec{x}) \to \eta\gamma_4\psi(-\vec{x})$, whereas in the q-number language where $\psi(\vec{x})$ is regarded as an operator, we have: $P\psi(\vec{x})P^{-1} = \eta\gamma_4\psi(-\vec{x})$. A plane wave wavefunction corresponding to a particle moving in the $+z$ direction and where the quantization axis is also along the $+z$ direction transforms as

$$\begin{pmatrix} 1 \\ 0 \\ \dfrac{p}{E+m} \\ 0 \end{pmatrix} \exp(ipz - iEt) \to \eta \begin{pmatrix} 1 \\ 0 \\ \dfrac{-p}{E+m} \\ 0 \end{pmatrix} \exp(i(-p)z - iEt)$$

The transformed wavefunction is still an eigenspinor of Σ_z but the momentum has reversed its sign. Thus the helicity $\mathscr{H} \to -\mathscr{H}$. This conforms to what we did before without using the Dirac Equation.

Since the adjoint spinor transforms under parity as

$$\bar{\psi} = \psi^\dagger\gamma_4 \to \eta^*(\gamma_4\psi)^\dagger\gamma_4 = \eta^*\bar{\psi}\gamma_4 \tag{3.58}$$

we have the transformation properties for charge and current

$$\rho = -ij_4 = \psi^\dagger\psi \xrightarrow{P} \rho(-\vec{x})$$
$$j_k = i\bar{\psi}\gamma_k\psi \xrightarrow{P} i|\eta|^2\bar{\psi}\gamma_4\gamma_k\gamma_4\psi = -j_k(-\vec{x}) \tag{3.59}$$

on the other hand, note

$$i\bar{\psi}\gamma_5\gamma_4\psi \rightarrow -i\bar{\psi}\gamma_5\gamma_4\psi$$
$$i\bar{\psi}\gamma_5\gamma_k\psi \rightarrow +i\bar{\psi}\gamma_5\gamma_k\psi$$

Thus the vector $i\bar{\psi}\gamma_\mu\psi$ and the axial vector $i\bar{\psi}\gamma_5\gamma_\mu\psi$ behave oppositely under parity.

Consider the non-derivative Yukawa coupling of a neutral scalar field to a fermion field. The scalar-scalar (s-s) coupling has the form: $\bar{\psi}\psi\phi$, where ϕ represents the scalar field. This can be made invariant under P.

$$\psi \xrightarrow{P} \eta_F\gamma_4\psi$$
$$\phi \xrightarrow{P} \eta_B\phi$$

if we choose $\eta_B = +1$. The pseudo-scalar-pseudo-scalar (p.s.-p.s.) coupling has the form: $i\bar{\psi}\gamma_5\psi\phi$. The i is required if we wish the interaction to be Hermitian, since

$$(\psi^\dagger\gamma_4\gamma_5\psi\phi)^\dagger = \psi^\dagger\gamma_5\gamma_4\psi\phi = -\bar{\psi}\gamma_5\psi\phi$$

This p.s.-p.s. coupling can be made invariant under P if we choose $\eta_B = -1$ since

$$i\bar{\psi}\gamma_5\psi\phi \xrightarrow{P} i\eta_B\bar{\psi}\gamma_4\gamma_5\gamma_4\psi\phi = -i\eta_B\bar{\psi}\gamma_5\psi\phi \tag{3.60}$$

If we consider both scalar and pseudo-scalar interactions together

$$g_1\bar{\psi}\psi\phi + ig_2\bar{\psi}\gamma_5\psi\phi$$

we see that there is no choice for η_B which makes this interaction invariant under P. If both g_1 and g_2 are nonvanishing, parity is not conserved. The non-relativistic limit (to lowest order) of this interaction has the form

$$g_1\phi\psi^{P\dagger}\psi^P + \frac{g_2}{2M}\psi^{P\dagger}\vec{\sigma}\psi^P\cdot\vec{\nabla}\phi$$

where ψ^P stands for appropriate Pauli two-component spinor. The first term gives rise to s-wave mesons and the second term to p-wave mesons.

Consider the coupling of a charged scalar field to a fermion field of the following type

$$i\bar{\psi}_p\gamma_5\psi_n\phi_{\pi+} + \text{H.c.}$$

This interaction can be made invariant under P if

$$\eta_p^*\eta_n\eta_{\pi+} = -1$$

Once we choose $\eta_p = \eta_n$ *by convention*, the π^+ field can be regarded as pseudo-scalar.

The relative parity of particle and anti-particle can be determined in the following manner. We have seen that γ_4 is the parity operator for the Dirac wavefunction. Let us consider how the positive and negative energy states transform under the parity operation. We do this for particles at rest in a definite spin state.

$$\begin{aligned}
\psi_{p_0>0} &= \begin{pmatrix}1\\0\\0\\0\end{pmatrix} \xrightarrow{P} \eta\gamma_4\begin{pmatrix}1\\0\\0\\0\end{pmatrix} = \eta\begin{pmatrix}1\\0\\0\\0\end{pmatrix} \\
\psi_{p_0<0} &= \begin{pmatrix}0\\0\\1\\0\end{pmatrix} \xrightarrow{P} \eta\gamma_4\begin{pmatrix}0\\0\\1\\0\end{pmatrix} = -\eta\begin{pmatrix}0\\0\\1\\0\end{pmatrix}
\end{aligned} \tag{3.61}$$

Thus if a particle in a positive energy state has even intrinsic parity, then a particle in a negative energy state has odd intrinsic parity. In the hole theory, however, a physical antiparticle corresponds to a hole or a vacant negative energy state in an otherwise completely filled sea of negative energy states. Thus, for example, we may regard the ${}^1S_0$ positronium annihilation process

$$e^- + e^+ \rightarrow 2\gamma$$
$$e^-(p_0 > 0) \rightarrow 2\gamma + e^-(p_0 < 0)$$

which corresponds to $\eta = (2\gamma \text{ parity}) \times (-\eta)$ as far as the equality between the parities of the initial and final state is concerned. The 2γ state must, therefore, have odd parity, and a wavefunction of the form $(\vec{\varepsilon}_1 \times \vec{\varepsilon}_2)\cdot\hat{k}$. The polarization directions should be perpendicular as in π^0 decay, as first emphasized by Wheeler (1946).

Note that the phase factor η plays no role and we can generally say that for any Dirac particle-antiparticle system the parity is $-(-1)^l$. This is to be contrasted with the boson case; $w = +(-1)^l$ for the $\pi^+\pi^-$ system. To sum up, the relative e^-e^+ parity is odd while the relative $\pi^+\pi^-$ parity is even.

The first conclusive experimental support of this point was presented by Wu and Shaknov (1950). They determined the polarization of the 2γ's from ${}^1S_0$ positronium decay by letting both γ rays Compton-scatter. Since the angular distribution of Compton scattering is strongly dependent on the initial photon polarization direction, they were able to ascertain that $\vec{\varepsilon}_1 \perp \vec{\varepsilon}_2$. Thus the prediction of the Dirac theory that the parity of particle and antiparticle are opposite is fulfilled. One cannot get this prediction from non-relativistic quantum mechanics, since the notion of the antiparticle is peculiar to the relativistic theory. Since this prediction is the result of a symmetry argument, it must be independent of the validity of perturbation theory. Detailed perturbation calculations on this annihilation process are worked out in Chapter 12 of Jauch and Rohrlich (1955) using the parity-conserving vector interaction $i\bar{\psi}\gamma_\mu\psi A_\mu$. Such calculations, of course, give $\vec{\varepsilon}^1 \perp \vec{\varepsilon}_2$ which we have derived from a more general consideration.

3.8. Beta Decay

Before Pauli's suggestion of the neutrino in 1930, the continuous nature of the electron energy spectrum in nuclear beta decay was a "puzzle" in the same sense as the τ-θ problem was a "puzzle" before 1957. Some people (e.g. N. Bohr) wondered if one would have to give up energy conservation. The gravity of the situation in those days is reflected in the following remark by Debye: "Oh, it is best not to think about it [continuous beta spectrum] at all—like the new taxes." An enlightening account of the early development of β-decay physics may be found in a contribution to the Memorial Volume to Wolfgang Pauli by C. S. Wu (1960).

With the help of Pauli's idea, Fermi (1934) made the first field-theoretical calculation of the β decay interaction. In analogy to the $j_\mu A_\mu$ electromagnetic interaction, Fermi used as the interaction density

$$H_{\text{int}} \propto (\bar{p}\gamma_\mu n)(\bar{e}\gamma_\mu \nu) + \text{H.c.} \tag{3.62}$$

(From now on we let the particle symbols p, n, etc., stand for the respective field operators whenever there is no possibility of confusion.) The nucleon is considered here as a source and sink of leptons (e, ν). The first term in (3.62), taken in lowest order, leads to a process in which the neutron and the neutrino are annihilated and the proton and the electron are created; since the field operator ν can create an antineutrino as well as annihilate a neutrino, the first term is responsible for a β^- process: $n \to p + e^- + \bar{\nu}$ as well as $\nu + n \to p + e^-$. The second term, H.c., which can be written explicitly as

$$\begin{aligned} \text{H.c.} &= (\bar{e}\gamma_\mu \nu)^\dagger(\bar{p}\gamma_\mu n)^\dagger \\ &= (\nu^\dagger \gamma_\mu \gamma_4 e)(n^\dagger \gamma_\mu \gamma_4 p) \\ &= (\bar{\nu}\gamma_\mu e)(\bar{n}\gamma_\mu p) \end{aligned} \tag{3.63}$$

can give rise to a β^+ process

$$p \to n + e^+ + \nu$$

and the antineutrino capture reaction

$$\bar{\nu} + p \to n + e^+$$

which was first observed by Cowan, Reines, and collaborators (Cowan *et al.*, 1956).

It is perhaps important to emphasize even at this early stage that the Fermi theory of β decay is essentially a phenomenological theory. In writing down the interaction (3.62) it has been assumed that the nucleons and leptons interact exactly at the same point. There is reason to believe that this assumption is incorrect; for instance, if we calculate the transition rate for $\bar{\nu} + p \to e^+ + n$, the total cross section exceeds the maximum value allowed by unitarity at $\bar{\nu}$ energies $\sim$300 Bev. However, since the de Broglie wavelength of the lepton in nuclear β decay is of the order of 10^{-11} cm, and is much larger than the range of the weak interactions which is at most $\sim 0.5 \times 10^{-13}$ cm, nuclear β decay, for all practical purposes, is rather insensitive to the "inner structure" of the weak interactions; so we might as well use the simple point interaction (3.63). In Chapter 11 we will briefly discuss the possibility that the Fermi-type four-fermion interaction might not be "fundamental."

The most general form of the β decay interaction based on the Fermi hypothesis was considered for over twenty years to be

$$H_{\text{int}} = \sum_i C_i(\bar{p}\Gamma_i n)(\bar{e}\Gamma_i \nu) + \text{H.c.} \tag{3.64}$$

where the Γ_i's are the various 4×4 matrices considered in Chapter 2.

TABLE 3.5.

Covariants in β Decay

$\Gamma_i =$	I	γ_μ	$\sigma_{\mu\nu}/\sqrt{2}$	$i\gamma_5\gamma_\mu$	γ_5
Name	S	V	T	A	P
	scalar	vector	tensor	axial-vector	pseudoscalar

The $\sqrt{2}$ in the denominator of Γ_T and i in Γ_A are chosen so that both the tensor interaction and the axial-vector interactions have the same non-relativistic limit as far as the nucleons are concerned. (Recall that $\sigma_{ij} = -\sigma_{ji}$ (six altogether) corresponds to $i\gamma_5\gamma_k$ (three altogether).)

For over twenty years an exhaustive study was made of many aspects of most of the known β decay reactions (e.g., energy spectrum, β-ν correlation, β-γ correlation, etc.). It remained for Lee and Yang (1956b) to point out that with the quantities that had been measured up to that time, it was impossible to distinguish the interaction (3.64) from the following parity-nonconserving one

$$H_{\text{int}} = \sum_i (\bar{p}\Gamma_i n)(\bar{e}\Gamma_i(C_i + C_i'\gamma_5)\nu) + \text{H.c.} \tag{3.65}$$

Since $\bar{e}\Gamma_i\nu$ and $\bar{e}\Gamma_i\gamma_5\nu$ behave oppositely under parity, this interaction leads to final states of opposite parity. Note that (3.65) conserves parity if either (a) all $C_i' = 0$. In this case we have P invariance with $\eta_p^*\eta_n\eta_e^*\eta_\nu = 1$ (even coupling) or (b) all $C_i = 0$. P invariance again holds with $\eta_p^*\eta_n\eta_e^*\eta_\nu = -1$ (odd coupling).

The physical meaning of C_i and C'_i can be made more transparent in the following way. We define

$$C_i^R = C_i - C'_i$$
$$C_i^L = C_i + C'_i$$

Then we have

$$(C_i + C'_i\gamma_5) = C_i^R\tfrac{1}{2}(1 - \gamma_5) + C_i^L\tfrac{1}{2}(1 + \gamma_5)$$

Since $\frac{1}{2}(1 \pm \gamma_5)$ are the projection operators for chirality, we see that $C_i^{R,L}$ are the amplitudes for annihilating right-, left-handed neutrinos. As we shall discuss in Chapter 5, $C_i^{R,L}$ can also be regarded as the amplitudes for creating left-, right-handed antineutrinos. If parity is conserved, $C'_i = 0$ or $C_i = 0$, hence $C_i^R = \pm C_i^L$, and we have as many right- as left-handed neutrinos.

As an example we consider the decay of a free neutron with a pure vector-type interaction. (The axial vector case, which is slightly more involved, is left as an exercise.) We have

$$H_{\text{int}} = (\bar{p}\gamma_\mu n)(\bar{e}\gamma_\mu(C_V + C'_V\gamma_5)\nu) + \text{H.c.} \tag{3.66}$$

In the non-relativistic limit $\left(\frac{v}{c}\right)_N \to 0$, the neucleon part gives 1 for $\mu = 4$ and 0 for $\mu = j$. The lepton part leads to

$$MM^\dagger = \bar{u}_e(\vec{p}_e)(C_V\gamma_4 + C'_V\gamma_4\gamma_5)u_v(\vec{p}_v)u_v(\vec{p}_V)\gamma_4(C_V^*\gamma_4 + C_V'^*\gamma_5\gamma_4)\gamma_4u_e(\vec{p}_e) \tag{3.67}$$

Summing over the electron and antineutrino spins we have

$$\tfrac{1}{2}\text{Tr}\left[\left(\frac{-i\gamma\cdot p_e + m_e}{2m_e}\right)(C_V\gamma_4 + C'_V\gamma_4\gamma_5)\left(\frac{-i\gamma\cdot p_\nu - m_\nu}{2m_\nu}\right)(C_V^*\gamma_4 + C_V'^*\gamma_4\gamma_5)\right]$$
$$= \frac{E_eE_\nu}{2m_em_\nu}(|C_V|^2 + |C'_V|^2)\left\{1 + \frac{\vec{p}_e\cdot\vec{p}_\nu}{E_eE_\nu} + \frac{-|C_V|^2 + |C'_V|^2}{|C_V|^2 + |C'_V|^2}\frac{m_em_\nu}{E_eE_\nu}\right\} \tag{3.68}$$

(The coefficients $E_eE_\nu/2m_em_\nu$ cancels with the wave function normalization factor m_em_ν/E_eE_ν. Thus even if $m_\nu = 0$, we are not in trouble.) If $m_\nu \approx 0$, which is the case experimentally, it is impossible to distinguish the following three cases

(a) $C'_V = 0$ (even coupling)

(b) $C_V = 0$ (odd coupling)

(c) $C_V \neq 0, C'_V \neq 0$ (parity nonconservation)

We simply need to replace $|C_V|^2$ in the old theory (even coupling case) with $|C_V|^2 + |C'_V|^2$ in the new theory. Thus measurement of the energy and angular distributions throws no light on the question of parity conservation in this interaction.

With $|C_V| = |C'_V|$, if we average over the dependence on $\vec{p}_e\cdot\vec{p}_\nu$ in the final distribution, the entire electron energy dependence comes from the phase space factor, which has the form $p_e E_e p_\nu E_\nu$ as shown in many text books. If $m_\nu = 0$, this is just $p_e E_e E_\nu^2$. Thus if one takes the experimental data and plots

$$\sqrt{\frac{W(E_e)}{p_e E_e}} \quad \text{vs.} \quad E_e$$

where $W(E_e)$ is the electron energy spectrum, a straight line corresponding to $E_\nu = E_e^{(\max)} - E_e$ should result provided that $m_\nu = 0$. Such a plot is known as a Kurie plot. In practice we must also take into account the final state Coloumb interaction between the electron and the proton. In particular, the region of minimum $\bar{\nu}$ energy (maximum electron energy) is most sensitive to the neutrino mass. The H^3 decay experiments show that, if $C_V = \pm C_V$ and $C'_A = \pm C_{A'}$ are assumed, then $m_\nu < 200$ ev.

The angular distribution for the vector interaction obtained above is of the form $1 + a\left(\frac{v}{c}\right)_e \cos\theta_{e\nu}$ where $a = +1$. If we calculate the corresponding distribution for the other interactions (S, T, A), we obtain the same form but with values shown in Table 3.6.

TABLE 3.6

The Electron-Neutrino Correlation Coefficient

	S	V	T	A
a	-1	$+1$	$+\frac{1}{3}$	$-\frac{1}{3}$

Thus one sees that for the V interaction the relativistic electron and the antineutrino are emitted preferentially in the same direction, but just the opposite is the case for the S interaction.

If both S and V are present, we obtain an additional term (the so-called Fierz interference term)

$$2\text{Re}(C_S C_V^* + C'_S C'_V{}^*)\frac{m_e}{E_e}$$

which would distort the linearity of the Kurie plot. No evidence for such an interference term has been reported.

It is evident that only certain quantities are sensitive to whether or not parity is conserved in the decay process. One of the sensitive quantities is the electron helicity $\mathscr{H}(e^-)$. We have seen in Chapter 2, Section 7 that the covariant spin projection operator in the direction of motion is $\frac{1}{2}(1 + i\gamma_5\gamma\cdot w)$ with $w_\mu = \left(\frac{E_e\hat{p}_e}{m_e}, \frac{i|\vec{p}_e|}{m_e}\right)$. With the use of this operator

$$\mathscr{H}(e^-) = \frac{\text{Tr}(i\gamma_5\gamma\cdot wMM^\dagger)}{\text{Tr}(MM^\dagger)}$$

where $MM^\dagger$ is given by (3.67). Explicit computations show

$$\mathscr{H}(e^-) = \frac{-2\,\mathrm{Re}\,(C_V C_V'^*)}{|C_V|^2 + |C_V'|^2} \frac{\left(\dfrac{|\vec{p}_e|}{E_e} + \hat{p}_e \cdot \vec{p}_\nu\right)}{1 + \dfrac{\vec{p}_e \cdot \vec{p}_\nu}{E_e E_\nu}} \tag{3.69}$$

Thus $\mathscr{H}(e^-)$ *is* sensitive to the question of whether or not both C_V and C_V' are nonvanishing.

If we do not observe the $\bar{\nu}$, we must average over the p_ν direction. Then

$$\mathscr{H}(e^-) = -\frac{2\,\mathrm{Re}\,(C_V C_V'^*)}{|C_V|^2 + |C_V'|^2} \frac{|\vec{p}_e|}{E_e} \tag{3.70}$$

So the maximum possible longitudinal polarization is v/c in magnitude, which is attained when $C_V = \pm C_V'$. Consider the case $C_V = C_V'$, which actually corresponds to reality. Then the relativistic electron is left-handed. This is quite reasonable since $\frac{1}{2}\bar{e}(1 - \gamma_5) = \frac{1}{2}((1 + \gamma_5)e)^\dagger\gamma_4$ creates left-handed electrons as $v/c \to 1$.

If we observe antineutrinos, the longitudinal polarization for the electron can be complete. Consider the case $C_V = C_V'$, and select those events in which the $\bar{\nu}$ and e^- are emitted in the same direction. Then (3.69) gives $\mathscr{H}(e^-) = -1$ independently of the electron energy. The reader may verify that this result is obvious from elementary angular momentum considerations.

PROBLEM 4. Given

$$\mathrm{H_{int}} = (\bar{p} i\gamma_5\gamma_\mu n)(\bar{e} i\gamma_5\gamma_\mu(C_A + C_A'\gamma_5)\nu) + \mathrm{H.c.}$$

compute the energy and angular distributions of e and ν

(a) when the neutron is unpolarized,
(b) when the neutron is 100% polarized.

CHAPTER 4

Time Reversal

4.1. Classical Physics

"Of time you would make a stream upon whose bank you would sit and watch its flowing." So sang the poet Kahlil Gibran. From our daily experience we are intuitively aware of the direction of time; in most natural processes—biological, chemical or physical—the concept of "time's arrow" seems self-evident. Yet Newtonian mechanics defines no sense of time's direction. If we assume that the forces on a particle are not dissipative and, further, are time-independent, then Newton's law is

$$m\frac{d^2\vec{x}}{dt^2} = \vec{F}(\vec{x}) \tag{4.1}$$

Clearly if $\vec{x}(t)$ is a possible trajectory, then likewise so is $\vec{x}(-t)$ since

$$m\frac{d^2\vec{x}}{d(-t)^2} = \vec{F}(\vec{x})$$

Consider a trajectory from $\vec{x}(\tau)$ (with $\tau < 0$) to $\vec{x}(0)$. At $t = 0$, "stop" and reverse the momentum and angular momentum; e.g. $\vec{p}'(t = 0) = -\vec{p}(t = 0)$. With the new initial conditions, let the system evolve in the same force field. The trajectory will be traced exactly backwards. As an example, imagine a motion picture of a falling body. If the film is run backwards, one cannot distinguish between this film and a motion picture of the same body thrown upwards.

In statistical mechanics, microscopic reversibility holds; that is, the collision shown in Fig. 4.1(a) is as likely as that shown in Fig. 4.1(b). We may recall that Boltzmann's H theorem (which essentially states that a nonequilibrium macroscopic state tends to approach a more "probable" state in equilibrium as time goes on) is actually derived on the basis of the principle of *microscopic* reversibility. As a simple illustrative example, suppose one had a box partitioned into two compartments with initially all of the N molecules confined in one of the compartments as in Fig. 4.2(a). Remove the partition and take motion pictures. This time, if the film is run backwards, we do see a difference. This is because the initial configuration was so well prepared that it occurs very seldom. Note that for small N we *cannot* tell as illustrated in Fig. 4.2(b).

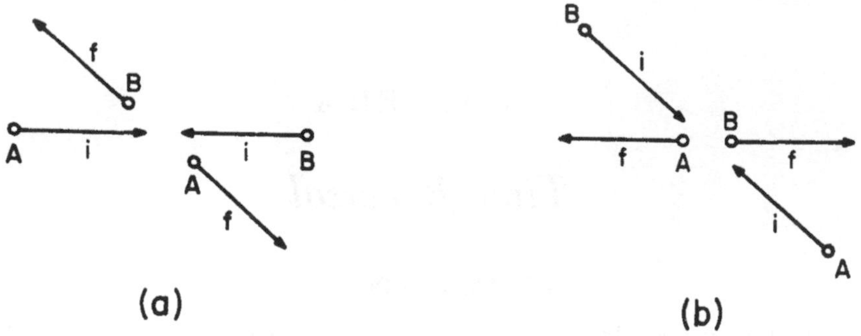

Figure 4.1. Microscopic reversibility

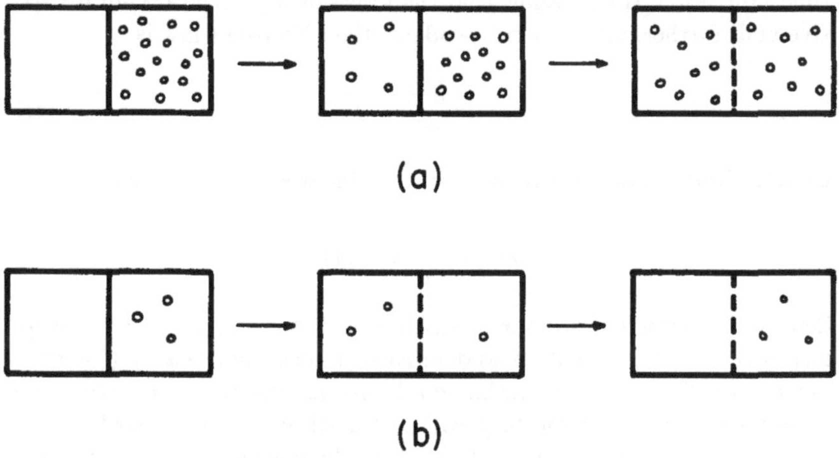

Figure 4.2. Molecules in a box

In general for diffusion-like processes, one does have a sense of direction of time. Consider the equation for heat conduction

$$\frac{\partial T}{\partial t} = \kappa \nabla^2 T$$

where κ and T stand for the diffusivity and temperature respectively. Suppose we start with some well-prepared distribution of temperature. As time goes on the temperature distribution becomes more and more uniform. For instance, if an instantaneous heat source is applied at $t = t_0$, $\vec{x} = \vec{x}_0$, the subsequent temperature distribution is given by

$$T \propto \frac{1}{[4\pi(t - t_0)]^{3/2}} \exp\left[-\frac{\kappa(\vec{x} - \vec{x}_0)^2}{4|t - t_0|}\right]$$

for infinite region problems. T is δ-function like as $t \to t = 0+$. At later times the temperature distribution peaked at $\vec{x} = \vec{x}_0$ becomes less and less sharp. As $t \to \infty$, the temperature distribution becomes completely uniform.

4.2. Wigner Time Reversal in Quantum Mechanics

The Schrödinger equation is

$$H\psi = i\frac{\partial\psi}{\partial t} \tag{4.4}$$

It has a diffusion-like appearance; $\psi(-t)$ does *not* satisfy (4.4). For instance, if $\psi \propto \exp[i(pz - Et)]$, the Schrödinger equation gives $\frac{p^2}{2m}\psi = E\psi$. But $\psi \propto \exp[i(pz + Et)]$ would give $\frac{p^2}{2m}\psi = -E\psi$, a contradiction. Wigner (1932, 1959) showed the way out of this dilemma. Let the time reversed state be $T\psi(t)$. We require the probability of finding a particle which will be the same; hence the norm is preserved.

$$\langle\psi|\psi\rangle = \langle T\psi|T\psi\rangle \tag{4.5}$$

This would be true if either

$$\langle\phi|\psi\rangle = \langle T\phi|T\psi\rangle \tag{4.6}$$

or

$$\langle\phi|\psi\rangle^* = \langle\psi|\phi\rangle = \langle T\phi|T\psi\rangle \tag{4.7}$$

In either case, the squares of scalar products are preserved. Since $T\psi(\vec{x}, t) = \psi(\vec{x}, -t)$ gets into trouble, we try

$$T\psi(\vec{x}, t) = \psi^*(\vec{x}, -t) \tag{4.8}$$

which implies (4.7). Equation (4.7) yields for any *Hermitian* operator Q that

$$\langle\psi|Q|\phi\rangle = \langle\phi|Q|\psi\rangle^* = \langle T\phi|TQT^{-1}|T\psi\rangle \tag{4.9}$$

Notice the difference between this expression and those for the other transformations we have studied. For instance, for parity we had

$$\langle\psi|Q|\phi\rangle = \langle P\psi|PQP^{-1}|P\phi\rangle$$

We now focus our attention on diagonal elements of Q. From (4.9) we have

$$\langle\psi|Q|\psi\rangle = \langle T\psi|TQT^{-1}|T\psi\rangle \tag{4.10}$$

This relation would be a trivial relation if T were unitary, but note that in our case, (4.7), T is not unitary. The diagonal elements of Hermitian operators may have classical analogs. We know from Ehrenfest's theorem that the expectation values of quantum mechanical observables approach classical quantities in certain limits. Now from the behavior of classical quantities under time reversal; e.g., $\vec{x} \xrightarrow{T} \vec{x}$, $\vec{p} \xrightarrow{T} -\vec{p}$ and $\vec{L} \xrightarrow{T} -\vec{L}$ we see that there are the two choices

$$TQT^{-1} = \pm Q\begin{cases}(+) \text{ for } \vec{x} \\ (-) \text{ for } \vec{p}, \vec{L}\end{cases} \tag{4.11}$$

corresponding to $\langle\psi|Q|\psi\rangle = \pm\langle T\psi|Q|T\psi\rangle$. For the operator $p = -i\vec{\nabla}$, $T(-i\vec{\nabla})T^{-1} = +i\vec{\nabla}$; hence $T \propto K$ where the transformation K is just that of complex conjugation on anything that stands to the right. $Ka = a^*K$ and

$$K\psi(\vec{x}) = \psi^*(\vec{x}). \tag{4.12}$$

Notice that K is not linear, i.e.

$$K(a_1\psi_1 + a_2\psi_2) = a_1^*\psi_1^* + a_2^*\psi_2^* \neq a_1\psi_1^* + a_2\psi_2^* \tag{4.13}$$

Wigner calls this property "anti-linear" and (4.7) is called the "anti-unitarity" condition.

We shall explicitly check the invariance of the commutation relations. First, notice that the classical (Poisson) bracket relation is invariant under T

$$[x_i, p_j]_{\text{classical}} = \sum_k \left(\frac{\partial x_i}{\partial x_k}\frac{\partial p_j}{\partial p_k} - \frac{\partial p_j}{\partial x_k}\frac{\partial x_i}{\partial p_k}\right) = \delta_{ij}$$

In quantum mechanics, we have

$$[x_i, p_j] = i\delta_{ij} \tag{4.14}$$

which is preserved since $x_i \xrightarrow{T} x_i$, $p_j \xrightarrow{T} -p_j$ and $i \xrightarrow{T} -i$. The orbital angular momenta commutators are also preserved since $\vec{L}_i \xrightarrow{T} -\vec{L}_i$ and $i \xrightarrow{T} -i$

$$[L_i, L_j] = iL_k, \qquad \{ijk\} = \{123\} \tag{4.15}$$

Although spin has no classical analog, its angular momentum properties suggest

$$T\vec{S}T^{-1} = -\vec{S} \tag{4.16}$$

If we assume $T = K$, then the effect on the $\vec{\sigma}$ matrices is

$$\begin{aligned} K\sigma_1K^{-1} &= \sigma_1 \\ K\sigma_2K^{-1} &= -\sigma_2 \\ K\sigma_3K^{-1} &= \sigma_3 \end{aligned} \tag{4.17}$$

Clearly $T = K$ is not sufficient here if (4.16) is to hold. But

$$T = \eta\sigma_2K, \qquad |\eta| = 1 \tag{4.18}$$

is appropriate.

More generally let

$$T = UK \tag{4.19}$$

Then (4.7) reads $\langle \phi^*/\psi^* \rangle = \langle \phi/\psi \rangle^* = \langle UK\phi/UK\psi \rangle = \langle U\phi^*/U\psi^* \rangle$. Since this must be true for all ϕ, ψ

$$U^\dagger U = I \tag{4.20}$$

U is unitary. Thus the time reversal transformation is the product of complex conjugation and a unitary transformation. Now if ψ satisfies the Schrödinger equation, (4.4), the condition for $\psi(t) = UK\psi(-t)$ to satisfy the same equation is clearly

$$K^{-1}U^{-1}HUK = H \tag{4.21}$$

This relation can be used to determine U explicitly in a given case.

It is to be emphasized that a particular form of U depends on the particular representation we use. The appearance of σ_2 in (4.18) is a direct consequence of the fact that we have used the "usual" representation of the Pauli matrices in which σ_2 is imaginary and σ_1 and σ_3 are real. In the momentum representation $T = K$ is not enough even for a spinless system. We must perform the transformation $\vec{p} \to -\vec{p}$ in addition to complex conjugation so that $x_k = i\dfrac{\partial}{\partial p_k} \to -i\dfrac{\partial}{\partial(-p_k)}$, $\phi(\vec{p}) \to \phi^*(-\vec{p})$, where $\phi(\vec{p})$ is a wave function in momentum space.

Using more or less heuristic arguments we have arrived at the "rule," $T = UK$. We could, of course, start with $T = UK$, and then *prove* (4.7), $\langle T\phi|T\psi \rangle = \langle \psi|\phi \rangle$. See Wigner (1932, 1959) and Wick (1958).

As a first example, consider the wave-function for a spinless particle traveling along $+z$.

$$\psi(z, t) \propto \exp\,[i(pz - Et)]$$

This satisfies the Schrödinger equation and so does

$$\psi^*(z, -t) \propto \exp\,[-i(pz + Et)] \tag{4.22}$$

However, (4.22) corresponds to a particle traveling along $-z$, as expected.

For spin one-half, the situation is slightly more complicated. Since $T = \eta\sigma_2 K$, we have

$$\begin{aligned} T\begin{pmatrix}1\\0\end{pmatrix} &= i\eta\begin{pmatrix}0\\1\end{pmatrix} \\ T\begin{pmatrix}0\\1\end{pmatrix} &= -i\eta\begin{pmatrix}1\\0\end{pmatrix} \end{aligned} \tag{4.23}$$

Equation (4.23) shows that the spin does indeed change direction. The phase η is essentially arbitrary and if the choice $\eta = 1$ is made, then

$$T\psi_{1/2}^{m} = i^{2m}\psi_{1/2}^{-m} \tag{4.24}$$

This relation can be made quite general. For orbital angular momentum we have

$$Y_{l,m} = (-1)^m\left[\frac{2l+1}{4\pi}\frac{(l-|m|)!}{(l+|m|)!}\right]^{1/2} P_l^m(\cos\phi)\exp(\text{im}\,\varphi) \quad (4.25)$$

for $m \geq 0$. When $m < 0$ the $(-1)^m$ factor is not included. This convention is standard (see p. 17 ff. of Condon and Shortley, 1951). Thus

$$Y_{l,m}^* = (-1)^m Y_{l,-m}$$

and

$$T\psi_l^m = i^{2m}\psi_l^{-m} \quad (4.26)$$

The case of arbitrary J can be treated by decomposing the wave-function into products, $Y_{l,m}\chi_{m_s}$ (Sachs, 1953, Appendix 3). Then $T = \sigma_2^{(1)}\ldots\sigma_2^{(N)}K$ and one finds that for appropriate choice of phase

$$T\psi_J^M = i^{2M}\psi_J^{-M} \quad (4.27)$$

While the phase factor i^{2m} is a matter of convention, there is no phase ambiguity for T^2, e.g. for Pauli two-component spinors,

$$T^2\begin{pmatrix}1\\0\end{pmatrix} = UKUK\begin{pmatrix}1\\0\end{pmatrix} = \sigma_2\sigma_2^*|\eta|^2\begin{pmatrix}1\\0\end{pmatrix} = -\begin{pmatrix}1\\0\end{pmatrix}.$$

Recall $|\eta|^2 = 1$. Generally, $T^2 = 1$ for integral J (which can be readily verified for $Y_{l,m}$) while $T^2 = -1$ for half-integral J. In fact one can show from the antiunitarity of the T operator that

$$\text{if } T^2 = \omega I, \text{ then } \omega = \pm 1 \quad (4.28)$$

PROOF: Since $UKUK = \omega$, then $U = \omega KU^\dagger K^{-1} = \omega U^T$ (recall $K^2 = 1$, $U^\dagger U = I$ and $KUK^{-1} = U^*$). Iterating $U = \omega^2 U^{T^T} = \omega^2 U$, then $\omega = \pm 1$.

Maxwell's equations are (in vacuum)

$$\begin{aligned}\vec{\nabla}\cdot\vec{E} &= \rho\\ \vec{\nabla}\cdot\vec{B} &= 0\\ \vec{\nabla}\times\vec{E} &= -\frac{\partial\vec{B}}{\partial t}\\ \vec{\nabla}\times\vec{B} &= \vec{j} + \frac{\partial\vec{E}}{\partial t}\end{aligned} \quad (4.29)$$

The electric charge is unchanged under time reversal while the current which is the product of charge times velocity changes sign: $\rho \xrightarrow{T} \rho$, $\vec{j} \xrightarrow{T} -\vec{j}$. Hence the equations will be invariant if

$$\begin{aligned}T\vec{E}(t) &= \vec{E}(-t)\\ T\vec{B}(t) &= -\vec{B}(-t)\end{aligned} \quad (4.30)$$

Then for the Poynting vector, $\vec{G} = \vec{E} \times \vec{B} \xrightarrow{T} -\vec{G}$, i.e. the energy flow is reversed. In terms of the potentials

$$\vec{E} = -\nabla\varphi + \frac{\partial\vec{A}}{\partial t}$$
$$\vec{B} = \vec{\nabla} \times \vec{A} \tag{4.31}$$

which implies

$$T\varphi(t) = \varphi(-t)$$
$$T\vec{A}(t) = -\vec{A}(-t) \tag{4.32}$$

This could also have been guessed from $\Box A_\mu = -j_\mu$. Note that $A_\mu \xrightarrow{T^2} +A_\mu$ as one expects for a spin one field.

Some interesting examples result from the stationary states of electrons in electro-magnetic fields. If there is an *external* magnetic field, $\vec{B}$, and ψ_J^M is a solution of the Schrödinger equation for this field, then $T\psi_J^M \propto \psi^{-M}_J$ is a solution of the equation with the same energy but for the field $-\vec{B}$.

A less trivial example is that of Kramers' degeneracy (Kramers, 1930). Suppose there is an external electric field, $\vec{E}$. Now if ψ is a solution of $H\psi = E\psi$, then $T\psi$ is also a solution with the same energy, $H(T\psi) = E(T\psi)$ in the *same* field $\vec{E}$. Suppose there is a degeneracy present in the case of zero field. The question here is whether the field, $\vec{E}$, could possibly result in the removal of the degeneracy. The degeneracy would be removed if $T\psi$ and ψ are essentially the same for each level (otherwise there would still be two states with the same energy), or, more precisely, if $T\psi = a\psi$, where a is a constant (also $|a|^2 = 1$ from normalization). But iterating this equation, $T^2\psi = |a|^2\psi = \psi$. However, we saw that $T^2 = +1$ is true for integral spin only. Thus a system of half-integral spin has an inherent two-fold degeneracy no matter how peculiar the field, $\vec{E}$, may be. This is a very far-reaching consequence of time reversal invariance.

Solid state physics has an important application of these considerations. In paramagnetic salts the electrons are in an inhomogeneous electric field due to the neighboring atoms. The $Y_{l,m}$ do not satisfy $T\psi = a\psi$. However, there are two superpositions of $Y_{l,m}$ and $Y_{l,-m}$ that do; namely, $P_l^m \cos m\phi$ and $P_l^m \sin m\phi$. Clearly, those levels that are split by $\vec{E}$ can not have m as a good quantum number. The spin wavefunction for an even number of electrons behaves similarly, e.g.

$$\psi = \frac{1}{\sqrt{2}}[\alpha(1)\alpha(2) \pm \beta(1)\beta(2)]$$

satisfies $T\psi = \mp\psi$. It is said that the magnetic quantum number m is "quenched" by $\vec{E}$ if the degeneracy is removed in such a way that the expectation value of m is zero for each level, e.g.

$$\langle P_l^m \cos m\phi | i \frac{\partial}{\partial\phi} | P_l^m \cos m\phi \rangle = 0$$

Curie's law for the paramagnetic susceptibility $\chi \propto \frac{1}{T}$, is dependent upon m being a good quantum number. Hence, deviations would be expected for ionic salts with m quenched (even number of electrons). Now in salts of Cu^{++} which has a configuration of one electron missing from the closed shell, Curie's law is followed. However, in Ni^{++} salts with two electrons missing from the closed shell, there are important deviations at low temperatures.

Historically, Kramers showed that if $\psi(\vec{x}, \ldots, \vec{x}_N; \vec{s}, \ldots, \vec{s}_N)$ is a solution in field $\vec{E}$, then $(-1)^N \psi^*(\vec{x}, \ldots, \vec{x}_N; -\vec{s}, \ldots, -\vec{s}_N)$ is also a solution in $\vec{E}$. Then he derived the degeneracy theorem. Two years later, Wigner showed that Kramers' transformation corresponded to time reversal. Thus in many ways the history of T (Kramers' theorem) is like the history of P (Laporte's rule).

One parenthetical remark: Equation (4.7), $\langle\phi|\psi\rangle^* = \langle\psi|\phi\rangle = \langle T\phi|T\psi\rangle$, suggests an alternative approach to time reversal due to Schwinger (1951): $|\psi\rangle \rightleftarrows \langle\psi|$ under time reversal. In the Schwinger procedure bras and kets are transformed into each other. The concept of the complex conjugation does not so directly appear. Instead the concept of transposition plays a crucial role. Since $(AB)^T = B^T A^T$, this led Pauli to make the following famous remark: "Schwinger taught us how to get more out of formulae by reading from right to left as well as from left to right." The commutation relation

$$[x_i, p_j] = i\delta_{ij}$$

is preserved since

$$[x_i, p_j] \to [(-p_j)^T, x_i^T]$$

without changing $i \to -i$. For Hermitian operators transposition and complex conjugation amount to the same thing, so there is no difference. For non-Hermitian operators the time reversed operator in the Wigner procedure and that in the Schwinger procedure differ by Hermitian conjugation. Needless to say, the two procedures are completely equivalent since "observables" are to be represented by Hermitian operators.

4.3. Scattering and Reaction Theory; Reciprocity Relation, K Matrix

We briefly discuss a formal theory of scattering. Our treatment is somewhat superficial; for a more complete discussion, see Lippmann and Schwinger (1950) and Gell-Mann and Goldberger (1953). A scattering process is viewed as beginning at time $t = -\infty$ with free states, proceeding through $t = 0$ and the interaction, and finally at $t = +\infty$ one again has free states. A free state of energy, E, satisfies the equation

$$(H_0 - E)\Phi = 0 \tag{4.33}$$

where H_0 denotes the free Hamiltonian in the Schrödinger representation. When the interaction V is present, the eigenstate of the total Hamiltonian satisfies

$$(H_0 - E)\Psi = -V\Psi \tag{4.34}$$

Notice that the *same* energy, E, is implied. This, of course, requires certain behavior of the two states. Denoting $(H_0 - E)^{-1}$ by $\dfrac{1}{H_0 - E}$ and subtracting (4.33) from (4.34), we obtain

$$\Psi = \Phi - \frac{1}{H_0 - E} V\Psi \tag{4.35}$$

which is to be understood as an integral equation for Ψ. More explicitly, (4.35) can be written as

$$\psi_E = \phi_E - \int dE' \frac{1}{E' - E} \phi_{E'} \langle \phi_{E'} | V | \psi_E \rangle \tag{4.36}$$

where ϕ_E is a suitably normalized eigenstate of the unperturbed Hamiltonian. Now the operator $\dfrac{1}{H_0 - E}$ is ambiguous at $H_0 = E$. There are three cases of physical interest

$$\Psi^{(\pm)} = \Phi + \frac{1}{E - H_0 \pm i\varepsilon} V\Psi^{(\pm)} \qquad \varepsilon > 0 \tag{4.37a}$$

$$\Psi^{(S)} = \Phi + \Pr \frac{1}{E - H_0} V\Psi^{(S)} \tag{4.37b}$$

The corresponding states have simple meanings. Φ corresponds to the free, unperturbed wave; $\Psi = \Phi$ if there were no interaction. $\Psi^{(-)}$ includes the ingoing waves; i.e. at $t = -\infty$ it included no outgoing waves. $\Psi^{(+)}$ includes the outgoing waves. $\Psi^{(S)}$ corresponds to stationary waves. These statements can be verified by setting $\phi_{E'} = u_{E'} \exp(-iE't)$ and inserting the appropriate ε factor in (4.36).

It is often convenient to use the Möller wave operator, $\Omega^{(\pm)}$, defined by

$$\Omega^{(\pm)}\Phi = \Psi^{(\pm)} \tag{4.38}$$

which, because of (4.37) satisfy the operator integral equation

$$\Omega^{(\pm)} = 1 + \frac{1}{E \pm i\varepsilon - H_0} V\Omega^{(\pm)} \tag{4.39}$$

Also convenient is the $T^{(\pm)}$ operator defined by

$$T^{(\pm)} = V\Omega^{(\pm)} \tag{4.40}$$

Its matrix elements (between free states) are

$$\mathrm{T}_{ba}^{(+)} = \langle \Phi_b | \mathrm{T}^{(+)} | \Phi_a \rangle \tag{4.41}$$

The integral equation is

$$\mathrm{T}^{(+)} = V + V \frac{1}{E + i\varepsilon - H_0} \mathrm{T}^{(+)} \tag{4.42}$$

Equation (4.42) has the iterated solution

$$\mathrm{T}^{(+)} = V + V \frac{1}{E + i\varepsilon - H_0} V + V \frac{1}{E + i\varepsilon - H_0} V \frac{1}{E + i\varepsilon - H_0} V + \dots \tag{4.43}$$

There is also a formal solution

$$\mathrm{T}^{(+)} = V + V \frac{1}{E + i\varepsilon - H_0 - V} V \tag{4.44}$$

which one can easily prove, using the identity

$$\frac{1}{A - B} = \frac{1}{A}\left(1 + B \frac{1}{A - B}\right)$$

In Chapter 3, from the perturbation expansion of the S matrix, we had

$$S = 1 - 2\pi i \delta(E_f - E_i)\mathrm{T}^{(+)}$$

By comparing (3.12) with (4.43) one sees that $\mathrm{T}^{(+)}$ defined in two different ways are indeed identical. In the interaction representation formalism, one would have only the operator $U(t, t_0)$ which would take the state from that at t_0 to that at t. In terms of U, one can prove (Gell-Mann and Goldberger, 1953)

$$\begin{aligned} \Omega^{(+)} &= U(0, -\infty) \\ \Omega^{(-)} &= U(0, +\infty) = U^\dagger(\infty, 0) \\ S &= U(\infty, -\infty) \end{aligned} \tag{4.45}$$

This leads to some other useful relations

$$S = \Omega^{(-)\dagger}\Omega^{(+)} \tag{4.46}$$

$$S_{fi} = \langle \Phi_f | S | \Phi_i \rangle = \langle \Psi_f^{(-)} | \Psi_i^{(+)} \rangle \tag{4.47}$$

To see what happens under time reversal, first notice

$$\mathrm{T}^{(+)\dagger} = V + V \frac{1}{E - i\varepsilon - H_0 - V} V \tag{4.48}$$

since E, H_0, V are Hermitian and $(A^{-1})^\dagger = (A^\dagger)^{-1}$. However, if H_0 and V are invariant under T, we have

$$T\mathrm{T}^{(+)}T^{-1} = V + V \frac{1}{E - i\varepsilon - H_0 - V} V = \mathrm{T}^{(-)} \tag{4.49}$$

thus

$$\mathcal{T}\mathrm{T}^{(+)}\mathcal{T}^{-1} = \mathrm{T}^{(+)\dagger} = \mathrm{T}^{(-)} \tag{4.50}$$

Contrast the above result with

$$P\mathrm{T}^{(+)}P^{-1} = \mathrm{T}^{(+)}$$

From (4.50) one gets the most general form of the reciprocity relation

$$\langle\Phi_f|\mathrm{T}^{(+)}|\Phi_i\rangle = \langle\mathcal{T}\mathrm{T}^{(+)}\mathcal{T}^{-1}\mathcal{T}\Phi_i|\mathcal{T}\Phi_f\rangle = \langle\mathcal{T}\Phi_i|\mathrm{T}^{(+)}|\mathcal{T}\Phi_f\rangle \tag{4.51}$$

where (4.7) has been used.

As an example, suppose

$$\begin{aligned}\Phi_i &= |\alpha, J, M\rangle \\ \Phi_f &= |\beta, J, M\rangle\end{aligned} \tag{4.52}$$

where α and β denote quantum numbers such as channel and parity. Since

$$\mathcal{T}\Phi_i = i^{2M}|\alpha, J, -M\rangle \tag{4.53}$$

we have

$$\begin{aligned}\langle\beta, J, M|\mathrm{T}^{(+)}|\alpha, J, M\rangle &= i^{2M-2M}\langle\alpha, J, -M|\mathrm{T}^{(+)}|\beta, J, -M\rangle \\ &= \langle\alpha, J, M|\mathrm{T}^{(+)}|\beta, J, M\rangle\end{aligned} \tag{4.54}$$

where the last equality is the result of rotational invariance. Thus, $\alpha, J \rightarrow \beta, J$ is equally as probable as $\beta, J \rightarrow \alpha, J$.

On the other hand, consider the plane wave states with spin

$$\begin{aligned}\Phi_i &= |\alpha, \vec{p}_i, m_i\rangle \\ \Phi_f &= |\beta, \vec{p}_f, m_f\rangle\end{aligned} \tag{4.55}$$

where m_i and m_f refer to the z-components of the initial and final spin.

$$\mathcal{T}\Phi_i = i^{2m_i}|\alpha, -\vec{p}_i, -m_i\rangle$$

Then (4.51) gives

$$\begin{aligned}&\langle\beta, \vec{p}_f, m_f|\mathrm{T}^{(+)}|\alpha, \vec{p}_i, m_i\rangle \\ &\qquad = i^{2m_i-2m_f}\langle\alpha, -\vec{p}_i, -m_i|\mathrm{T}^{(+)}|\beta, -\vec{p}_f, -m_f\rangle\end{aligned} \tag{4.56}$$

The factor $i^{2m_i-2m_f} = \pm 1$ since there can be no transitions between half-integral and integral J systems. If parity is conserved (4.56) becomes

$$\langle\beta, \vec{p}_f, m_f|\mathrm{T}^{(+)}|\alpha, \vec{p}_i, m_i\rangle = \pm\langle\alpha, \vec{p}_i, -m_i|\mathrm{T}^{(+)}|\beta, \vec{p}_f, -m_f\rangle \tag{4.57}$$

This is not the same as detailed balance since the spins are reversed. If the spins are summed over, we have (Coester 1951)

$$\begin{aligned}&\sum |\langle\beta, \vec{p}_f, m_f|\mathrm{T}^{(+)}|\alpha, \vec{p}_i, m_i\rangle|^2 \\ &\qquad = \sum |\langle\alpha, \vec{p}_i, m_i|\mathrm{T}^{(+)}|\beta, \vec{p}_f, m_f\rangle|^2\end{aligned} \tag{4.58}$$

Heitler calls this "semi-detailed balance."

There is a classical analog to this situation (Blatt and Weisskopf, 1952). Boltzmann originally pointed out that there is no detailed balance if the objects have a non-spherical shape. This point is illustrated in Fig. 4.3. The reciprocal collision occurs with the same probability as the original collision while the "detailed balancing" collision need not.

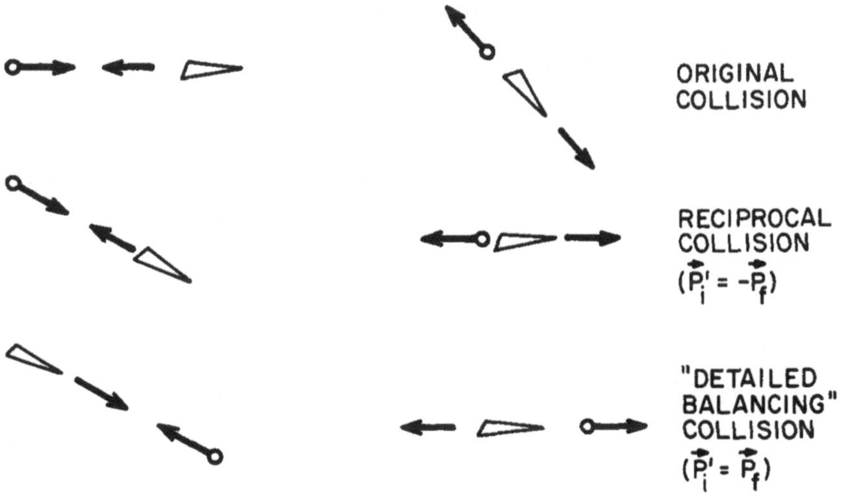

Figure 4.3. Reciprocal collision and "detailed balancing" collision for objects with non-spherical shapes

Marshak (1951) and Cheston (1951) proposed a method for determining the π^+ spin by applying the principle of detailed balancing to the reactions

$$p + p \rightleftarrows \pi^+ + d$$

The differential cross sections for unpolarized initial beams are given by

$$\frac{d\sigma}{d\Omega}(pp \to \pi d) = \frac{2\pi}{v_{pp}} \frac{1}{4} \sum_{\text{spin}} |T_{ab}^{(+)}|^2 \frac{p_\pi^2}{(2\pi)^3 v_{\pi d}}$$

$$\frac{d\sigma}{d\Omega}(\pi d \to pp) = \frac{2\pi}{v_{\pi d}} \frac{1}{3(2S_\pi + 1)} \sum_{\text{spin}} |T_{ab}^{(+)}|^2 \frac{p_p^2}{(2\pi)^3 v_{pp}} \tag{4.59}$$

where $v_{\pi d}(v_{pp})$ refers to the relative velocity of the πd (pp) system. This leads to

$$\frac{d\sigma}{d\Omega}(pp \to \pi d) = \frac{3}{4}(2S_\pi + 1) \frac{p_\pi^2}{p_p^2} \frac{d\sigma}{d\Omega}(\pi d \to pp)$$

for the differential cross sections, and

$$\sigma(pp \to \pi d) = \tfrac{3}{2}(2S_\pi + 1) \frac{p_\pi^2}{p_p^2} \sigma(\pi d \to pp)$$

for the total cross sections. The two processes are to be compared at the same c.m. energies. The reaction $pp \to \pi d$ was studied at Berkeley (Cartwright *et al.*, 1951), and $\pi d \to pp$ at Columbia and Rochester (Durbin, Loar, and Steinberger, 1951; Clark, Roberts, and Wilson, 1951). In both cases the angular distributions are of the form $0.1 + \cos^2\theta$ at 21 to 25 Mev above the πd threshold, and from the relative magnitude $2S_\pi + 1 = 1$ or $S_\pi = 0$ was deduced.

The importance of the density of final states can be illustrated in a picturesque fashion. Parking a car is difficult but leaving a parking space is easy. (cf. Fig. 4.4.)

Figure 4.4. Available phase space in parking a car and in leaving a parking space

In low energy nuclear physics, the reaction $N^{14}(d, \alpha)C^{12}$ and $C^{12}(\alpha, d)N^{14}$ have been studied (Bodansky *et al.*, 1959). As is shown in Fig. 4.5, the average difference is less than 6% in agreement with time reversal invariance.

It is important to notice that detailed balance is possible even if T invariance is violated; i.e., T invariance is a sufficient but not a necessary condition. Let us suppose that the interaction is weak; then $T^{(+)} \approx V$ (see (4.43)); thus, the Hermiticity of V is sufficient,

$$\langle \Phi_f | V | \Phi_i \rangle = \langle \Phi_i | V | \Phi_f \rangle^*$$

For other cases where detailed balance throws no light on T invariance, see Henley and Jacobsohn (1959).

Spin zero-spin $\frac{1}{2}$ scattering, e.g., π-p or p-α is also interesting. Consider (4.56) when $m_i = m_f$. It must hold whether or not the quantization axis is along the beam direction. We can write down for the elements of the transition matrix, M,

$$\begin{aligned}\langle \vec{p}_f, \vec{\sigma}\cdot\hat{\omega} &= 1 | M(\vec{\sigma}, \vec{p}_i, \vec{p}_f) | \vec{p}_i, \vec{\sigma}\cdot\hat{\omega} = 1 \rangle \\ &= \langle -\vec{p}_i, \vec{\sigma}\cdot\hat{\omega} = -1 | M(\vec{\sigma}, \vec{p}_i, \vec{p}_f) | -\vec{p}_f, \vec{\sigma}\cdot\hat{\omega} = -1 \rangle \qquad (4.60)\end{aligned}$$

where $\hat{\omega}$ is a unit vector in *any* direction. Now from rotational invariance M has the form

$$M = a + b\vec{\sigma}\cdot\hat{n} + c\vec{\sigma}\cdot\hat{p}_i + d\vec{\sigma}\cdot\hat{p}_f$$

$$\hat{n} = \frac{|\vec{p}_i \times \vec{p}_f|}{\vec{p}_i \times \vec{p}_f} \qquad (4.61)$$

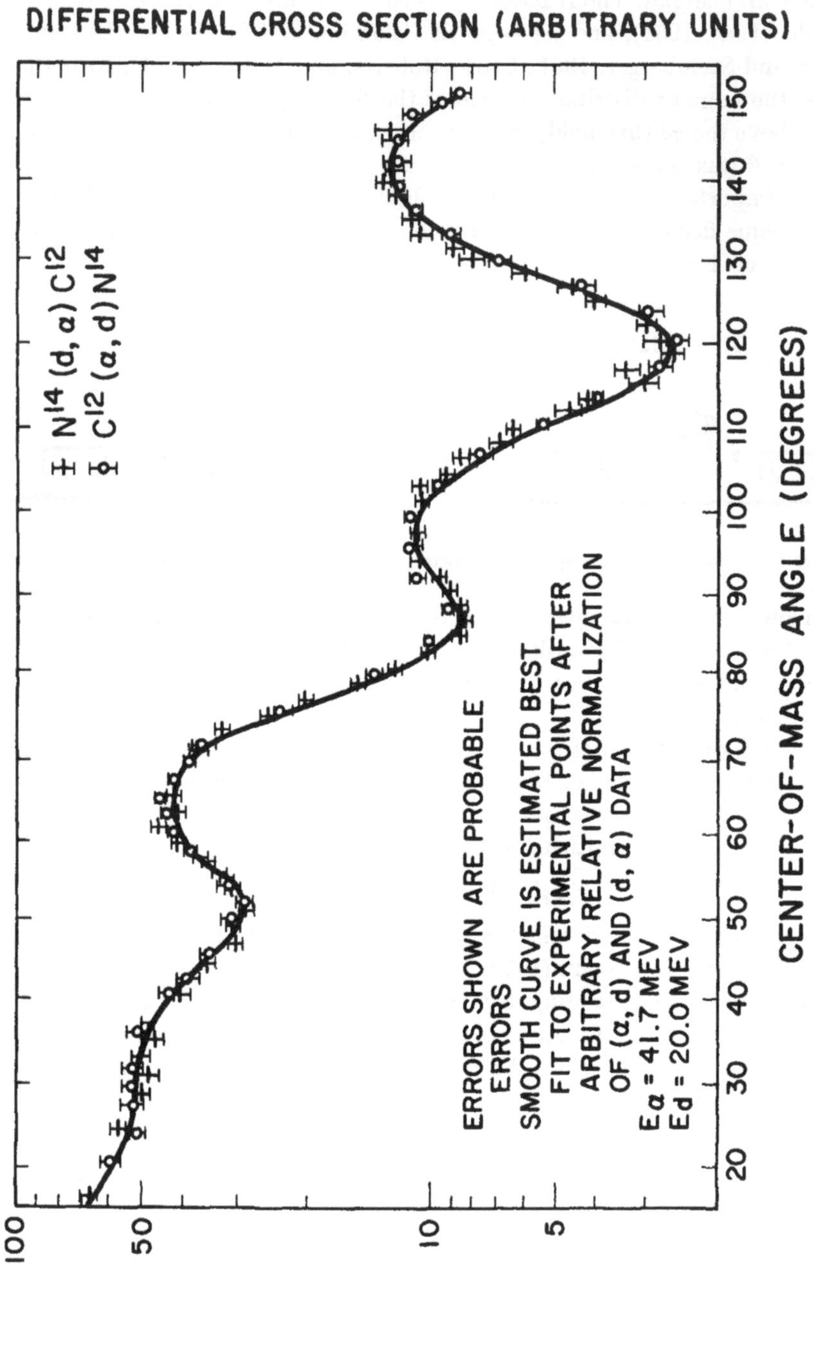

Figure 4.5. Comparison between $N^{14}(d, \alpha)C^{12}$ and $C^{12}(\alpha, d)N^{14}$

If T invariance holds, M must be unchanged under the substitutions $\vec{\sigma} \to -\vec{\sigma}$ and $\vec{p}_i \to -\vec{p}_f$, hence $c = d$. However, if parity is conserved, c and d cannot appear at all. The presence of the b term is compatible with T invariance, so polarization effects shed no further light on T invariance if parity is conserved.

In spin $\frac{1}{2}$-spin $\frac{1}{2}$ scattering (e.g., pp) things are more complicated (Wolfenstein and Ashkin, 1952; Dalitz 1952). Let $\hat{N}$ be a unit vector along $\vec{p}_i \times \vec{p}_f$, $\hat{P}$ be a unit vector along $\vec{p}_i + \vec{p}_f$, and $\hat{K}$ be a unit vector along $\vec{p}_f - \vec{p}_i$. Note that $\hat{K}$, $\hat{P}$ and $\hat{N}$ form a set of orthornormal vectors if $\vec{p}_i$ and $\vec{p}_f$ are measured in the center-of-mass system. The transition matrix, which is now a 4 by 4 matrix in the composite spin space, has the form

$$M = a + b(\sigma_N^{(1)} + \sigma_N^{(2)}) + c\sigma_N^{(1)}\sigma_N^{(2)} + (d + e)\sigma_P^{(1)}\sigma_P^{(2)} \\ + (d - e)\sigma_K^{(1)}\sigma_K^{(2)} + f(\sigma_K^{(1)}\sigma_P^{(2)} + \sigma_P^{(1)}\sigma_K^{(2)}) \tag{4.62}$$

where $\sigma_N^{(1)}$ stands for $\vec{\sigma}^{(1)} \cdot \hat{N}$ etc. If T invariance hold, M must be unchanged under the substitutions $\vec{\sigma}^{(1)} \to -\vec{\sigma}^{(1)}$, $\vec{\sigma}^{(2)} \to -\vec{\sigma}^{(2)}$, $\hat{N} \to -\hat{N}$, $\hat{P} \to -\hat{P}$, and $\hat{K} \to \hat{K}$. Thus the f term would be present only if T invariance were violated. The first quantity of interest is the polarization of initially unpolarized protons produced in scattering from an unpolarized target,

$$P(\theta) = \frac{\mathrm{Tr}\,(MM^\dagger \vec{\sigma}^{(1)} \cdot \hat{N})}{\mathrm{Tr}\,(MM^\dagger)} \tag{4.63}$$

On the other hand, one may start with a proton polarized 100% along $\hat{\omega}$ (in practice, along the normal to the scattering plane in the previous scattering) and study the azimuthal angular distribution produced by the scattering for some fixed θ. Since the density matrix for a 100% polarized beam is $\frac{1}{2}(1 + \vec{\sigma}^{(1)} \cdot \hat{\omega})$, we obtain the azimuthal distribution

$$1 + \alpha(\theta)\hat{N} \cdot \hat{\omega} = \frac{\mathrm{Tr}\,(M(1 + \vec{\sigma}^{(1)} \cdot \hat{\omega})M^\dagger)}{\mathrm{Tr}\,(MM^\dagger)} \tag{4.64}$$

where we have used the fact that $\mathrm{Tr}(M\vec{\sigma}^{(1)}M^\dagger)$ is necessarily in the $\hat{N}$ direction (which is obvious from parity conservation). $\mathrm{Tr}\,(M\vec{\sigma}^{(1)}M^\dagger) \neq 0$ implies that there is a left-right asymmetry, e.g. for fixed θ different counting rates at $\phi = 90^0$ ($\hat{\omega}$ and $\hat{N}$ antiparallel) and -90^0 ($\hat{\omega}$ and $\hat{N}$ parallel). See Fig. 4.6 for geometry.

Explicit computations show

$$\tfrac{1}{4}\,\mathrm{Tr}\,(MM^\dagger\vec{\sigma}^{(1)}) = [2\,\mathrm{Re}\,\{(a + c)b^*\} + 4\,\mathrm{Im}\,(f^*e)]\hat{N}$$

$$\tfrac{1}{4}\,\mathrm{Tr}\,(M\vec{\sigma}^{(1)}M^\dagger) = [2\,\mathrm{Re}\,\{(a + c)b^*\} - 4\,\mathrm{Im}\,(f^*e)]\hat{N} \tag{4.65}$$

(Actually for pp scattering the ab^* term must be absent if the effect of the Pauli principle is properly taken into account. See Wolfenstein (1954).) We have seen that if time reversal invariance holds, then f must vanish identically. Then Tr $(MM^\dagger\vec{\sigma}^{(1)})$ and Tr $(M\vec{\sigma}^{(1)}M^\dagger)$ must be equal by T invariance. Thus the polarization-asymmetry equality

$$P(\theta) = a(\theta) \tag{4.66}$$

is a consequence of time reversal invariance. (Note that for a spin $\frac{1}{2}$-spin 0 scattering the equality relation follows immediately from parity conservation.) Experimentally, in pp scattering

$$P(\theta) - a(\theta) = -0.014 \pm 0.014$$

for $\theta = 30^0$ in center-of-mass energies of 180 to 200 Mev where $P(\theta)$ and $a(\theta)$ are of the order of 25% (Abashian and Hafner, 1958). The polarization asymmetry equality was also verified in p-Ni, p-Fe, and p-Co scattering (Rosen and Brolley, 1959).

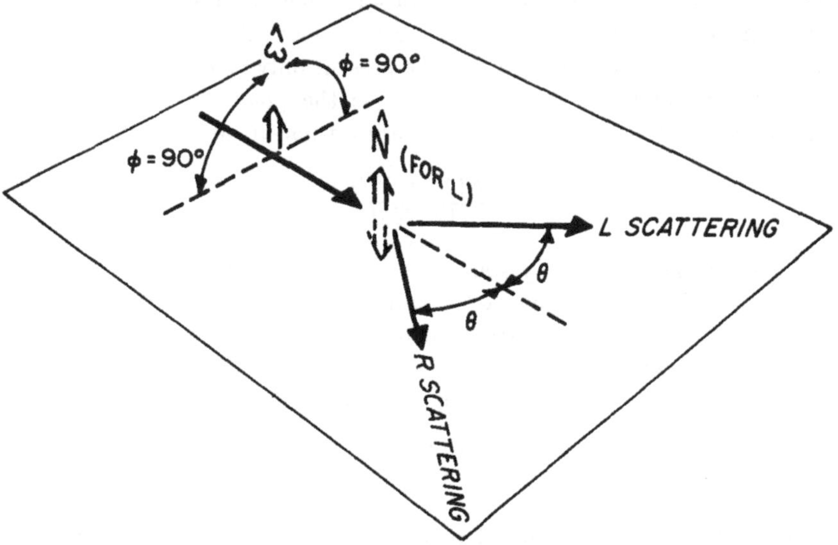

Figure 4.6. Left-right asymmetry in the scattering of a polarized particle

We may mention in passing that, even with $f = 0$, the pp scattering amplitude contains five complex amplitudes to be determined. In other words, we have ten real functions of θ. Of these the common phase can be determined from interference of the nuclear scattering with Coulomb scattering. Besides the differential cross section and the polarization (asymmetry) we still need seven experiments. In triple scattering experiments in which one analyzes the polarization after the second scattering, we can determine

$$R_{ij} = \tfrac{1}{4}\,\mathrm{Tr}\,(M\sigma_i^{(1)}M^\dagger\sigma_j^{(1)}) \tag{4.67}$$

where $i, j = 1, 2, 3$ refer to $\hat{N}$, $\hat{P}$, $\hat{K}$. Thus R_{NN}, R_{KK}, R_{PP}, and R_{PK} (R_{NP} is identically zero) are in principle measurable. We still need three more. These would be provided by correlation experiments in which the polarization of the scattered particle and that of the recoil particle are measured in coincidence. From such experiments one determines

$$C_{ij} = \tfrac{1}{4}\,\mathrm{Tr}\,(MM^{\dagger}\sigma_i^{(1)}\sigma_j^{(2)}) \tag{4.68}$$

In practice some of these experiments are exceedingly difficult to perform. At Berkeley the so-called D, R, and A parameters (which are related to our R_{ij}) as well as $d\sigma/d\Omega$ and $P(\theta)$ have been measured at $E_{\mathrm{lab}}^{(\mathrm{K.E.})} = 310$ Mev. Since the various amplitudes are functions of phase shifts, the values at one angle are not independent of the values at other angles. This means that it is possible to carry out a phase-shift analysis to determine M without performing all possible experiments. For further reading consult a review article by Wolfenstein (1956) and an experimental paper by Chamberlain *et al.* (1957).

The K matrix is another useful concept in scattering theory. K satisfies the equation

$$K = V + V\,\mathrm{Pr}\,\frac{1}{E - H_0}\,K \tag{4.69}$$

From now on the Pr (Principal value) can be omitted and considered as understood. In contrast to $\Psi^{(\pm)} = \Omega^{(\pm)}\Phi$, $K\Phi$ represents a stationary, standing wave. The formal solution is

$$K = V + V\,\frac{1}{E - H_0 - V}\,V \tag{4.70}$$

Before proceeding, some useful formulae will be recorded

$$\lim_{\epsilon\to 0}\frac{1}{\pi}\int_{-\infty}^{\infty}\frac{x}{x^2 + \varepsilon^2}f(x)\,dx = \mathrm{Pr}\int_{-\infty}^{\infty}\frac{f(x)}{x}\,dx \tag{4.71}$$

$$\lim_{\epsilon\to 0}\frac{1}{\pi}\int_{-\infty}^{\infty}\frac{\varepsilon}{x^2 + \varepsilon^2}f(x)\,dx = f(0) \tag{4.72}$$

and using (4.71) and (4.72)

$$\frac{1}{E \pm i\varepsilon - H_0} = \mathrm{Pr}\,\frac{1}{E - H_0} \mp i\pi\delta(E - H_0) \tag{4.73}$$

The relation between K and $T^{(+)}$ is

$$K = \mathrm{T}^{(+)} + i\pi K\delta(E - H_0)\mathrm{T}^{(+)} \tag{4.74}$$

To prove this, note

$$\mathrm{T}^{(+)} + i\pi K\delta(E - H_0)\mathrm{T}^{(+)}$$
$$= V + V\frac{1}{E - H_0}\mathrm{T}^{(+)} - i\pi V\delta(E - H_0)\mathrm{T}^{(+)} + i\pi K\delta(E - H_0)\mathrm{T}^{(+)}$$
$$= V + V\frac{1}{E - H_0}[\mathrm{T}^{(+)} + i\pi K\delta(E - H_0)\mathrm{T}^{(+)}]$$
$$- i\pi\underbrace{\left(V + V\frac{1}{E - H_0}K - K\right)}_{\text{zero}}\delta(E - H_0)\mathrm{T}^{(+)}$$

Now K and $\mathrm{T}^{(+)} + i\pi K\delta(E - H_0)\mathrm{T}^{(+)}$ satisfy the same inhomogeneous integral equation with the same boundary conditions; hence they are equal. In practice, in studying the energy dependence of the cross sections, it is often convenient to separate out the energy dependence expected from "kinematical factors." Since

$$\sum_m \langle n|\delta(E - H_0)|m\rangle\langle m|\mathrm{T}^{(+)}|j\rangle = \sum_m \rho_n(E)\langle n|m\rangle\langle m|\mathrm{T}^{(+)}|j\rangle$$

where $\rho_n(E)$ denotes the phase-space density for states of the system $|n\rangle$, (4.74) can be written as

$$K = \mathrm{T}^{(+)} + i\pi K\rho\mathrm{T}^{(+)} \tag{4.75}$$

where ρ denotes the matrix of phase space densities. If we are interested in the K and $\mathrm{T}^{(+)}$ matrices on the energy shell i.e. in those elements of the K and $\mathrm{T}^{(+)}$ matrices which are diagonal in the energy variable), we may make the following change of notation

$$\bar{K} = \pi\rho^{1/2}K\rho^{1/2}, \qquad \bar{\mathrm{T}}^{(+)} = \pi\rho^{1/2}\mathrm{T}^{(+)}\rho^{1/2} \tag{4.76}$$

Then (4.74) becomes

$$\bar{K} = \bar{\mathrm{T}}^{(+)} + i\bar{K}\bar{\mathrm{T}}^{(+)} \tag{4.77}$$

where $\bar{K}$ and $\bar{\mathrm{T}}^{(+)}$ are defined only on the energy shell.

The $\bar{K}$ matrix has the expansion (for given J)

$$\bar{K}_J = \sum_\alpha \sum_{\alpha'} \sum_M \langle\alpha|\bar{K}_J|\alpha'\rangle|\alpha, J, M\rangle\langle\alpha', J, M|$$

Note that $\bar{K}_J$ does not depend on M because of rotational invariance. The label α denotes channel, parity, isospin etc. Under time reversal

$$T\bar{K}_J T^{-1} = \sum_\alpha \sum_{\alpha'} \sum_M \langle\alpha|\bar{K}_J|\alpha'\rangle^* i^{2M-2M}|\alpha, J, -M\rangle\langle\alpha', J, -M| \tag{4.78}$$

but M is summed over. From (4.70) it is clear that if V and H_0 are invariant under T, then K is also. From (4.78) the elements of $\bar{K}_J$ are real if $\bar{K}_J$ and $T\bar{K}_J T^{-1}$ are to be equal. But K is also Hermitian from (4.70) since H_0 and V are Hermitian, hence also symmetric. If we know $\bar{K}$, we know $\bar{\mathrm{T}}^{(+)}$; it is gratifying that the real and symmetric $\bar{K}$ matrix completely characterizes the various reactions whose channels are open. In contrast, the elements of $\bar{T}^{(+)}$ are in general complex.

As a very simple example of the use of the K matrix formalism, consider elastic scattering when there is no other open channel. Denote an element of $\bar{K}$ by κ_A where A may stand for J, parity, etc. Define δ_A so that

$$\kappa_A = -\tan \delta_A \tag{4.79}$$

From (4.76), (4.79) implies that an element of $\bar{T}^{(+)}$ denoted by τ_A has the form

$$\tau_A = \frac{-\tan \delta_A}{1 - i \tan \delta_A} = -\exp(i\delta_A) \sin \delta_A \tag{4.80}$$

We now see that the meaning of δ_A is just that of the conventional phase shift. Notice that if the interaction is weak and repulsive, $V > 0$, then from (4.43) $\tau_A > 0$, and (4.80 implies, as expected, that $\delta_A < 0$. Suppose the elements of the K matrix (not $\bar{K}$ matrix) are slowly varying functions of energy. Then the entire energy dependence comes from the phase-space ρ factor in (4.76). Since ρ is proportional to q, (4.79) now reads

$$\frac{q}{\text{const.}} = -\tan \delta_A$$

or

$$q \operatorname{ctn} \delta_A = -\text{const.} \tag{4.81}$$

This is the well-known effective range formula in the zero-range approximation.

In the case of photoproduction we must consider several reactions

$$\gamma + p \leftrightarrow \begin{cases} \pi^+ + n \\ \pi^0 + p \end{cases} \qquad \text{(medium weak)}$$

$$\gamma + p \leftrightarrow \gamma + p \qquad \text{(weak)}$$

$$\pi^+ + n \leftrightarrow \begin{cases} \pi^0 + p \\ \pi^+ + n \end{cases} \qquad \text{(strong)}$$

Lumping together the two pion-nucleon channels together for simplicity (actually, we must analyze the pion-nucleon system in terms of two isospin channels, $T = \frac{1}{2}$ and $T = \frac{3}{2}$), we have

$$\kappa_A = \begin{matrix} & \gamma\text{-}p & \pi\text{-}N \\ \gamma\text{-}p & 0 & -\gamma \\ \pi\text{-}N & -\gamma & -\tan\delta \end{matrix} \qquad \text{with } \gamma \text{ real} \tag{4.82}$$

where we have ignored γ - p scattering.

Also

$$(1 + i\kappa_A)^{-1} = \frac{1}{1 - i \tan \delta} \begin{pmatrix} 1 - i \tan \delta & i\gamma \\ i\gamma & 1 \end{pmatrix}$$

Hence $\tau_A = (1 + i\kappa_A)^{-1}\kappa_A$

$$= -\begin{pmatrix} 0 & \gamma \exp(i\delta)\cos\delta \\ \gamma\exp(i\delta)\cos\delta & \exp(i\delta)\sin\delta \end{pmatrix} \qquad (4.83)$$

where terms $0(\gamma^2)$ have been ignored. Thus time reversal invariance implies that the phase of the photoproduction amplitude equals the phase of the corresponding scattering amplitude in the final state of the πN system as pointed out by Aizu (1954), Watson (1954), and Fermi (1955). (See also Gell-Mann and Watson (1954) whose derivation we have reproduced here).

Now to apply these considerations, assume that only $E1 \to s_{1/2}$ (with amplitude a) and $M1 \to p_{3/2}$ (with amplitude b) contribute. Using the production matrix discussed in Chapter 3, Section 4, we have

$$\frac{d\sigma}{d\Omega} = |a|^2 + \frac{|b|^2}{2}(2 + 3\sin^2\theta) - 2\,\mathrm{Re}\,(ab^*)\cos\theta \qquad (4.84)$$

The angular distribution for $\gamma + p \to \pi^+ + n$ is shown in Fig. 4.7. The pion-nucleon system exhibits a resonant behavior in the $p_{3/2}$ state. This means that the phase of b goes through 90°. Meanwhile the s wave phase shifts in πN scattering are small so that a is almost real. Then the forward-backward asymmetry in the photoproduction angular distribution characterized by the coefficient Re (ab^*) must change sign as the phase of b passes 90°. This is indeed observed.

It is interesting to note that essentially the same final state theorem holds for the reaction

$$\bar{K} + N \to \Sigma + \pi \quad \text{(exothermic)}$$
$$\pi + \Sigma \to \pi + \Sigma$$

at zero kinetic energy of the $\bar{K}N$ *system* even if $K^- + p \to K^- + p$ is "strong." This means that the phase of $\pi\Sigma$ scattering at the $\bar{K}N$ threshold can be determined from the phase of the $\bar{K}N \to \Sigma\pi$ absorption process. This is because the element of κ corresponding to the $\bar{K}N \to \bar{K}N$ channel vanishes not because the interaction is weak but because the phase space factor is zero at the $\bar{K}N$ threshold (Dalitz and Tuan, 1960).

The relation between the S and K operators is easily found from

$$S = 1 - 2\pi i\delta(E - H_0)\mathrm{T}^{(+)} \qquad (3.12)$$

and

$$\mathrm{T}^{(+)} = (1 + i\pi\delta(E - H_0)K)^{-1}K$$

We have

$$S = \frac{1 - i\pi\delta(E - H_0)K}{1 + i\pi\delta(E - H_0)K} \qquad (4.85)$$

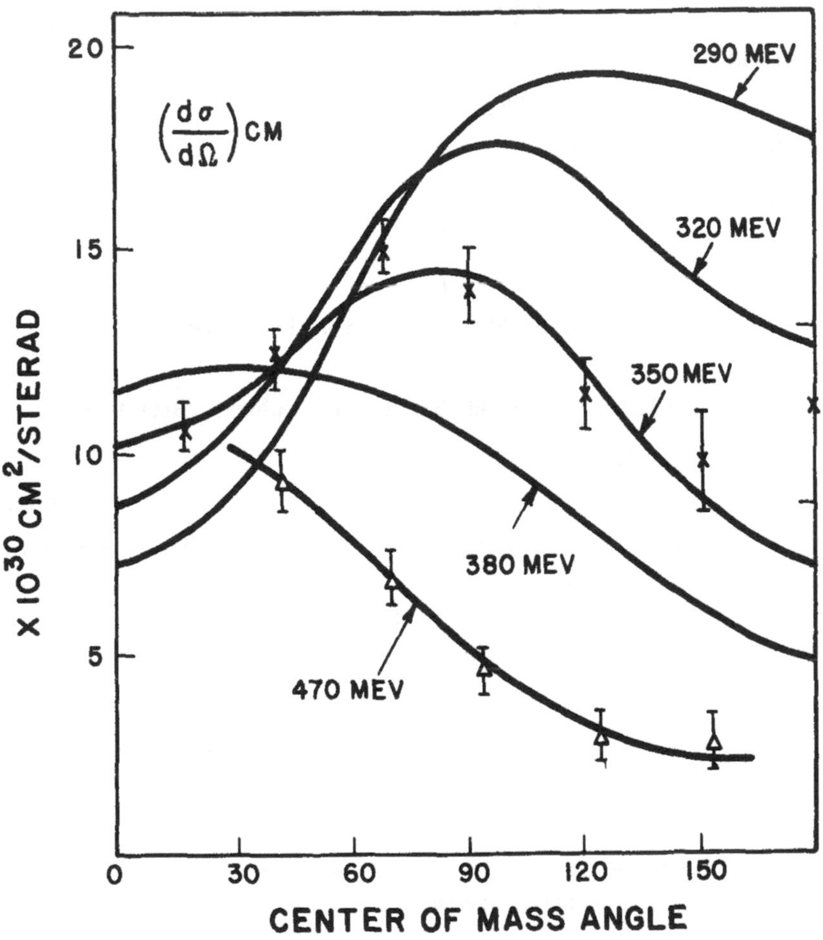

Figure 4.7. Angular distribution for $\gamma + p \rightarrow \pi^+ + n$

The Hermiticity of K makes the unitarity of S evident

$$S^\dagger = \frac{1 + i\pi\delta(E - H_0)K}{1 - i\pi\delta(E - H_0)K} = S^{-1} \tag{4.86}$$

Under time reversal

$$TST^{-1} = \frac{1 + i\pi\delta(E - H_0)K}{1 - i\pi\delta(E - H_0)K} = S^\dagger = S^{-1} \tag{4.87}$$

which was intuitively expected. Also we could use (3.12) backwards to find $T\mathrm{T}^{(+)}T^{-1}$,

$$S^\dagger = 1 + 2\pi i\delta(E - H_0)\mathrm{T}^{(+)\dagger} = TST^{-1} = 1 + 2\pi i\delta(E - H_0)T\mathrm{T}^{(+)}T^{-1} \tag{4.88}$$

hence

$$\mathcal{T}\mathrm{T}^{(+)}\mathcal{T}^{-1} = \mathrm{T}^{(+)\dagger} \tag{4.50}$$

from which follows

$$\langle \Phi_f | \mathrm{T}^{(+)} | \Phi_i \rangle = \langle \mathcal{T}\Phi_i | \mathrm{T}^{(+)} | \mathcal{T}\Phi_f \rangle \tag{4.51}$$

the reciprocity relation.

PROBLEM 5.

$$\mathcal{T}\Omega^{(+)}\mathcal{T}^{-1} = ?$$

$$\mathcal{T}\Omega^{(-)}\mathcal{T}^{-1} = ?$$

Show $\langle \Phi_f | S | \Phi_i \rangle = \langle \mathcal{T}\Phi_i | S | \mathcal{T}\Phi_f \rangle$ by using the above relations.

4.4. Time Reversal in Field Theory; "Diracology," Beta Decay

The Klein-Gordon equation

$$(\square^2 - m^2)\phi = 0 \tag{4.89}$$

involves a real differential operator. Hence one could equally well have either $\phi(\vec{x}, x_0) \xrightarrow{T} \phi^*(\vec{x}, -x_0)$ or $\phi(\vec{x}, x_0) \xrightarrow{T} \phi(\vec{x}, -x_0)$. But here we will require that the c-number wave function ϕ_c transform similarly to the Schrödinger wave function of the non-relativistic theory

$$\phi_c \xrightarrow{T} \phi_c^* \tag{4.90}$$

And the necessary connection between ϕ_c and the q-number field operator ϕ is

$$\phi_c = \langle \Phi_0 | \phi | a^\dagger(\vec{p}) \Phi_0 \rangle \tag{4.91}$$

which represents a one-particle state with momentum $\vec{p}$ ($|\Phi_0\rangle$ is the vacuum state). In the usual expansion

$$\phi = \frac{1}{(2\pi)^{3/2}} \int \frac{d^3p}{\sqrt{2p_0}}$$
$$\times \{a(\vec{p}) \exp[i(\vec{p}\cdot\vec{x} - p_0 x_0) + a^\dagger(\vec{p}) \exp[-i(\vec{p}\cdot\vec{x} - p_0 x_0)]\}$$
$$[a(\vec{p}), a^\dagger(\vec{p}\,')] = \delta^{(3)}(\vec{p} - \vec{p}\,')$$
$$a(\vec{p})|\Phi_0\rangle = 0 \tag{4.92}$$

Using (4.92) one can readily check that ϕ_c in (4.91) is indeed the usual plane wave solution. Using (4.90) and the relation (4.91) the transformation properties of ϕ can be determined,

$$\langle \mathcal{T}\Phi_0 | \phi | \mathcal{T}a^\dagger \Phi_0 \rangle = \langle \Phi_0 | \phi | a^\dagger \Phi_0 \rangle^* \tag{4.93}$$

where the left side is the wave function in the time reversed state. But

$$\langle \Phi_0 | \phi | a^\dagger \Phi_0 \rangle^* = \langle \mathcal{T}\Phi_0 | \mathcal{T}\phi a^\dagger \Phi_0 \rangle$$
$$= \langle \mathcal{T}\Phi_0 | \mathcal{T}\phi \mathcal{T}^{-1} | \mathcal{T}a^\dagger \Phi_0 \rangle \tag{4.94}$$

Thus

$$T\phi T^{-1} = \phi(\vec{x}, -x_0) \tag{4.95}$$

which should be compared to the c-number relation (4.90). Notice that this is consistent with invariance of the commutation relations,

$$\left[\phi(x), \frac{\partial\phi(x')}{\partial x_0}\right]_{x_0=x_0'} = i\delta^{(3)}(\vec{x} - \vec{x}') \tag{4.96}$$

which are transformed by $x_0 \to -x_0$, $i \to -i$. The transformation (4.95) can be achieved by transforming the creation and annihilation operators in the following manner,

$$\begin{aligned} Ta(\vec{p})T^{-1} &= a(-\vec{p}) \\ Ta^{\dagger}(\vec{p})T^{-1} &= a^{\dagger}(-\vec{p}) \end{aligned} \tag{4.97}$$

(within an irrelevant phase factor) just as one would expect.

In summary

$$\begin{aligned} \phi_c(\vec{x}, x_0) &\to \eta_T\phi_c^*(\vec{x}, -x_0), \quad \text{c-number} \\ T\phi(\vec{x}, x_0)T^{-1} &= \eta_T\phi(\vec{x}, -x_0), \quad \text{q-number} \end{aligned} \tag{4.98}$$

whereas under parity

$$\begin{aligned} \phi_c(\vec{x}, x_0) &\to \eta_P\phi_c(-\vec{x}, x_0), \quad \text{c-number} \\ P\phi(\vec{x}, x_0)P^{-1} &= \eta_P\phi(-\vec{x}, x_0), \quad \text{q-number} \end{aligned}$$

Before we discuss time reversal in the Dirac theory, first some mathematical results will be summarized (Pauli, 1936). A special form of Schur's lemma is: If A is a 4×4 matrix with $A\gamma_\mu A^{-1} = \gamma_\mu$ for every $\mu(=1, \ldots, 4)$, then A is a constant times the unit matrix. Pauli's fundamental theorem is also a special form of a general theorem in group theory. Given a set $\{\gamma_\mu\}$ and another set $\{\gamma_\mu'\}$ with $\{\gamma_\mu, \gamma_\nu\} = 2\delta_{\mu\nu}$ and $\{\gamma_\mu', \gamma_\nu'\} = 2\delta_{\mu\nu}$, there exists a nonsingular 4×4 matrix, S, such that $\gamma_\mu' = S\gamma_\mu S^{-1}$ and S is unique up to a constant. Only the uniqueness proof is included here.

PROOF:

Let $S_1\gamma_\mu S_1^{-1} = S_2\gamma_\mu S_2^{-1} = \gamma_\mu'$

Multiply from the left by S_1^{-1} and from the right by S_2.

Then $\gamma_\mu S_1^{-1}S_2 = S_1^{-1}S_2\gamma_\mu$, hence, by Schur's lemma

$$S_1^{-1}S_2 = \text{const.} \times I$$

Thus $S_2 = \text{const.} \times S_1$ Q.E.D.

The Dirac equation which the field operator ψ (in q number theory) must satisfy

$$\left(\gamma_j \frac{\partial}{\partial x_j} + \gamma_4 \frac{\partial}{\partial(ix_0)} + m\right)\psi = 0$$

can be made invariant under time reversal. In the time reversed state the equation reads

$$\left(\gamma_j^* \frac{\partial}{\partial x_j} + \gamma_4^* \frac{1}{(-i)} \frac{\partial}{\partial(-x_0)} + m\right)\psi' = 0 \tag{4.99}$$

In the usual way we assume

$$\psi' = \eta B \psi$$

where B is a 4×4 matrix. Then

$$B^{-1}\gamma_\mu^* B = \gamma_\mu \tag{4.100}$$

for the Dirac equation to be invariant. Note that $\{\gamma_\mu^*, \gamma_\nu^*\} = 2\delta_{\mu\nu}$; by Pauli's fundamental theorem, B exists and is unique up to a constant. B must satisfy other conditions, the invariance of $\rho = \psi^\dagger\psi = \bar{\psi}\gamma_4\psi$; thus $B^\dagger B = I$, B is unitary. Also B is antisymmetric if Hermitian γ matrices have been used. *Proof*: since $\gamma_\mu^* = \gamma_\mu^T = B\gamma_\mu B^{-1}$, $(B^{-1})^T\gamma_\mu^T B^T = \gamma_\mu$; iterating yields $\gamma_\mu = (B^{-1}B^T)^{-1}\gamma_\mu B^{-1}B^T$. By Schur's lemma $B^T = cB$; c is a constant. But iterating $B = c^2 B$, hence $c = \pm 1$. However, if $c = +1$, $B^T = B$, one can find a contradiction. $B\sigma_{\mu\nu}$ $(= -iB\gamma_\mu\gamma_\nu, \mu \neq \nu)$ is an antisymmetric matrix; $(-iB\gamma_\mu\gamma_\nu)^T = -i\gamma_\nu^T\gamma_\mu^T B^T = -iB\gamma_\nu\gamma_\mu = iB\gamma_\mu\gamma_\nu$ using $B^T = B$ and $B^{-1}\gamma_\mu^T B = \gamma_\mu$ twice. Similarly $(iB\gamma_5\gamma_\mu)^T = -iB\gamma_5\gamma_\mu$; hence if $B^T = +B$, there would exist ten linearly independent antisymmetric 4×4 matrices. Since this is impossible (there can only be six), $B^T = -B$. Notice also that $BB^* = -1$ since $B^* = (B^T)^\dagger = (-B)^\dagger = -B^{-1}$. This implies that $T^2\psi T^{-2} = T\eta B\psi T^{-1} = |\eta|^2 BB^*\psi = -\psi$, as expected for a half-integral J system.

In specific representations, B will have different forms. In the Dirac-Pauli representation γ_2 and γ_4 are real while γ_1 and γ_3 are imaginary. In order to have $B\gamma_{2,4}B^{-1} = \gamma_{2,4}$ and $B\gamma_{1,3}B^{-1} = -\gamma_{1,3}$ we must choose $B = \gamma_1\gamma_3$ $(= -i\Sigma_2)$ up to a phase factor. Notice that this satisfies $B^T = -B$. Just as in the Klein-Gordon case, $T\psi T^{-1} = \eta B\psi(\vec{x}, -x_0)$ in the q-number theory while $\psi_c \rightarrow B\eta\psi_c^*(\vec{x}, -x_0)$ in the c-number theory. Then the Dirac-Pauli wave function for a single particle with spin up

$$\psi_\uparrow = \begin{pmatrix} 1 \\ 0 \\ \dfrac{p_3}{E+m} \\ \dfrac{p_1 + ip_2}{E+m} \end{pmatrix} \exp\left[i(\vec{p}\cdot\vec{x} - p_0x_0)\right]$$

is transformed into

$$-i\eta \begin{pmatrix} 0 & -i & 0 & 0 \\ i & 0 & 0 & 0 \\ 0 & 0 & 0 & -i \\ 0 & 0 & i & 0 \end{pmatrix} \begin{pmatrix} 1 \\ 0 \\ \dfrac{p_3}{E+m} \\ \dfrac{p_1 - ip_2}{E+m} \end{pmatrix} \exp\left[-i(\vec{p}\cdot\vec{x} + p_0x_0)\right]$$

$$= \eta \begin{pmatrix} 0 \\ 1 \\ \dfrac{-(p_1 - ip_2)}{E + m} \\ \dfrac{p_3}{E + m} \end{pmatrix} \exp\left[-i(\vec{p}\cdot\vec{x} + p_0 x_0)\right]$$

This should be compared with the spin down solution

$$\psi_\downarrow = \begin{pmatrix} 0 \\ 1 \\ \dfrac{p_1 - ip_2}{E + m} \\ \dfrac{-p_3}{E + m} \end{pmatrix} \exp\left[i(\vec{p}\cdot\vec{x} - p_0 x_0)\right]$$

Physically the changes have been spin up $\xrightarrow{T}$ spin down, $\vec{p} \xrightarrow{T} -\vec{p}$, as expected.

$B = \gamma_1\gamma_3$ is true only in the Dirac Pauli representation but, of course, $B^T = -B$ is always true. This is different from the case of parity for which $P\psi P^{-1} = \eta\gamma_4\psi$ in any representation. That $B = \gamma_1\gamma_3$ is not universal should really be expected since there is no reason for nature to prefer the 1 or 3 axis in time reflection.

It is useful to discuss the notion of "equivalent" representations. Suppose we know ψ in one representation for which $\{\gamma_\mu, \gamma_\nu\} = 2\delta_{\mu\nu}$ and we want the form of ψ' in another representation for which $\{\gamma'_\mu, \gamma'_\nu\} = 2\delta_{\mu\nu}$ where $\gamma'_\mu = S\gamma_\mu S^{-1}$. Without guessing, we would try $\psi' = U\psi$. Since $\psi^\dagger\psi$ and $\psi'^\dagger\psi'$ must be equal, U must be unitary. Since $\bar{\psi}\gamma_\mu\psi$ must be equal to $\bar{\psi}'\gamma'_\mu\psi'$, we have $U^{-1}\gamma'_\mu U = \gamma_\mu$. Thus $U = S$ (up to a phase factor) will do.

The Majorana representation $\{\gamma_\mu^{(M)}\}$ is often useful: $\gamma_1^{(M)} = \gamma_4^{(D-P)}$ $\gamma_2^{(M)} = \gamma_2^{(D-P)}$, $\gamma_3^{(M)} = \gamma_5^{(D-P)}$, $\gamma_4^{(M)} = \gamma_1^{(D-P)}$ and $\gamma_5^{(M)} = \gamma_3^{(D-P)}$. The Dirac equation, $\left(\gamma_\mu \dfrac{\partial}{\partial x_\mu} + m\right)\psi = 0$ is purely real in this representation. Thus if ψ is a solution, ψ^* is also. The charge conjugation formalism becomes very simple in this representation, as we shall discuss in Chapter 5. However, remember that in the Dirac-Pauli representation the upper two components are large in the non-relativistic limit; the limit yields a Pauli spinor times a Schrödinger wave function. In the Majorana representation the large and small components are mixed up.

In the Weyl representation $\gamma_k^{(W)} = -\gamma_k^{(D-P)}$, $\gamma_4^{(W)} = -\gamma_5^{(D-P)}$ and $\gamma_5^{(W)} - \gamma_7^{(D-P)}$. This representation is useful in the extreme relativistic limit since

$$\gamma_\mu p_\mu = i\begin{pmatrix} 0 & \vec{\sigma}\cdot\vec{p} + p_0 \\ -\vec{\sigma}\cdot\vec{p} + p_0 & 0 \end{pmatrix} \tag{4.101}$$

Thus whenever $m \ll p_0, |\vec{p}|$ one has two *uncoupled* two component equations. The Weyl representation is used in the treatments of electron-proton scattering at high energies (Yennie, Ravenhall, and Wilson, 1954) and for the description of the neutrino. Chirality (γ_5) is diagonal,

$$\gamma_5 = -\begin{pmatrix} I & 0 \\ 0 & -I \end{pmatrix}$$

PROBLEM 6. (Trivial):

(1) Find S such that

$$\gamma_\mu^{(M)} = S\gamma_\mu^{(D-P)}S^{-1}$$
$$\gamma_\mu^{(W)} = S\gamma_\mu^{(D-P)}S^{-1}$$

(2) Write down the analogs of

$$\begin{pmatrix} 1 \\ 0 \\ \dfrac{p_3}{E+m} \\ \dfrac{p_1 + ip_2}{E+m} \end{pmatrix} \quad \text{and} \quad \begin{pmatrix} 0 \\ 1 \\ \dfrac{p_1 - ip_2}{E+m} \\ \dfrac{-p_3}{E+m} \end{pmatrix}$$

in the Majorana and Weyl representations.

(3) Find B and C such that

$$B\gamma_\mu B^{-1} = \gamma_\mu^* = \gamma_\mu^T$$
$$C^{-1}\gamma_\mu C = -\gamma_\mu^* = -\gamma_\mu^T$$

in the Majorana and Weyl representations.

From now on, particular representations will not be used but only the defining equation

$$B\gamma_\mu B^{-1} = \gamma_\mu^* = \gamma_\mu^T \tag{4.100$'$}$$

To examine the charge current, first, how does $\bar{\psi}$ transform?

$$\bar{\psi} \xrightarrow{T} \psi^\dagger B^\dagger \gamma_4^* B B^{-1} = \bar{\psi} B^{-1} \tag{4.102}$$

Then

$$\bar{\psi}\gamma_\mu\psi \xrightarrow{T} \bar{\psi}B^{-1}\gamma_\mu^* B\psi = \bar{\psi}\gamma_\mu\psi$$

or

$$j_k = i\bar{\psi}\gamma_k\psi \xrightarrow{T} -i\bar{\psi}\gamma_k\psi = -j_k \tag{4.103}$$
$$\rho = -ij_4 = \bar{\psi}\gamma_4\psi \xrightarrow{T} \bar{\psi}\gamma_4\psi = \rho$$

Recall also that $\vec{A} \xrightarrow{T} -\vec{A}$ and $A_0 \xrightarrow{T} A_0$, in fact, the Fourier components $(\vec{\varepsilon}_1 + i\vec{\varepsilon}_2)\exp[i(\vec{p}\cdot\vec{x} - p_0x_0)]$, $m = 1$ and $(\vec{\varepsilon}_1 - i\vec{\varepsilon}_2)\exp[-i(\vec{p}\cdot\vec{x} + p_0x_0)]$, $m = -1$ transform into each other. Hence the spin reverses its direction. But as for helicity, $\mathscr{H}_\gamma \to \mathscr{H}_\gamma$, since $\vec{p} \xrightarrow{T} -\vec{p}$. also.

Similarly, for the Klein-Gordon case

$$j_k = ie\left[\left(\frac{\partial\phi^\dagger}{\partial x_k} - ieA_k\phi^\dagger\right)\phi - \phi^\dagger\left(\frac{\partial\phi}{\partial x_k} + ieA_k\phi\right)\right] \xrightarrow{T} -j_k$$

$$\rho = -ij_4 = e\left[\left(\frac{\partial\phi^\dagger}{\partial(ix_0)} + eA_0\phi^\dagger\right)\phi - \phi^\dagger\left(\frac{\partial\phi}{\partial(ix_0)} - eA_0\phi\right)\right] \xrightarrow{T} \rho \tag{4.104}$$

as $\phi \xrightarrow{T} \eta\phi$.

The moment interactions can also be examined. The Pauli moment (anomalous magnetic moment) interaction that reduces to $\vec{\sigma}\cdot\vec{B}$ in the non-relativistic limit is

$$H_{\text{int}} = \frac{\mu^{(m)}}{2}\bar{\psi}\sigma_{\mu\nu}\psi F_{\mu\nu} \tag{4.105}$$

$\mu^{(m)}$ must be purely real from the Hermiticity requirement as the reader may verify. Under time reversal,

$$T\frac{\mu^{(m)}}{2}\bar{\psi}\sigma_{\mu\nu}\psi F_{\mu\nu}T^{-1} = -\frac{\mu^{(m)}}{2}\bar{\psi}B^{-1}\sigma_{\mu\nu}^{*}B\psi F_{\mu\nu}$$

$$= \frac{\mu^{(m)}}{2}\bar{\psi}\sigma_{\mu\nu}\psi F_{\mu\nu} \tag{4.106}$$

where we have used

$$TF_{\mu\nu}T^{-1} = -F_{\mu\nu}$$

for both $F_{ij} = B_k$ and $F_{4i} = iE_i$, and

$$B^{-1}\sigma_{\mu\nu}^{*}B = B^{-1}\frac{\gamma_\mu^{*}\gamma_\nu^{*}}{2i}B = -\sigma_{\mu\nu} \quad (\mu \neq \nu)$$

Hence the Pauli moment interaction is invariant under T. On the other hand, for the electric moment interaction, which can be obtained from the Pauli moment interaction by interchanging the electric and magnetic field $F_{\mu} \to \frac{i}{2}\varepsilon_{\mu\nu\lambda\sigma}F_{\lambda\sigma}$, we have (note $\gamma_5\sigma_{\mu\nu} = -\frac{1}{2}\varepsilon_{\mu\nu\lambda\sigma}\sigma_{\lambda\sigma}$)

$$T\left(-\frac{i}{2}\mu^{(e)}\bar{\psi}\gamma_5\sigma_{\mu\nu}\psi F_{\mu\nu}\right)T^{-1} = \frac{i}{2}\mu^{(e)}\bar{\psi}\gamma_5\sigma_{\mu\nu}\psi F_{\mu\nu} \tag{4.107}$$

where we have used $B^{-1}\gamma_5^{*}\sigma_{\mu\nu}^{*}B = -\gamma_5\sigma_{\mu\nu}$. Thus unless $\mu^{(e)}$ is purely imaginary, the presence of an electric moment interaction implies non-variance under time reversal. However, the Hermiticity of the Hamiltonian requires that $\mu^{(e)}$ be real. Therefore the electric moment interaction would violate both P and T invariance. This could also have been guessed from the non-relativistic limit

$$-\frac{i}{2}\mu^{(e)}\bar{\psi}\gamma_5\sigma_{\mu\nu}\psi F_{\mu\nu} \overset{\text{N.R.}}{\sim} \mu^{(e)}\vec{\sigma}\cdot\vec{E}$$

and $\vec{\sigma} \xrightarrow{T} -\vec{\sigma}$ and $\vec{E} \xrightarrow{T} \vec{E}$. In Chapter 3 we have already discussed the upper limits on the electric dipole moment of the neutron. The limits on the electric dipole moments of other fermions are summarized in Table 4.1.

TABLE 4.1

Upper Limits on the Electric Dipole Moments of Fermions

n	$e \times 5 \times 10^{-20}$ cm	(Smith, Purcell, and Ramsey, 1957)
p	$e \times 1.3 \times 10^{-13}$ cm	(Sternheimer, 1959)
μ	$e \times 2 \times 10^{-16}$ cm	(Berley and Gidal, 1960)
e	$e \times 3 \times 10^{-15}$ cm	(Nelson *et al.*, 1959)

Of course, in non-diagonal transitions an electric moment can be present; e.g., $\Sigma^0 \to \Lambda^0 + \gamma$ if the $\Lambda\Sigma$ parity were odd. The interaction would be represented phenomenologically by

$$H_{\text{int}} = i\mu^{(e)}_{\Lambda\Sigma}\bar{\psi}_\Lambda\gamma_5\sigma_{\mu\nu}\psi_\Sigma F_{\mu\nu} + \text{H.c.} \tag{4.108}$$

We need only choose η's so that $\eta_\Sigma = -\eta_\Lambda = \eta$. Then $\psi_\Sigma \xrightarrow{T} \eta B\psi_\Sigma$, $\psi_\Lambda \xrightarrow{T} -\eta B\psi_\Lambda$, and the interaction is invariant under T with $\mu^{(e)}$ real.

The non-derivative Yukawa coupling of a neutral meson ($\phi^\dagger = \phi$) to a spinor field can be generally written as

$$H_{\text{int}} = \phi(g_s\bar{\psi}\psi + ig_p\bar{\psi}\gamma_5\psi) \tag{4.109}$$

With this form, Hermiticity requires that g_s and g_p be real. Under T,

$$H_{\text{int}} \xrightarrow{T} \eta\phi(g_s\bar{\psi}\psi - ig_p\bar{\psi}\gamma_5\psi) \tag{4.110}$$

with $\phi \xrightarrow{T} \eta\phi$. Since ϕ is Hermitian, $\eta = \pm 1$. Thus scalar and pseudoscalar coupling cannot be mixed. This means that for the non-derivative Yukawa coupling of a neutral spin-zero boson field to a spinor field, time reversal invariance implies parity conservation. This theorem can be generalized to the charge-independent pion-nucleon interaction (Feinberg, 1957; Gupta, 1957; Solov'ev, 1957).

We now turn our attention to beta decay. The time reversal properties of the most general four-fermion point interaction are

$$\begin{aligned} T\{\sum_i (\bar{p}\Gamma_i n)(\bar{e}C_i\Gamma_i + C_i'\Gamma_i\gamma_5)\nu\}T^{-1} \\ = \eta_p^*\eta_n\eta_e^*\eta_\nu \sum_i (\bar{p}B^{-1}\Gamma_i^* Bn)\{(\bar{e}B^{-1}(C_i^*\Gamma_i^* + C_i'^*\Gamma_i^*\gamma_5^*)B\nu\} \\ = \eta_p^*\eta_n\eta_e^*\eta_\nu \sum_i (\bar{p}\Gamma_i n)\{\bar{e}(C_i^*\Gamma_i + C_i'^*\Gamma_i\gamma_5)\nu\} \end{aligned} \tag{4.111}$$

The relations $B^{-1}\Gamma_i^* B = \pm\Gamma_i$ and $B^{-1}\Gamma_i^*\gamma_5 B = \pm\Gamma_i\gamma_5$ have been used. The sign depends on whether the Γ_i contain i or not; i.e., 1, γ_μ, γ_5 have the (+) sign and $\sigma_{\mu\nu}$, $i\gamma_5\gamma_\mu$ have (−). If all the coupling constants are relatively real, we can choose $\eta_p^*\eta_n\eta_e^*\eta_\nu$ in such a way that the β interaction is invariant under time reversal; i.e. that $TH(\vec{x})T^{-1} = H(\vec{x})$ in the Schrödinger representation and $TH(\vec{x}, x_0)T^{-1} = H(\vec{x}, -x_0)$ in the interaction representation. That time reversal invariance imposes the reality condition on the coupling constants was first noted by Biedenbarn and Rose (1951).

In the decay of the free neutron (also see Problem 4), if one assumes $C_V = C_V'$, $C_A = C_A'$ and all the other constants are zero (which coupling constant combination will be justified in Chapter 7), then the angular distribution for polarized neutrons (omitting solid angles and phase space) is

$$\begin{aligned} w = (|C_V|^2 + 3|C_A|^2)\Bigg[1 &+ \frac{|C_V|^2 - |C_A|^2}{|C_V|^2 + 3|C_A|^2}\,\vec{p}_e\cdot\vec{p}_\nu \\ &+ \langle\vec{\sigma}_n\rangle\cdot\left(\frac{-2|C_A|^2 - 2Re(C_V C_A^*)}{|C_V|^2 + 3|C_A|^2}\,\frac{\vec{p}_e}{E_e}\right. \\ &+ \frac{2|C_A|^2 - 2Re(C_V C_A^*)}{|C_V|^2 + 3|C_A|^2}\,\frac{\vec{p}_\nu}{E_\nu} \\ &\left.+ \frac{2\,\mathrm{Im}\,(C_V C_A^*)}{|C_V|^2 + 3|C_A|^2}\,\frac{\vec{p}_e\times\vec{p}_\nu}{E_e E_\nu}\right)\Bigg] \end{aligned} \tag{4.112}$$

As measured at Chalk River by Robson, the electron-neutrino angular correlation coefficient ($\vec{p}_e\cdot\vec{p}_\nu$ term) is approximately zero; hence $|C_V| \approx |C_A|$. The electron asymmetry coefficient($\langle\vec{\sigma}_n\rangle\cdot\vec{p}_e$ term) was measured at Argonne, and it was $(-0.09 \pm 0.03)\,\frac{v_e}{c}$. The neutrino asymmetry coefficient, ($\langle\vec{\sigma}_n\rangle\cdot\vec{p}_\nu$ term) also measured at Argonne was 0.88 ± 0.15. Thus $C_V \approx -C_A$. The $\langle\vec{\sigma}_n\rangle\cdot\vec{p}_e$ and $\langle\vec{\sigma}_n\rangle\cdot\vec{p}_\nu$ terms violate parity but not time reversal. The $\langle\vec{\sigma}_n\rangle\cdot(\vec{p}_\nu\times\vec{p}_e)$ term (which would be present only if $\mathrm{Im}\,(C_V C_A^*) \neq 0$) is more interesting since it violates time reversal but not parity. Notice that

$$\begin{aligned} \langle\vec{\sigma}_n\rangle\cdot(\vec{p}_e\times\vec{p}_\nu) &= \langle\sigma_n\rangle\cdot[\vec{p}_e\times(-\vec{p}_e - \vec{p}_{\text{recoil}})] \\ &= -\vec{p}_{\text{recoil}}\cdot(\langle\vec{\sigma}_n\rangle\times\vec{p}_e). \end{aligned}$$

Thus the coefficient can be determined by counting the recoil protons above and below the $\langle\vec{\sigma}_n\rangle\times\vec{p}_e$ plane (see Fig. 4.8). The two numbers should be the same if T invariance holds. Experimentally, it was found that

$\dfrac{2\,\mathrm{Im}\,(C_V C_A^*)}{|C_V|^2 + 3|C_A|^2} = -0.04 \pm 0.07$ by Burgy *et al.* (1958b). Writing $C_A = \dfrac{|C_A|}{|C_V|}\,C_V \exp(i\lambda)$ where for $\lambda = 0$ or π, T invariance holds, the experimental value above implies $|\pi - \lambda| < 8°$. This experiment was suggested by Jackson, Treiman, and Wyld (1957).

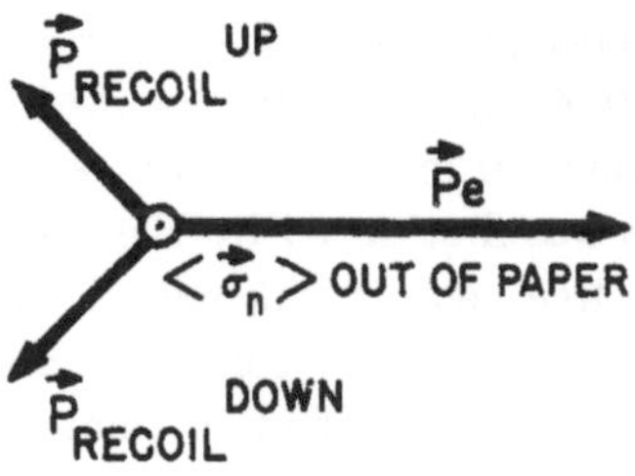

Figure 4.8. Test of time reversal invariance in neutron decay

4.5. Rule for Testing Time Reversal Invariance in Decay Processes

In decay processes, it is difficult to apply reciprocity since one must compare $n \rightarrow p + e^- + \bar{\nu}$ with $\bar{\nu} + e^- + p \rightarrow n$. The latter has not been seen in nature. But there is a rule that works solely because the interaction is weak (in the sense that first order perturbation theory is adequate). This rule, which we will derive, is not related to the reciprocity relation discussed earlier. The reciprocity relation works regardless of the strength of the interaction. The following rule is applicable *only* if the primary decay interaction is weak. This point is often obscure in the literature.

Case 1. If there is no final state interaction; e.g., $\mu^- + p \rightarrow \nu + n$; $K^+ \rightarrow \mu^+ + \nu$. If the decaying state is $|D_m\rangle$ the final state is $|\psi_f\rangle = H_W|D_m\rangle$ (since the $T^{(+)}$ matrix is essentially the interaction Hamiltonian itself). Treatment to lowest order in the decay interaction is justified because the decay interaction is weak. Suppose an observable, Q, is measured in the final state. There will be two kinds; e.g., $\vec{p}_e \cdot \vec{p}_\nu$, $\vec{\sigma} \cdot \vec{p}_e$ where $TQ_+T^{-1} = Q_+$ and $\vec{\sigma} \cdot (\vec{p}_g \times \vec{p}_\nu)$, $(\vec{\sigma}_1 \times \vec{\sigma}_2) \cdot \vec{p}$ where $TQ_-T^{-1} = -Q$. The final state matrix element is

$$\begin{aligned} \langle \psi_f | Q_\pm | \psi_f \rangle &= \langle TQ_\pm T^{-1} TH_W T^{-1} TD_m | TH_W T^{-1} TD_m \rangle \\ &= i^{2m-2m} \langle H_W D_{-m} | \pm Q_\pm | H_W D_{-m} \rangle \\ &= \langle H_W D_m | \pm Q_\pm | H_W D_m \rangle \end{aligned} \tag{4.113}$$

(where the last equality follows from rotational invariance). Thus, if the system is T invariant, then $\langle Q_- \rangle = 0$; i.e. observable quantities will not depend on terms like $\vec{\sigma} \cdot (\vec{p}_e \times \vec{p}_\nu)$. For instance, in $K^+_{\mu 3} \rightarrow \pi^0 + \mu^+ + \nu$ one can look at the polarization of the μ along the normal to the decay plane. If the term depending on $\vec{\sigma}_\mu \cdot (\vec{p}_\pi \times \vec{p}_\nu)$ is non-vanishing, then time reversal invariance does not hold.

Case 2. Suppose there is a final state interaction; i.e. the products of the primary decay interact further. In the decay of the neutron, the Coulomb attraction of the proton and electron was negligible. However, in $Co^{60} \to Ni^{60} + e^- + \bar{\nu}$ ($Z = 28$), it is appreciable. Also in $\Lambda \to p + \pi^-$ the π-p scattering affects the final state. Now this time

$$|\psi_f\rangle = U(\infty, 0)|H_W D_m\rangle \tag{4.114}$$

$U(\infty, 0) = U^\dagger(0, \infty) = \Omega^{(-)\dagger}$ where $\Omega^{(-)}$ satisfies

$$\Omega^{(-)} = 1 + \frac{1}{E - H_0 - i\varepsilon} V\Omega^{(-)}$$

Here V is the part of the Hamiltonian responsible for the *stronger* interaction in the final state. For instance, it would be the Coulomb potential in β decay and the πN Hamiltonian in Λ decay. If V is very small, then $\Omega^{(-)\dagger} \approx 1$ and Case 2 reduces to Case 1. $|\psi_f\rangle$ can be expanded in terms of the eigenfunctions of the non-interacting Hamiltonian $|w_{(\text{parity})}, J, M\rangle$.

$$|\psi_f\rangle = \Sigma|w, J, M\rangle\langle w, J, M|\Omega^{(-)\dagger}H_W|D_m\rangle \tag{4.115}$$

In general, when there is a final state interaction, we must project the state obtained by the primary decay interaction onto an incoming state of the final system. This is often a source of confusion. However, it arises in a natural manner in the general theory of scattering (see Gell-Mann and Goldberger, 1953).

$\Omega^{(-)}$ is unitary and

$$\Omega^{(-)}|w, J, M\rangle = \exp(-i\delta_{wJM})|w, J, M\rangle \tag{4.116}$$

Scattering with $\Omega^{(-)}$cannot change w, J, M which are good quantum numbers, and δ_{wJM} can be identified with the ordinary phase shift from the asymptotic form of the incoming wave function at large distances.

$$|\psi_f\rangle = \sum |w, J, M\rangle\langle w, J, M|H_W|D_m\rangle \exp(-i\delta_{wJM}) \tag{4.117}$$

If H_W is invariant under T, the expectation value, $\langle Q\rangle$, is

$$\begin{aligned}\langle\psi_f|Q|\psi_f\rangle &= \sum\sum \langle w', J', M'|Q_\pm|w, J, M\rangle\langle w, J, M|H_W|D_m\rangle \\ &\quad\times \langle w', J', M'|H_W|D_m\rangle^* \exp[i(\delta_{w'J'M'} - \delta_{wJM})] \\ &= \sum\sum \langle w, J, -M|\pm Q_\pm|w', J', -M'\rangle\langle D_{-m}|H_W|w, J, -M\rangle \\ &\quad\times \langle D_{-m}|H_W|w'J', -M'\rangle^* \exp[i(\delta_{w'J'M'} - \delta_{wJM})]\end{aligned} \tag{4.118}$$

Equation (4.118) can be rewritten by using rotational invariance and noting $\cos(\delta' - \delta)$ is even and $\sin(\delta' - \delta)$ is odd under the interchange of w, J, M and w', J', M'. In short, we can write $\pm\langle Q_\pm\rangle \cos(\delta' - \delta)$ for $\langle Q_\pm\rangle \cos(\delta' - \delta)$ and $\mp\langle Q_\pm\rangle \sin(\delta' - \delta)$ for $\langle Q_\pm\rangle \sin(\delta' - \delta)$. And the general rule (no accidental cancellations) is that those with $(-)$ signs, $\langle Q_-\rangle \cos(\delta' - \delta)$ and $\langle Q_+\rangle \sin(\delta' - \delta)$ are *not* compatible with T invariance. Of course, when $\delta' - \delta = 0$, Case 2 reduces to Case 1.

This situation may be contrasted with the case of parity where, say, $\langle \vec{J} \cdot \vec{p} \rangle$ always violates P regardless of the strength of the primary interaction and whether or not there is a final state interaction.

In the β decay of nuclei that are not too heavy, $Z \ll 137$, $\sin(\delta' - \delta) \propto Z/137$ and $\cos(\delta' - \delta) \approx 1$, one can look at the coefficient of $\langle \vec{J} \cdot \vec{p}_e \rangle$ e.g., in Co^{60} decay. If there were no final state interaction, the observation of such terms would shed no light on T invariance, the coefficient would not contain $\mathrm{Im}\,(C_i C_j^*)$. Lee and Yang (1956b) calculated this term and obtained

$$\frac{1}{|C_T|^2 + |C_A|^2 + |C_T'|^2 + |C_A'|^2} \times \left[Re(C_T C_T'^* - C_A C_A'^*) - \frac{Zm_e}{137 p_e} \mathrm{Im}\,(C_A C_T'^* - C_A' C_T^*) \right] \frac{v_e}{c} \tag{4.119}$$

If both A and T were present, the energy dependence of this term would tell us something about time reversal. Note that $\mathrm{Im}\,(C_A C_T'^* - C_A' C_T^*)$ appears together with a parameter that characterizes the final state interaction, namely $Z/137$.

In the decay of the $\Sigma^- \to n + \pi^-$, suppose the initial Σ^- is polarized. Measure the neutron polarization, $\langle \vec{\sigma}_n \rangle \cdot (\langle \vec{\sigma}_\Sigma \rangle \times \vec{p}_n)$. In Problem 3, we saw that the coefficient was proportional to $\mathrm{Im}\,(a_s a_p^*)$. If the neutron and π did not scatter after they were created, a_s and a_p would be relatively real, provided T invariance is good. But since they do interact, the relative phase of a_s and a_p is determined by the phase shift of the final state scattering. This can be proved, using the same argument as we used for photoproduction. In the Σ^- decay we know this phase shift, since from charge independence, the π^--n scattering should be the same as the π^+-p scattering. At 190 $\frac{\text{Mev}}{c}$ the s-wave phase shift is $-14°$ and that for the p-wave $-3°$. Thus, if T invariance is good, the phase difference between a_s and a_p should be at most 17°.

CHAPTER 5

Charge Conjugation

5.1. General Considerations

In the old days, before the advent of modern physics, the asymmetry between positive and negative charge was considered to be fundamental. If one rubs a glass rod with silk, the rod is left with a net positive charge, while a hard-rubber rod rubbed with cat's fur retains a net negative charge. It seems, at least in this example, that the two kinds of charge have intrinsically different origins. Yet it should be remembered that the laws of electromagnetism are symmetric between positive and negative charge: two positive charges repel each other in the same way as do two negative charges. One can, in fact, easily see that Maxwell's equations are invariant with respect to a change in sign of the charge density ρ. When we interchange $\rho \rightleftarrows -\rho$ we must also interchange $\vec{j} \rightleftarrows -\vec{j}$, $\vec{E} \rightleftarrows -\vec{E}$, and $\vec{B} \rightleftarrows -\vec{B}$. We suspect that there may be something of fundamental significance in this symmetry.

In a sense one can say that the Dirac theory predicted the existence of a particle with charge opposite to that of the electron. In his hole theory, a hole in the normally-filled negative energy states acted in every respect (but charge) like an ordinary particle—its charge was opposite, however.

The relativistic scalar wave equation was first proposed in 1926 and 1927 by Klein and Gordon. The expression which satisfies the continuity equation in this case is $i\left(\phi^* \dfrac{\partial \phi}{\partial x_0} - \dfrac{\partial \phi^*}{\partial x_0}\phi\right)$. As is quite evident from the form of this expression, an inherent difficulty arises when one tries to interpret it as a probability density. It can be negative. It remained for Pauli and Weisskopf to show in 1934 that this expression should be interpreted as the charge density and that one can associate positively and negatively charged particles with the *same* field ϕ. The subsequent discoveries of the various antiparticles were a triumph of relativistic quantum mechanics. Note that the notion of antiparticle is alien to non-relativistic quantum mechanics.

We postulate the existence of a unitary operator C with the property that

$$C|\Psi(Q, \vec{p}, \vec{s})\rangle = |\Psi(-Q, \vec{p}, \vec{s})\rangle \tag{5.1}$$

Note that the space properties of the state $|\Psi\rangle$ remain unchanged.

$|\Psi\rangle$ is here intended as a single particle state with charge Q, momentum $\vec{p}$, and spin $\vec{s}$. For the moment we accept the validity of this postulate. Later on we will show that the existence of the unitary operator C is compatible with the rest of quantum field theory (e.g. with the equation of motion and the commutation relations). Now we will derive several conclusions with only the assumption that C is unitary.

Suppose

$$CHC^{-1} = H \tag{5.2}$$

then it follows that for the S matrix

$$CSC^{-1} = S \tag{5.3}$$

using the same argument as we used for the parity operator (see Chapter 3, Section 1). Remember that the time-reversal operator T was not unitary and therefore $TST^{-1} = S^{-1} = S^{\dagger}$.

Consider the example of π^+-p scattering.

$$\begin{aligned}\langle \pi^+ p \text{ final } |S| \pi^+ p \text{ initial}\rangle &= \langle \pi^+ p \text{ final } |C^{-1}SC| \pi^+ p \text{ initial}\rangle \\ &= \langle \pi^- \bar{p} \text{ final } |S| \pi^- \bar{p} \text{ initial}\rangle \end{aligned} \tag{5.4}$$

This tells us that π^+-p scattering should be identical in every respect to π^--$\bar{p}$ scattering provided that the strong interactions responsible for π^+-p scattering are invariant under charge conjugation.

Note added in proof. In practice it is difficult to perform π^--$\bar{p}$ scattering. Instead we may study (Pais, 1959)

$$\bar{p} + p \rightarrow \pi^+ + \text{other particles}$$
$$\bar{p} + p \rightarrow \pi^- + \text{other particles}$$

in flight. Xuong, Lynch, and Hinrichs (1961) and Maglić, Kalbfleisch, and Stevenson (1961) measured the energy spectra and the angular distributions of π^+ and π^- created by an antiproton beam of $p_{\text{lab}} \sim 1.6$ BeV/c. To the extent that the two processes are charge conjugate to each other, we must have, in the c.m. system,

$$\left(\frac{d\sigma(\theta)}{d\Omega}\right)_{\pi^+} = \left(\frac{d\sigma(\pi - \theta)}{d\Omega}\right)_{\pi^-}$$

(Note that we compare a configuration in which a π^+ is emitted at an angle θ with respect to the incident $\bar{p}$ direction with a configuration in which a π^- is emitted at an angle $\pi - \theta$ with respect to the $\bar{p}$ direction.) Equality relations of this kind have been shown to be satisfied within experimental errors.

As an example from weak interactions, consider muon decay, $\mu^{\pm} \rightarrow e^{\pm} + \nu + \bar{\nu}$. Let H_W be the weak interaction Hamiltonian responsible for the μ decay. Since the interaction is weak, the transition amplitude into a particular helicity state of e^-, say left, is $\langle e_L^- \nu\bar{\nu}|H_W|\mu^-\rangle$. If H_W is invariant under C, then

$$\langle e_L^- \nu\bar{\nu}|H_W|\mu^-\rangle = \langle e_L^+ \bar{\nu}\nu|H_W|\mu^+\rangle \tag{5.5}$$

and similarly for e_R. Thus C invariance implies

$$\mathscr{H}(e^-) = \mathscr{H}(e^+) \tag{5.6}$$

This must be true even if the initial μ is unpolarized, and the ν, $\bar{\nu}$ momenta are not observed. These helicities were first measured at Liverpool (Culligan *et al.*, 1957) and Berkeley (Macq, Crowe, and Haddock, 1958) by looking at the circular polarization of bremsstrahlung produced by e^+ and e^- from muon decay. The total transmission cross section of photons through magnetized iron depends on the photon helicity. The results were $\mathscr{H}(e^+) \approx 1$ and $\mathscr{H}(e^-) \approx -1$ from unpolarized muons at rest. This shows that charge conjugation invariance is violated in muon decay independent of the details of any theory.

5.2. Charge Conjugation Parity and Selection Rules

We may ask what kinds of states are eigenstates of C. Let

$$Q^{(\mathrm{op})}|\Psi(Q)\rangle = Q|\Psi(Q)\rangle \tag{5.7}$$

where $Q^{(\mathrm{op})}$ stands for the electric charge operator; Q, for the electric charge. When we operate with C, we have

$$CQ^{(\mathrm{op})}|\Psi(Q)\rangle = QC|\Psi(Q)\rangle = Q|\Psi(-Q)\rangle \tag{5.8}$$

and in opposite order:

$$Q^{(\mathrm{op})}C|\Psi(Q)\rangle = Q^{(\mathrm{op})}|\Psi(-Q)\rangle = -Q|\Psi(-Q)\rangle \tag{5.9}$$

Thus we have the important result that

$$[C, Q^{(\mathrm{op})}] \neq 0 \quad \text{if } Q \neq 0 \tag{5.10}$$

or that states with non-zero charge cannot be eigenstates of C. *Some* states with zero charge *can*, however, be eigenstates of C. Actually C has other properties than we have thus far shown. Among several other attributes which C changes there are baryon number and strangeness. The π^0, photon, positronium with definite (l, s) are eigenstates of C, while the neutron, K^0 created in π^--p collision and the hydrogen atom cannot be. Since $C^2 = +1$, the eigenvalues of C are ± 1. C is sometimes called the "particle-antiparticle conjugation" operator.

Let us assume that the electromagnetic interaction is invariant under C. Then, since $Cj_\mu C^{-1} = -j_\mu$ (both charge and current change sign), it follows that $CA_\mu C^{-1} = -A_\mu$. Since A_μ can be expanded in terms of the creation and annihilation operators

$$A_\mu = \frac{1}{(2\pi)^{3/2}} \int \frac{d^3k}{\sqrt{2k_0}} \{a_\mu(\vec{k}) \exp(ik\cdot x) + a_\mu^\dagger(\vec{k}) \exp(-ik\cdot x)\} \tag{5.11}$$

and C does not affect the space parts; then we must have

$$C\begin{Bmatrix} a_\mu(\vec{k}) \\ a_\mu^\dagger(\vec{k}) \end{Bmatrix}C^{-1} = -\begin{Bmatrix} a_\mu(\vec{k}) \\ a_\mu^\dagger(\vec{k}) \end{Bmatrix} \tag{5.12}$$

We assume now that the vacuum state $|\Phi_0\rangle$ has the property

$$C|\Phi_0\rangle = |\Phi_0\rangle \tag{5.13}$$

Then the one photon state $a_\mu^\dagger|\Phi_0\rangle$ has odd "charge-conjugation parity." It follows that all states with an odd number of photons in any space configuration (the spin and momentum are not important here) have $C = -1$ and that all states with an even number of photons in any space configuration have $C = +1$.

Furry's theorem tells us that any matrix element with an odd number of external photons (and with no other external particles) must be zero if H is invariant under C (Furry, 1937). Let N be the total number of external photons. Assume that N is odd. If we initially have n photons, then the initial state has $C = (-1)^n$. The final state has $(N - n)$ photons, thus $C = (-1)^{N-n} = -(-1)^n$. We see that such a matrix element would connect states with $C = +1$, with states of $C = -1$.

As an example, we can consider electrodynamics with $H_{\text{int}} = ie\bar{\psi}\gamma_\mu\psi A_\mu$ and calculate with the use of the Feynman rules. Thus for the triangle diagram shown in Fig. 5.1, the matrix element has the form

$$M \propto \iint d^4q_1 d^4q_2 \operatorname{Tr}\left(\frac{1}{iq_1\cdot\gamma + m}\,\varepsilon_\mu^{(2)}\gamma_\mu\,\frac{1}{iq_2\cdot\gamma + m}\,\varepsilon_\lambda^{(3)}\gamma_\lambda \times \frac{1}{-i\,q_3\cdot\gamma + m}\,\varepsilon_\sigma^{(1)}\gamma_\sigma\right) \tag{5.14}$$

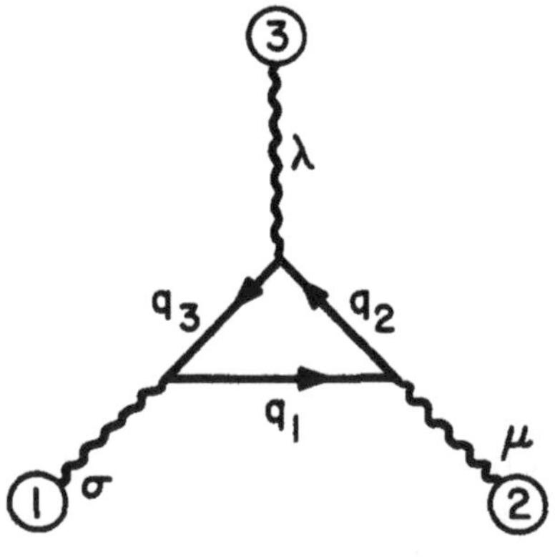

Figure 5.1. Example of Furry's theorem

We note that a change of representation $\gamma_\mu \to -\gamma_\mu$ (this is a valid step because the commutation rules for the γ's still hold) introduces a factor

$(-1)^3$ in M. Thus $\int\int \mathrm{Tr}$ must equal zero. (The $iq\cdot\gamma$ terms have no effect here since we integrate over all the internal momenta.) Thus, we see that Furry's theorem does hold for this example.

Feynman's "physical" explanation for this is that, for every loop diagram with electron line going clockwise, there exists another loop diagram with *positron* line going also clockwise (or electron line going counterclockwise) (Feynman, 1949, especially Footnote 9). Since the "potential" is opposite in sign for positron and electron and we must add amplitudes corresponding to all possible "paths"; for the case of three (or any odd number of $\varepsilon\cdot\gamma$ vertices, we get $(-e)(-e)(-e) + (+e)(+e)(+e) = 0$. However, the advantage of the symmetry approach is that we need not restrict ourselves to a particular form of interaction such as $ie\bar{\psi}\gamma_\mu\psi A_\mu$.

We have thus far limited our discussion to electromagnetic interactions, but one can, quite generally, speak of other interactions. For example, consider the known π^0 decay reaction.

$$\pi^0 \to 2\gamma$$

The charge-conjugation parity of the final state is $+1$. Thus if the interaction responsible for π^0 decay is invariant under C, then the C of the π^0 must also be $+1$. Note that this argument is independent of the specific mechanism; e.g., $\pi^0 \to p\bar{p}$ or other baryon pairs $\to 2\gamma$. In fact we do not know the detailed mechanism of this decay.

If the interaction is charge-conjugation invariant, then the decay $\pi^0 \to 3\gamma$ is *strictly forbidden*. In the absence of charge-conjugation invariance, this decay would be expected to occur about 1/137th of the time. This branching ratio has never been looked for experimentally with accuracy. Such an experiment might be a nice way to check the validity of C invariance for hyperon interactions if hyperon-pion interactions have something to do with π^0 decay.

In the case of positronium, neither the positron nor the electron alone is an eigenstate of C. However, we shall see that the e^--e^+ system in a definite (l, s) state is an eigenstate of C. We regard e^+ and e^- as identical fermions differing only in their charge labels. Since we are dealing with fermions, the wavefunction must acquire a minus sign under "total" exchange of particles, which consists of changing the Q, $\vec{r}$, and s labels. (Actually we are assuming here a kind of "generalized" Pauli principle. This assumption can be justified since the creation operator for the electron anticommutes with the creation operator for the positron as we shall see later.) The following factors are seen to appear under particle exchange.

space exchange	$(-1)^l$
spin exchange	$(-1)^{s+1}$
charge conjugation	C

($s = 1$ for triplet spin state)
($s = 0$ for singlet spin state)

Thus we have the relation

$$(-1)^l(-1)^{s+1}C = -1$$

or

$$C = (-1)^{l+s} \tag{5.15}$$

Let us now see what restrictions C invariance imposes on the decay into photons. If n is the number of photons in the final state, we see that the conservation of charge conjugation parity requires

$$(-1)^{l+s} = (-1)^n$$

Then the singlet S state ${}^1S_0$, $l = s = 0$ is allowed to decay into 2γ's. Similarly for the triplet S state ${}^3S_1$, $l = 0$, $s = 1$, the 2γ decay is forbidden but the 3γ decay occurs, which we could have seen from angular momentum considerations alone (c.f., Chapter 2, Section 4). What we could not have seen, however, is that ${}^1S_0 \to 3\gamma$ is strictly forbidden and not just suppressed by $\frac{1}{137}$. Historically people did perturbation calculations to show that the transition ${}^1S_0 \to 3\gamma$ is forbidden up to $\left(\frac{1}{137}\right)^3$. Wolfenstein and Ravenhall (1952) and Michel (1953) showed that the transition of ${}^1S_0$ into an odd number of photons is absolutely forbidden if C invariance holds, independently of perturbation theory.

The proton-antiproton system is entirely analogous to positronium. Thus we have $C = (-1)^{l+s}$. One can ask if a π^0 can be regarded as a bound state of proton and antiproton (Fermi and Yang, 1949). This is not necessarily such a meaningful question when the binding energy is so enormous (i.e., $2M_p \gg \mu_\pi$) but we can certainly consider the virtual process $\pi^0 \to p + \bar{p}$. Since angular momentum is conserved in a virtual process and the π^0 is a spin-zero particle, the only angular momentum states possible for the $p\bar{p}$ system are ${}^3P_0$ and ${}^1S_0$. In both cases $C = +1$. Thus the charge conjugation parity of the π^0 must be even. This, of course, agrees with the empirical observation: $\pi^0 \to 2\gamma$. We note also that $C = 1$ for the π^0 has nothing to do with the transformation properties of the π^0 field under space inversion. We have ${}^3P_0$ for a hypothetical scalar π^0 and ${}^1S_0$ for the pseudo-scalar π^0. $C = 1$ in both cases.

Now consider the π^+-π^- system. Since this system is a zero charge eigenstate of $Q^{(\mathrm{op})}$, it can also be an eigenstate of C. The wave function must be even under total interchange. Using the same arguments as for the e^+-e^- system, we see that $C = (-1)^l$.

We can consider the C parity of the π^0-π^0 system from two points of view. The first method is to note that for a total exchange, $C = (-1)^l$. However, since we have a system of two identical bosons, l cannot be odd-valued. Thus C is always $+1$. The alternative point of view is to use our knowledge that the charge-conjugation parity of a single π^0 is $+1$ for any spatial configuration. In fact the charge-conjugation parity of any system composed of π^0's *only* is $+1$. Note that this is quite different from the neutral system composed of π^+ and π^-. In the latter case, we have to know the spatial configuration of the various particles (e.g. relative angular momenta) to obtain C.

5.3. Charge Conjugation in the Dirac Theory

The existence of the positron was first inferred from the hole interpretation of the negative energy solutions of the Dirac equation (Dirac, 1931). It was, however, not immediately obvious that the hole must have the same mass as the electron. Dirac even thought that the hole should be identified with the proton. Weyl (1931) showed that the hole must have the same mass. Oppenheimer (1930) calculated the annihilation rate expected if the electron and proton were particle and antiparticle. If this were the case, the hydrogen atom would undergo self-annihilation in about 10^{-10} seconds. Thus the existence of the negative energy solutions was considered by many to be an undesirable feature of the Dirac theory (e.g. Pauli, 1933). Then the positron was discovered by Anderson (1933). And more than twenty years later, a Berkeley group established the existence of an antiproton (Chamberlain *et al.* 1955).

In the hole theory the e^- and e^+ appear asymmetrically. The physical vacuum is described as an infinite sea of negative energy electrons. Majorana (1937) and Kramers (1937) showed that the Dirac theory can actually be cast in a completely symmetric form.

We assume that the charge conjugate wave function ψ^c (which would "look like" the positive energy wave function) can be obtained from the negative energy wave function ψ in the following manner.

$$\psi^c = 4 \times 4 \text{ matrix } \psi^*$$

The appearance of $*$ is reasonable if we are to obtain $\exp(i\vec{p}\cdot\vec{x} - ip_0x_0)$ from $\exp(-i\vec{p}\cdot\vec{x} + ip_0x_0)$. Since for covariant calculations it is easier to work with $\bar{\psi}$, we write

$$\psi^c = \eta C\bar{\psi}^T \qquad |\eta|^2 = 1 \tag{5.16}$$

C here is a 4×4 matrix, *not* the charge conjugation operator. (We let the same symbol stand for both.)

The Dirac free particle operator $(\gamma_\mu \dfrac{\partial}{\partial x_\mu} + m)$ is associated with the space-time properties of the fermion and thus should not be changed under the charge conjugation operation. We should not star the γ matrices either since the charge conjugation operation is unitary and not antiunitary. The invariance of the Dirac equation holds if

$$\left[\gamma_\mu\left(\frac{\partial}{\partial x_\mu} + ieA_\mu\right) + m\right]C\bar{\psi}^T = 0 \tag{5.17}$$

follows from the original Dirac equation

$$\left[\gamma_\mu\left(\frac{\partial}{\partial x_\mu} - ieA_\mu\right) + m\right]\psi = 0$$

where A_μ is the external electromagnetic field. Note that $\psi^c = \eta C\bar{\psi}^T$ and ψ satisfy the same equation except for the sign of the $ieA_\mu\gamma_\mu$ term. Let us first consider the "adjoint" equation

$$-\frac{\partial}{\partial x_\mu}\bar{\psi}\gamma_\mu - ieA_\mu\bar{\psi}\gamma_\mu + m\bar{\psi} = 0$$

which can be obtained from the Hermitian conjugate of the original Dirac equation

$$\vec{\nabla}\psi^\dagger\cdot\vec{\gamma} - \frac{\partial}{\partial x_4}\psi^\dagger\gamma_4 + m\psi^\dagger + ieA^\dagger\psi^\dagger\gamma_\mu = 0$$

by multiplying γ_4 from the right. Actually the equation to which (5.17) should be compared is the transpose of the adjoint equation

$$\left[-\gamma_\mu^T\left(\frac{\partial}{\partial x_\mu} + ieA_\mu\right) + m\right]\bar{\psi}^T = 0 \tag{5.18}$$

In (5.17), multiplying C^{-1} from the left, we note that (5.17) is equivalent to (5.18) if there exists a C such that

$$C^{-1}\gamma_\mu C = -\gamma_\mu^T \tag{5.19}$$

Since the commutation rules still hold $\{(-\gamma_\mu^T), (-\gamma_\nu^T)\} = 2\delta_{\mu\nu}$, then by Pauli's fundamental theorem, C exists. Recall that B was defined in such a way that

$$B\gamma_\mu B^{-1} = \gamma_\mu^* = \gamma_\mu^T$$

The product CB has the property

$$CB\gamma_\mu(CB)^{-1} = C\gamma_\mu^T C^{-1} = -\gamma_\mu \tag{5.20}$$

and we can identify (up to a constant) $CB = \gamma_5$. As in the case of the B matrix, the nonsingular matrix C can be chosen unitary. A second property of C is that it is antisymmetric. We can offer two proofs of this property.

Proof 1. This one is exactly analogous to the one given in Chapter 4, Section 4 for $B^T = -B$. We first show that $C = \pm C^T$. Then we show that there would be too many (namely, 10) independent antisymmetric matrices if $C = +C^T$.

Proof 2. First note that $C\gamma_5^T C^{-1} = C\gamma_4^T\gamma_3^T\gamma_2^T\gamma_1^T C^{-1} = \gamma_4\gamma_3\gamma_2\gamma_1 = \gamma_5$, or $\gamma_5^T = C^{-1}\gamma_5 C$. Now from $CB = \gamma_5$ we write

$$C^T = (\gamma_5 B^{-1})^T = (-B^{-1})C^{-1}\gamma_5 C = -C \qquad \text{Q.E.D.}$$

Now we may ask, what form does C have in the Dirac-Pauli representation? We have previously seen that in this representation $B = \gamma_3\gamma_1$. Thus, in order to have $CB = \gamma_5$, we must have,

$$C = \gamma_4\gamma_2 \qquad \text{(D.P. only)} \tag{5.21}$$

Now the relation $C\gamma_\mu^T C^{-1} = -\gamma_\mu$ has the appearance

$$\gamma_4\gamma_2\gamma_\mu^T\gamma_2\gamma_4 = -\gamma_\mu$$

which is satisfied since, as we remember, $\gamma_\mu^T = -\gamma_\mu$ for $\mu = 1, 3$ and $\gamma_\mu^T = \gamma_\mu$ for $\mu = 2, 4$.

Needless to say, a particular form of C depends on the particular representation we use. According to Problem 5, C is γ_4 within a phase factor in the Majorana representation in which γ_k are purely real and γ_4 purely imaginary. Then

$$\psi^c = \eta\gamma_4\gamma_4^T(\psi^\dagger)^T = -\eta\psi^* \qquad \text{(Maj. only)} \tag{5.22}$$

The invariance of the Dirac equation under charge conjugation in this representation is nothing more than the statement that if ψ is a solution to

$$\left[\gamma_\mu\left(\frac{\partial}{\partial x_\mu} - ieA_\mu\right) + m\right]\psi = 0$$

then ψ^* is a solution to

$$\left[\gamma_\mu\left(\frac{\partial}{\partial x_\mu} - ieA_\mu\right) + m\right]^*\psi^* = 0$$

since

$$\left[\gamma_\mu\left(\frac{\partial}{\partial x_\mu} - ieA_\mu\right)\right]^* = \gamma_\mu\left(\frac{\partial}{\partial x_\mu} + ieA_\mu\right) \tag{5.23}$$

In fact, this is how Majorana (1937) demonstrated that the Dirac equation is invariant under charge conjugation.

Since positron $\rightarrow$ electron under charge conjugation, we would like to have the electromagnetic current, j_μ transform as $j_\mu \xrightarrow{C} -j_\mu$. We already know that $\psi \xrightarrow{C} \eta C\bar{\psi}^T$. The adjoint wave function transforms as

$$\bar{\psi} \rightarrow \eta^*(C\bar{\psi}^T)^\dagger\gamma_4 = \eta^*(C\gamma_4^T C^{-1}C(\psi^\dagger)^T)^\dagger\gamma_4 = -\eta^*\psi^T C^{-1} \tag{5.24}$$

where the unitarity of C has been used.

Thus we see

$$\begin{aligned} j_\mu = i\bar{\psi}\gamma_\mu\psi &\xrightarrow{C} -i\psi^T C^{-1}\gamma_\mu C\bar{\psi}^T \\ &= i(\bar{\psi}\gamma_\mu\psi)^T = i\bar{\psi}\gamma_\mu\psi \end{aligned}$$

There appears to be something wrong. It is clear that we do not have $j_\mu \xrightarrow{C} -j_\mu$. In fact, we could have anticipated this difficulty by looking at the fourth component $\rho = \bar{\psi}\gamma_4\psi = \psi^\dagger\psi$ which is positive definite and thus cannot have its sign reversed by any transformation.

A further unsatisfactory feature of the formalism is that the total energy

$$P_0 = -\int d^3x T_{44} = i\int d^3x \bar{\psi}\gamma_4 \frac{\partial\psi}{\partial x_0} \tag{5.25}$$

can assume negative values. This is admissible in the hole theory but inadmissible if we want to have a theory symmetric in particle and antiparticle. In this case the physical electron and positron must both have positive energies: "The vacuum is the state of lowest energy."

Pauli (1940) showed that the positive definite character of charge and the indefinite character of energy are characteristic of a spin $\frac{1}{2}$ integer field without quantization. Our goal here will be to display the theory in which the charge density ρ can be either positive or negative but in which P_0 can only be positive.

5.4. Quantization and the Spin-Statistics Connection

The reason why the hole theory (which is essentially a single particle theory) works so well is that electrons obey the exclusion principle. The concept of a hole makes sense only when one is not allowed to put as many electrons as one wants in the same state. If this were not the case, electrons would not stay in positive energy states forever, but would undergo transitions to negative energy states.

If we abandon the exclusion principle, we arrive at Klein's paradox (Klein, 1929). Assume a potential of the form

$$\text{Region I:}\quad E - V = \sqrt{|\vec{p}|^2 + m^2} > m$$

$$\text{Region II:}\quad m > E - V > -m$$

$$\text{Region III:}\quad -m > E - V = -\sqrt{|\vec{p}|^2 + m^2}$$

where m is the rest mass of the electron. Note that region II is not an accessible region classically since $\vec{p}$ would be imaginary. Quantum-mechanically, however, electrons can undergo transitions from I to III through II by means of the well-known "tunnel effect" just as in the α decay case. Being a "downhill" transition, this would imply that eventually all electrons would find themselves in the negative energy states. Hence the paradox! Dirac's solution to this paradox was, of course, to invoke the exclusion principle and to define the physical vacuum as the state in which all negative energy states are filled.

Jordan and Wigner (1928) first developed a formalism of second quantization which takes into account the exclusion principle behavior of the Dirac particles. In this theory the exclusion principle is incorporated in the very beginning. We consider an operator a and its Hermitian conjugate $a^\dagger$. Their algebraic properties are defined by the relations

$$a^2 = (a^\dagger)^2 = 0 \tag{5.26}$$
$$\{a, a^\dagger\} = aa^\dagger + a^\dagger a = 1$$

a and $a^\dagger$ are interpreted as annihilation and creation operators respectively. Now we define the number operator $N = a^\dagger a$. It follows that N has the properties

$$1 - N = 1 - a^\dagger a = aa^\dagger \tag{5.27}$$
$$N(1 - N) = a^\dagger a a a^\dagger = 0$$

Thus the eigenvalues of N are 0 and 1. This is precisely what we want for the exclusion principle.

More generally, for each spin and momentum state we have

$$\{a^{(r)}(\vec{p}), a^{(s)\dagger}(\vec{q})\} = \delta_{\vec{p}\vec{q}}\delta_{rs} \tag{5.28}$$
$$\{a, a\} = \{a^\dagger, a^\dagger\} = 0$$

If we apply $a^{(r)\dagger}(\vec{p})$ twice to the vacuum state $|\Phi_0\rangle$ we get zero

$$a^{(r)\dagger}(\vec{p})a^{(r)\dagger}(\vec{p})|\Phi_0\rangle = 0 \tag{5.29}$$

In other words, two electrons cannot be created in the same momentum and spin state.

Now, let us expand the Dirac field operator ψ in terms of the $a^{(r)}(\vec{q})$'s and the plane-wave solutions of the Dirac equation

$$\psi = \frac{1}{(2\pi)^{3/2}} \int d^3q \sqrt{\frac{m}{|q_0|}} \sum_{r=1}^{4} u^{(r)}(\vec{q})a^{(r)}(\vec{q}) \exp(iq\cdot x)$$

$$\text{with } r = 1, 2 \qquad q_0 = \sqrt{|\vec{q}|^2 + m^2} > 0$$
$$r = 3, 4 \qquad q_0 = -\sqrt{|\vec{q}|^2 + m^2} < 0 \tag{5.30}$$

We rewrite the expansion in the form:

$$\psi = \frac{1}{(2\pi)^{3/2}} \int d^3q \sqrt{\frac{m}{|q_0|}} \left\{ \sum_{r=1}^{2} u^{(r)}(\vec{q})a^{(r)}(\vec{q}) \exp(iq\cdot x) + \sum_{r=3}^{4} u^{(r)}(-\vec{q})a^{(r)}(-\vec{q}) \exp(-iq\cdot x) \right\} \tag{5.31}$$

where, in this last expression q_0 is always positive. We now define $a^{(r)}(\vec{q}) = a^{(s)}(\vec{q})$ for $r = s = 1, 2$ and $a^{(r)}(-\vec{q}) = b^{(s)\dagger}(\vec{q})$ with $s = 2$ when $r = 3$ and $s = 1$ when $r = 4$. The motivation for this is that we will later identify $b^\dagger$ as the creation operator for a positron—remember that the annihilation of a spin up negative energy electron with momentum $\vec{q}$ must correspond to the creation of a spin down positive energy positron with momentum $-\vec{q}$. We redefine also $v^{(s)}(\vec{q}) = u^{(r)}(-\vec{q})$ where $s = 2, 1$ for $r = 3, 4$ respectively and $u^{(s)}(\vec{q}) = u^{(r)}(\vec{q})$ where $s = r = 1, 2$. Thus we can further rewrite

$$\psi = \frac{1}{(2\pi)^{3/2}} \int d^3q \sqrt{\frac{m}{|q_0|}} \sum_{s=1}^{2} \{u^{(s)}(\vec{q})a^{(s)}(\vec{q}) \exp{(iq\cdot x)} + v^{(s)}(\vec{q})b^{(s)\dagger}(\vec{q}) \exp{(-iq\cdot x)}\} \tag{5.32}$$

The adjoint spinor has the form:

$$\bar{\psi} = \frac{1}{(2\pi)^{3/2}} \int d^3q \sqrt{\frac{m}{|q_0|}} \sum_{s=1}^{2} \{\bar{v}^{(s)}(\vec{q})b^{(s)}(\vec{q}) \exp{(iq\cdot x)} + \bar{u}^{(s)}(\vec{q})a^{(s)\dagger}(\vec{q}) \exp{(-iq\cdot x)}\} \tag{5.33}$$

We thus have for the charge conjugate spinor:

$$\psi^c = \eta_c C\bar{\psi}^T = \frac{\eta_c}{(2\pi)^{3/2}} \int d^3q \sqrt{\frac{m}{|q_0|}} \sum_{s=1}^{2} \{C\bar{v}^{(s)T}(\vec{q})b^{(s)}(\vec{q}) \exp{(iq\cdot x)} + C\bar{u}^{(s)T}(\vec{q})a^{(s)\dagger}(\vec{q}) \exp{(-iq\cdot x)}\} \tag{5.34}$$

A symmetry requirement is that $u^{(s)c} = C\bar{v}^{(s)T}$ and $v^{(s)c} = C\bar{u}^{(s)T}$ satisfy the same equation as $u^{(s)}$ and $v^{(s)}$ respectively. In particular, we will show here that $u^{(1)c} = Cv^{(1)T}$ has the same form as $u^{(1)}$. We have (apart from normalization factors)

$$v^{(1)}(\vec{p}) = u^{(4)}(-\vec{p}) = \begin{pmatrix} \dfrac{p_1 - ip_2}{|p_0| + m} \\ \dfrac{-p_3}{|p_0| + m} \\ 0 \\ 1 \end{pmatrix}$$

in the Dirac-Pauli representation. Therefore

$$C\bar{v}^{(1)T}(\vec{p}) = -\gamma_2 u^{(4)*}(-\vec{p}) = \begin{pmatrix} 1 \\ 0 \\ \dfrac{p_3}{|p_0| + m} \\ \dfrac{p_1 + ip_2}{|p_0| + m} \end{pmatrix} \tag{5.35}$$

Thus we see that ψ and ψ^c "look" the same except for the interchange of $a^{(s)}$ and $b^{(s)}$. $\psi \rightleftarrows \psi^c$ can be accomplished by $a^{(s)}(\vec{p}) \rightleftarrows b^{(s)}(\vec{p})$, $a^{(s)\dagger}(\vec{p}) \rightleftarrows b^{(s)\dagger}(\vec{p})$. We identify $b^{(s)}(\vec{p})$ as a positron annihilation operator and $b^{(s)\dagger}(\vec{p})$ as a positron creation operator. It is to be noted that $b^{(s)}$ and $b^{(s)\dagger}$ satisfy the same commutation relations as $a^{(s)}$ and $a^{(s)\dagger}$, and that $a^{(s)}$ and $b^{(s)}$ anti-commute.

With this new notation, we can now define the number operator for positrons as well as the one for electrons:

$$\begin{aligned} N^{(s,+)}(\vec{p}) &= b^{(s)\dagger}(\vec{p})b^{(s)}(\vec{p}) \qquad \text{for positrons} \\ N^{(s,-)}(\vec{p}) &= a^{(s)\dagger}(\vec{p})a^{(s)}(\vec{p}) \qquad \text{for electrons} \end{aligned} \tag{5.36}$$

It can easily be shown that another property of the field operators is expressed in their commutation relations (for equal times):

$$\{\psi_\alpha(\vec{x}, x_0), \bar{\psi}_\beta(\vec{x}', x_0)\} = (\gamma_4)_{\alpha\beta}\delta^{(3)}(\vec{x} - \vec{x}') \tag{5.37}$$

A discussion of the covariant commutation relations (at arbitrary x_0 and x_0') appears later in this chapter.

It was recognized as early as 1934 by Heisenberg (1934) that unless matrix elements were properly symmetrized in the field operators, certain undesirable features of the theory still remained. Heisenberg's "rule" for proper symmetrization to remove these undesirable features is:

$$\begin{aligned} \bar{\psi}\Omega\psi \to :\bar{\psi}\Omega\psi: &\equiv \tfrac{1}{2}\sum(\bar{\psi}_\alpha\Omega_{\alpha\beta}\psi_\beta - \psi_\beta\Omega_{\alpha\beta}\bar{\psi}_\alpha) \\ &= \tfrac{1}{2}(\bar{\psi}\Omega\psi - \psi^T\Omega^T\bar{\psi}^T) \end{aligned} \tag{5.38}$$

where the notation : : is termed as an antisymmetrized product. The difference between $\bar{\psi}\Omega\psi$ and $:\bar{\psi}\Omega\psi:$ is a *c*-number, and this is sufficient to remove undesired singularities that appear when the product is not properly symmetrized. Note that

$$\langle\Phi_0|j_\mu|\Phi_0\rangle = +\infty$$

if we take

$$j_\mu = i\bar{\psi}\gamma_\mu\psi$$

(as proved in Schweber, Bethe, and de Hoffman (1955) p. 211). This is clearly unacceptable since the properties of the vacuum depend on the sign of the electric charge. Physically this shows that the contributions from the infinite sea of negative energy electrons have not been eliminated. However, if we use instead

$$j_\mu = :i\bar{\psi}\gamma_\mu\psi: \text{ we get } \langle\Phi_0|j_\mu|\Phi_0\rangle = 0$$

Similarly, a much more satisfactory expression for the charge results if we use the antisymmetrized product in the calculation,

$$\begin{aligned} Q &= e\int d^3x :\bar{\psi}\gamma_4\psi: \\ &= \frac{e}{2}\int d^3x(\psi^\dagger\psi - \psi\psi^\dagger) \\ &= \frac{e}{2}\sum_{n=1}^{2}\int d^3p\{a^{(r)\dagger}(\vec{p})a^{(r)}(\vec{p}) + b^{(r)}(\vec{p})b^{(r)\dagger}(\vec{p}) \\ &\qquad\qquad - a^{(r)}(\vec{p})a^{(r)\dagger}(\vec{p}) - b^{(r)\dagger}(\vec{p})b^{(r)}(\vec{p})\} \qquad (5.39) \\ &= \frac{e}{2}\sum_{r=1}^{2}\int d^3p[(2a^{(r)\dagger}(\vec{p})a^{(r)}(\vec{p}) - 1) - (2b^{(r)\dagger}(\vec{p})b^{(r)}(\vec{p}) - 1)] \\ &= e\sum_{r=1}^{2}\int d^3p[N^{(r,-)}(\vec{p}) - N^{(r,+)}(\vec{p})] \end{aligned}$$

where $u_r^\dagger(\vec{p})u_{r'}(\vec{p}) = \delta_{rr'}p_0/m$ and $u_r^\dagger(\vec{p})v_{r'}(-\vec{p}) = 0$ have been used. Thus Q can now assume either positive or negative values depending on how many electrons and positrons are present.

The calculation of the field energy using the antisymmetrized product furnishes the following results:

$$P_0 = \int d^3x \frac{i}{2}\left(\psi^\dagger \frac{\partial \psi}{\partial x_0} - \frac{\partial \psi}{\partial x_0}\psi^\dagger\right)$$
$$= \sum_{r=1}^{2} \int d^3p[N^{(r,+)}(\vec{p}) + N^{(r,-)}(\vec{p}) - 1]p_0 \tag{5.40}$$

where $p_0 > 0$ in the last expression. This expression satisfies the requirement that the vacuum characterized by $N^{(+)} = N^{(-)} = 0$ be the lowest state of energy. However, it is still somewhat unsatisfactory in that we must redefine the energy scale such that the "-1" term drops out. Pauli feels that this points out a basic incompleteness in the present theory: "One begins to doubt all volume integrals."

Let us now consider the abnormal cases of Bose-Einstein quantization of a spin $\frac{1}{2}$ field and Fermi-Dirac quantization of a spin 0 field. We shall see that these procedures lead to entirely unacceptable features.

First, let us quantize a spin $\frac{1}{2}$ field according to the commutation $[a, a^\dagger] = 1$. This normally implies Bose-Einstein statistics and hence no exclusion principle. We find in this case

$$Q \propto N^{(+)} + N^{(-)}$$
$$P_0 \propto N^{(+)} - N^{(-)} + \text{constant}$$

with the positive-definiteness belonging to the first quantity instead of to the second, as we would like. Actually Pauli showed that $N^{(+)}$ and $N^{(-)}$ can be redefined to give

$$Q \propto N^{(+)} - N^{(-)}$$
$$P_0 \propto N^{+} + N^{(-)}$$

but then we get a contradiction with the positive definite metric (i.e., $\langle\psi|\psi\rangle < 0$). This last point is clearly inadmissible since we have no way to interpret negative probability in quantum mechanics.

The Feynman rules for calculating matrix elements are developed from a rather intuitive approach and are equivalent to the results of second quantized field theory (Feynman, 1949). One of his rules is that the exclusion principle be used for electrons. When he calculates the probability of the vacuum remaining as the vacuum (Feynman diagrams with no external lines) without using the exclusion principle, he gets the meaningless result: Probability > 1. This point is, of course, related to Pauli's contradiction with the positive definite metric.

Before considering the results of abnormal quantization of the spin 0 field, we note that the indefinite character of Q and positive definite character of P_0 are satisfied even before second quantization, since

$$\rho = i\left(\phi^\dagger \frac{\partial \phi}{\partial x_0} - \frac{\partial \phi^\dagger}{\partial x_0}\phi\right)$$

$$-T_{44} = \tfrac{1}{2}\left(\vec{\nabla}\phi^\dagger \cdot \vec{\nabla}\phi + \frac{\partial \phi^\dagger}{\partial x_0}\frac{\partial \phi}{\partial x_0} + \mu^2 \phi^\dagger \phi\right) \tag{5.41}$$

Under charge conjugation

$$\begin{aligned} \phi &\to \eta_c \phi^\dagger \\ \phi^\dagger &\to \eta_c^* \phi \end{aligned} \tag{5.42}$$

and ρ reverses sign whereas T_{44} is unchanged. Thus it seems that one does not learn anything here about whether the ϕ field should be quantized using the Bose-Einstein or Fermi-Dirac statistics. However, one gets into a different kind of trouble if the spin 0 field is quantized according to Fermi-Dirac statistics.

For simplicity, let us consider the neutral Hermitian spin 0 field

$$\phi = \frac{1}{(2\pi)^{3/2}} \int \frac{d^3p}{\sqrt{2p_0}} [a(\vec{p}) \exp(ip\cdot x) + a^\dagger(\vec{p}) \exp(-ip\cdot x)] \quad \text{with } p_0 > 0$$

Case 1: (Normal quantization according to Bose-Einstein statistics). For this case we use the following commutation relations and number operator N:

$$\begin{aligned} [a(\vec{p}), a^\dagger(\vec{p}\,')] &= \delta_{\vec{p}\vec{p}\,'} \\ [a(\vec{p}), a(\vec{p}\,')] &= [a^\dagger(\vec{p}), a^\dagger(\vec{p}\,')] = 0 \\ N(\vec{p}) &= a^\dagger(\vec{p})a(\vec{p}) \end{aligned} \tag{5.43}$$

As is well known, this choice of commutators corresponds to the physical possibility of having as many particles in the same state as we wish. The statement that $a^\dagger(\vec{p})$ and $a^\dagger(\vec{p}\,')$ commute means, in wave function language, that the two-particle amplitude $f(\vec{p}, \vec{p}\,')$ is symmetric under interchange of $\vec{p}$ and $\vec{p}\,'$ (i.e., Bose-Einstein statistics). (For a discussion of the connection between the wave-function formalism and the field operator formalism for many-particle systems see e.g. p. 144 ff. of Schweber, Bethe, and de Hoffman, 1955.)

$$\begin{aligned} [\phi(x), \phi(x')] &= \frac{1}{2(2\pi)^3} \int \frac{d^3p}{\sqrt{p_0}} \int \frac{d^3p'}{\sqrt{p_0'}} \{[a(\vec{p}), a^\dagger(\vec{p}\,')] \exp(ip\cdot x - ip'\cdot x) \\ &\quad + [a^\dagger(\vec{p}), a(\vec{p}\,')] \exp(ip'\cdot x' - ip\cdot x)]\} \\ &= -\frac{1}{2(2\pi)^3} \int_{p_0>0} \frac{d^3p}{p_0} \{\exp[ip\cdot(x-x')] \\ &\quad - \exp[-ip\cdot(x-x')]\} \\ &= \frac{-i}{(2\pi)^3} \int_{p_0>0} \frac{d^3p}{p_0} \exp[i\vec{p}\cdot(\vec{x}-\vec{x}\,')) \sin(p_0(x_0 - x_0'))] \\ &\equiv i\Delta(x - x') \end{aligned} \tag{5.44}$$

where Δ has the properties

$$\Delta(x) = 0 \qquad \text{for } x^2 > 0 \text{ (spacelike)}$$

$$\Delta(\vec{x}, x_0) = \Delta(-\vec{x}, x_0) = -\Delta(\vec{x}, -x_0) \tag{5.45}$$

$$\left(\frac{\partial \Delta}{\partial x_0}\right)_{x_0=0} = -\delta^{(3)}(\vec{x})$$

An explicit form for Δ is

$$\Delta(x) = \frac{1}{4\pi}\frac{1}{r}\frac{\partial}{\partial r}F(r, x_0) \tag{5.46}$$

where

$$F = \begin{cases} J_0(\mu\sqrt{x_0^2 - r^2}) & x_0 > r \\ 0 & \text{for } -r < x_0 < r \text{ (outside the light cone)} \\ -J_0(\mu\sqrt{x_0^2 - r^2}) & x_0 < -r \end{cases}$$

as shown in many books (e.g. Schiff, 1949, p. 357).

The behavior of Δ is seen to satisfy the causality requirement of special relativity that measurements at two points separated by a space-like distance should not influence each other.

Case 2: ("Abnormal" quantization according to Fermi-Dirac statistics). For this case we make use of the anticommutators:

$$\{a(\vec{p}), a^\dagger(\vec{p}\,')\} = \delta_{\vec{p}\vec{p}\,'}$$

$$\{a(\vec{p}), a(\vec{p}\,')\} = \{a^\dagger(p), a^\dagger(p')\} = 0$$

Exactly in the same manner as in Case 1, we calculate the anticommutator of the field operators $\phi(x)$ and $\phi(x')$:

$$\{\phi(x), \phi(x')\} = \frac{1}{(2\pi)^3}\int_{p_0>0}\frac{d^3p}{p_0}\exp\left[i\vec{p}\cdot(\vec{x} - \vec{x}\,')\right]\cos\left[p_0(x_0 - x_0')\right]$$

$$\equiv \Delta^{(1)}(x - x')$$

An explicit form for $\Delta^{(1)}$ is

$$\Delta^{(1)}(x) = \frac{1}{4\pi}\frac{1}{r}\frac{\partial}{\partial r}F_1(r, x_0)$$

where

$$F_1 = \begin{cases} N_0(\mu\sqrt{x_0^2 - r^2}) & \text{for } |x_0| > r \\ -iH_0^{(1)}(i\mu\sqrt{r^2 - x_0^2}) & \text{for } r > |x_0| \end{cases}$$

We see that $\Delta^{(1)}$ does not vanish for $(x - x')^2 > 0$ outside the light cone, but falls off exponentially. Difficulty with causality is apparent even with the { } bracket since, if $\{\phi(x), \phi(x')\}$ did not vanish outside the light cone, observable quantities constructed bilinearly from ϕ would not commute for space-like separation.

Back to the Fermi-Dirac case for completeness, we note that we can express the anticommutator between the spinor field operators ψ and $\bar{\psi}$ in terms of Δ (not $\Delta^{(1)}$)

$$\{\psi_\alpha(x), \bar{\psi}_\beta(x')\} = i\left(-\gamma_\mu \frac{\partial}{\partial x_\mu} + m\right)_{\alpha\beta} \Delta(x - x') \equiv -iS_{\alpha\beta}(x - x') \quad (5.47)$$

For equal times this expression takes the simpler form which we have seen earlier in this chapter (see (5.37)).

PROBLEM 7. Show that

$$[\phi(x), \phi^\dagger(x')] = i\Delta(x - x')$$

and

$$\{\psi_\alpha(x), \bar{\psi}_\beta(x')\} = -iS_{\alpha\beta}(x - x')$$

are invariant under P, T and C.

We have shown that quantization of a free spin $\frac{1}{2}$ field according to Bose-Einstein statistics leads to a contradiction with either the postulate that the physical vacuum is the state of lowest energy of the postulate concerning the positive definite metric, and that quantization of a free neutral scalar field, according to Fermi-Dirac statistics, leads to a contradiction with causality. These are two special cases of a very general theorem in Quantum Field Theory.

In this general theorem, it can be shown that the following assumptions lead to the conclusions that 1/2 integral spin fields obey Fermi-Dirac statistics and that integral spin fields obey Bose-Einstein statistics.

(a) Lorentz Invariance.

(b) The vacuum is the state of lowest energy.

(c) Microcausality. The appropriate commutor or anticommutator is zero for space-like separation.

(d) Positive definite metric.

The theorem was first proved for the free field cases by Pauli (1940). More recent work using the Heisenberg representation has been done by Lüders and Zumino (1958) and by Burgoyne (1958). For a discussion of "abnormal" cases see Pauli (1950).

5.5. Parity, Chirality, and Helicity under Charge Conjugation

We now examine the behavior of the charge conjugate field ψ^c under the parity, chirality, and helicity operations respectively.

Under the parity operation the Dirac field operator ψ transforms as

$$P\psi P^{-1} = \eta_P \gamma_4 \psi$$

Now we ask how ψ^c transforms under parity

$$\begin{aligned}\psi^c = \eta_C C\bar{\psi}^T &\xrightarrow{P} \eta_C\eta_P^* C((\gamma_4\psi)^\dagger\gamma_4)^T \\ &= \eta_C\eta_P^* C\gamma_4^T C^{-1}C\bar{\psi}^T \\ &= -\eta_P^*\gamma_4\psi^c\end{aligned} \tag{5.48}$$

Thus we see that the intrinsic parity of the anti-particle is opposite to that of the particle. We had previously obtained this same result in Chapter 3, Section 7 using the hole theory. However, the present approach is seen to be less clumsy.

Analogous to the chirality-defining expression for the C-number wave function $\gamma_5\psi = \lambda\psi$, where λ is the chirality, is the expression for the wave function for the charge conjugate state ψ^c:

$$\gamma_5\psi^c = \lambda^c\psi^c \tag{5.49}$$

Thus we can solve for λ^c:

$$\begin{aligned}\gamma_5\eta_C C\bar{\psi}^T &= \eta_C CC^{-1}\gamma_5 C\bar{\psi}^T \\ &= \eta_C C(\bar{\psi}\gamma_5)^T \\ &= \eta_C C(-\gamma_4\gamma_5\psi)^{\dagger T} \\ &= -\lambda\psi^c \\ \therefore\ \lambda^c &= -\lambda\end{aligned} \tag{5.50}$$

provided that λ is real. We have the result that the chirality operation anticommutes with the charge-conjugation operation. Note that chirality also anticommutes with the parity operation (since $\gamma_4\gamma_5 = -\gamma_5\gamma_4$). Suppose we have a rule that says $(1 + \gamma_5)$ is inserted in front of every Fermion field in writing down a weak interaction Hamiltonian. Then we see that this rule is completely equivalent to the rule: Insert $(1 - \gamma_5)$ in front of every anti-Fermion field in the Hamiltonian. This rule will be useful in Chapter 7.

In a similar way we get at the behavior of the charge conjugate field ψ^c under the helicity operation. We again work in the c-number formalism. The helicity $\mathscr{H}$ is defined for the particle field by $\vec{\Sigma}\cdot\hat{p}\psi = \mathscr{H}\psi$. For the charge conjugate field we have:

$$\begin{aligned}\vec{\Sigma}\cdot\hat{p}\psi^c &= \vec{\Sigma}\cdot\hat{p}\eta_C C\bar{\psi}^T \\ &= \eta_C C(\bar{\psi}\vec{\Sigma})^T\cdot\hat{p} \\ &= \eta_C C(\gamma_4\vec{\Sigma}\psi)^{\dagger T}\cdot\hat{p} \\ &= \mathscr{H}\psi^C\end{aligned} \tag{5.51}$$

Thus we see that helicity does not change under charge conjugation. This is, of course, quite reasonable, since

$$\vec{\sigma} \xrightarrow{C} \vec{\sigma}$$
$$\vec{p} \xrightarrow{C} \vec{p} \tag{5.52}$$

5.6. Behavior of Bilinear Covariants under Charge Conjugation

The behavior of the bilinear covariants under charge conjugation is easily determined in the following manner (Furry, 1937; Pais and Jost, 1952). Since $\psi \to \eta_C C\bar{\psi}^T$, $\bar{\psi} \to \eta_C^* \psi^T C^{-1}$, we have (see (5.38))

$$:\bar{\psi}\Gamma_i\psi: \xrightarrow{C} -:\psi^T C^{-1}\Gamma_i C\bar{\psi}^T = :\omega_i\psi^T\Gamma_i^T\bar{\psi}^T = :\omega_i\bar{\psi}\Gamma_i\psi: \tag{5.53}$$

where ω_i is defined by

$$C^{-1}\Gamma_i C = \omega_i\Gamma_i^T \tag{5.54}$$

Thus to see how the bilinear covariants transform under charge conjugation, it is sufficient to examine whether

$$C^{-1}\Gamma_i C = \Gamma_i^T \quad \text{or} \quad -\Gamma_i^T$$

We have the following five cases:

(1) $\Gamma_i = 1 \qquad \omega_i = 1$

(2) $\Gamma_i = \gamma_\mu \qquad \omega_i = -1$

since $C^{-1}\gamma_\mu C = -\gamma_\mu^T$ by definition

(3) $\Gamma_i = \sigma_{\mu\nu} \quad (\mu \neq \nu) \qquad \omega_i = -1$

since $C^{-1}\gamma_\mu\gamma_\nu C = \gamma_\mu^T\gamma_\nu^T = -(\gamma_\nu\gamma_\mu)^T$

(4) $\Gamma_i = i\gamma_5\gamma_\mu \qquad \omega_i = +1$

since $C^{-1}\gamma_5\gamma_\mu C = -\gamma_5^T\gamma_\mu^T = (\gamma_5\gamma_\mu)^T$

(5) $\Gamma_i = \gamma_5 \qquad \omega_i = +1$

since $C^{-1}\gamma_5 C = \gamma_5^T$

Thus, to sum up

$$\begin{aligned} \omega_i &= +1 \qquad \text{for } S, A, P \\ \omega_i &= -1 \qquad \text{for } V, T \end{aligned} \tag{5.55}$$

Now we are in a position to examine the over all charge-conjugation invariance properties of several familiar interaction Hamiltonians.

Examples:

(a) "Minimal" electromagnetic coupling:

$$ie\bar{\psi}\gamma_\mu\psi A_\mu$$

Each vector part transforms as

$$:\bar{\psi}\gamma_\mu\psi: \xrightarrow{C} -:\bar{\psi}\gamma_\mu\psi:$$

$$A_\mu \xrightarrow{C} -A_\mu$$

Thus this Hamiltonian is invariant under charge conjugation.

(b) Pauli Moment: $\frac{\mu_m}{2} F_{\mu\nu}\bar{\psi}\sigma_{\mu\nu}\psi$

$$:\bar{\psi}\sigma_{\mu\nu}\psi: \xrightarrow{C} -\bar{\psi}\sigma_{\mu\nu}\psi$$

$$F_{\mu\nu} \xrightarrow{C} -F_{\mu\nu}$$

This Hamiltonian is also invariant under charge conjugation.

(c) Electric dipole moment interaction:

$$\frac{i\mu_e}{2} F_{\mu\nu}\bar{\psi}\gamma_5\sigma_{\mu\nu}\psi$$

It is seen that this interaction is also invariant. However, it is interesting to note that the term *violates* parity and time reversal invariance.

(d) Non-derivative type Yukawa interaction (with neutral boson). The interaction is $g_1\bar{\psi}\psi\phi + ig_2\bar{\psi}\gamma_5\psi\phi$ with g_1 and g_2 real.

If both g_1 and g_2 are non-vanishing, this Hamiltonian violates P and T invariance. Under charge conjugation, both $\bar{\psi}\psi$ and $i\bar{\psi}\gamma_5\psi$ are even. Thus it is seen that invariance under C is guaranteed even if both g_1 and g_2 are non-vanishing provided that η_C is chosen to be $+1$ for the neutral scalar field ϕ so that $\phi \xrightarrow{C} \phi$. However, invariance under CP will then be violated.

It is interesting to note parenthetically at this point that for these four interactions considered here which are invariant under C, either P and T are violated together (as in (c) and (d)) or neither is violated (as in (a) and (b)). In other words, when T invariance is violated, the product CP is also violated. We shall see in a later chapter that these statements follow more generally from the CPT theorem.

(e) Pseudoscalar-pseudovector Yukawa interaction (with neutral boson):

$$\frac{F}{\mu}\bar{\psi}\gamma_5\gamma_\mu\psi\frac{\partial\phi}{\partial x_\mu}$$

Since $\gamma_5\gamma_\mu$ is even under charge conjugation, this interaction is C invariant if we choose $\eta_C = +1$ for the boson field so that $\phi \xrightarrow{C} \phi$. In a relativistic field theory in which pions and nucleons are assumed to be "fundamental," the π^0 is coupled to the proton via a Yukawa-type interaction. However, we cannot tell whether this coupling is *ps-ps* or *ps-pv* (or possibly both) since both require the π^0 to be even under charge conjugation (as noted earlier in this chapter, a C invariant π^0 decay interaction and the observation of $\pi^0 \to 2\gamma$ require that the π^0 be even under charge conjugation).

(f) Axial vector and tensor Interaction:

$$g_1\phi_\mu(i\bar{\psi}\gamma_5\gamma_\mu\psi) + g_2\left(\frac{\partial\phi_\nu}{\partial x_\mu} - \frac{\partial\phi_\nu}{\partial x_\mu}\right)(\bar{\psi}\gamma_5\sigma_{\mu\nu}\psi)$$

where ϕ_μ is some spin 1 field. If both g_1 and g_2 are non-vanishing, this interaction violates C invariance (and also T invariance) since axial vector and tensor terms behave oppositely under C.

5.7. The Breakdown of Charge Conjugation Invariance in Weak Interactions

In Section 1 we remarked that in the μ^+ decay $\mathscr{H}(e^+) \approx 1$ while in the μ^- decay $\mathscr{H}(e^-) \approx -1$. A particular Hamiltonian that can give this effect (and also the other muon decay parameters correctly) is

$$H_{\text{int}} = G(\bar{e}\gamma_\lambda(1 + \gamma_5)\nu)(\bar{\nu}\gamma_\lambda(1 + \gamma_5)\mu) + \text{H.C.} \tag{5.56}$$

where e, μ etc. mean ψ_e, ψ_μ etc. It has been assumed that e^-, μ^-, and ν are "particles" while e^+, μ^+ and $\bar{\nu}$ are "anti-particles." H.C. can be written explicitly as follows

$$\begin{aligned}\text{H.C.} &= G^*[\mu^\dagger\gamma_4\gamma_4(1 + \gamma_5)\gamma_\lambda\gamma_4\nu][\nu^\dagger\gamma_4\gamma_4(1 + \gamma_5)\gamma_\lambda\gamma_4 e] \\ &= G^*[\bar{\mu}\gamma_\lambda(1 + \gamma_5)\nu][\bar{\nu}\gamma_\lambda(1 + \gamma_5)e]\end{aligned} \tag{5.57}$$

We will show that H_{int} violates C in two ways:

(1) Recall from Chapter 2, that as $v/c \to 1$

$$\gamma_5\begin{pmatrix} I \\ \vec{\sigma}\cdot\hat{p}\end{pmatrix} = -\vec{\Sigma}\cdot\hat{p}\begin{pmatrix} I \\ \vec{\sigma}\cdot\hat{p}\end{pmatrix}$$

and (5.58)

$$\gamma_5\begin{pmatrix} -\vec{\sigma}\cdot\hat{p} \\ I\end{pmatrix} = \vec{\Sigma}\cdot\hat{p}\begin{pmatrix} -\vec{\sigma}\cdot\hat{p} \\ I\end{pmatrix}$$

in the Dirac-Pauli representation.

Thus $\frac{1}{2}(1 + \gamma_5)e$ is the wave function for a left-handed e^- for $p_0 > 0$ and a right-handed e^- if $p_0 < 0$. In a quantized theory $\frac{1}{2}(1 + \gamma_5)e$ annihilates e_L^- with $p_0 > 0$ (as $v/c \to 1$) and annihilates e_R^- with $p_0 < 0$ or equivalently creates e_R^+ with $p_0 > 0$. Notice that a hole has the *same* helicity as the corresponding positive energy anti-particle because the absence of spin up is spin down and also the absence of momentum $\vec{p}$ is momentum $-\vec{p}$. Similarly $\frac{1}{2}\bar{e}(1 - \gamma_5) = \frac{1}{2}[(1 + \gamma_5)e]^\dagger\gamma_4$ creates $e_L^-(p_0 > 0)$ or annihilates $e_R^+(p_0 > 0)$. Thus H_{int} creates e_L^- in μ^- decay (described by the G term) and creates e_R^+ in μ^+ decay (G^* term) since $v/c \approx 1$ for nearly the entire electron spectrum (the mean energy is $\sim$40 Mev.).

(2) Write H.C. in (5.56) terms of the charge conjugate spinors. First note that we can write $\psi = \eta C\bar{\psi}^{cT}$ and $\bar{\psi} = -\eta^*\psi^{cT}C^{-1}$. Assuming that kinematically independent spinor fields such as μ and ν anticommute and using the Heisenberg rule for bilinear products of two different fields, we have

$$\begin{aligned} :\bar{\mu}\gamma_\lambda(1+\gamma_5)\nu: &= -:\eta_\mu^*\eta_\nu\mu^{cT}[-\gamma_\lambda^T + (\gamma_\lambda\gamma_5)^T]\bar{\nu}^{cT}: \\ &= -\eta_\mu^*\eta_\nu:(\bar{\nu}^c\gamma_\lambda(1-\gamma_5)\mu^c): \end{aligned} \tag{5.59}$$

so

$$\text{H.C.} = G^*\eta_\mu^*\eta_\nu\eta_\nu^*\eta_e[\bar{\nu}^c\gamma_\lambda(1-\gamma_5)\mu^c][\bar{e}^c\gamma_\lambda(1-\gamma_5)\nu^c] \tag{5.60}$$

If charge conjugation invariance holds, (5.60) must be identical to $G[\bar{\nu}\gamma_\lambda(1+\gamma_5)\mu][\bar{e}\gamma_\lambda(1+\gamma_5)\nu]$ except for the superscripts c. This is clearly not the case. In fact, there is no choice of phases, η, which makes the G^* term identical to the G term after we interchange $\mu \rightleftarrows \mu^c$, $e \rightleftarrows e^c$, and $\nu \rightleftarrows \nu^c$. So one sees that the conclusion is the same as in (1).

In beta decay, one cannot (yet) study the decay of anti-neutrons and anti-nuclei. Yet it can be shown that recent β decay experiments have unambiguously demonstrated that C invariance is violated if the β decay Hamiltonian is to be Hermitian. Let

$$H_{\text{int}} = \sum_i [\bar{p}\Gamma_i n][\bar{e}(C_i\Gamma_i + C_i'\Gamma_i\gamma_5)\nu] + \text{H.C.} \tag{5.61}$$

Under charge conjugation

$$\begin{aligned} \bar{p}\Gamma_i n &\xrightarrow{C} \omega_i\eta_p^*\eta_n\bar{n}\Gamma_i p \\ \bar{e}\Gamma_i\nu &\xrightarrow{C} \omega_i\eta_e^*\eta_\nu\bar{\nu}\Gamma_i e \\ \bar{e}\Gamma_i\gamma_5\nu &\xrightarrow{C} \omega_i\eta_e^*\eta_\nu\bar{\nu}\gamma_5\Gamma_i e \end{aligned} \tag{5.62}$$

where $C^{-1}\Gamma_i\gamma_5 C = \omega_i(\gamma_5\Gamma_i)^T$ was used. Meanwhile under Hermitian conjugation,

$$\begin{aligned} \bar{p}\Gamma_i n &\xrightarrow{\text{H.C.}} \pm\bar{n}\Gamma_i p \\ \bar{e}\Gamma_i\nu &\xrightarrow{\text{H.C.}} \pm\bar{\nu}\Gamma_i e \\ \bar{e}\Gamma_i\gamma_5\nu &\xrightarrow{\text{H.C.}} \mp\bar{\nu}\gamma_5\Gamma_i e \end{aligned} \tag{5.63}$$

where the upper (lower) sign is to be chosen if γ_4 commutes (anticommutes) with Γ_i. Now the H.C. (C_i^*, $C_i'^*$ terms) can give rise to beta decay in the charge conjugate system, e.g., $\bar{n} \to e^+ + \nu + \bar{p}$. Thus, if charge conjugation invariance holds, the charge conjugate of $(\bar{p}\Gamma_i n)(\bar{e}C_i\Gamma_i\nu)$ etc. must be equal to the Hermitian conjugate of the same term. This means

$$\begin{aligned} &\sum_i \{C_i^*(\bar{n}\Gamma_i p)(\bar{\nu}\Gamma_i e) - C_i'^*(\bar{n}\Gamma_i p)(\bar{\nu}\gamma_5\Gamma_i e)\} \\ &\qquad = \eta_p^*\eta_n\eta_e^*\eta_\nu \sum_i \{C_i(\bar{n}\Gamma_i p)(\bar{\nu}\Gamma_i e) + C_i'(\bar{n}\Gamma_i p)(\bar{\nu}\gamma_5\Gamma_i e)\} \end{aligned} \tag{5.64}$$

since the ω_i and $\pm$ cancel. Thus C invariance implies that all C_i's have the same phase, all C_i''s have the same phase, but that C_i and C_i' be 90° out of phase; only then can the η's be chosen in such a way that invariance is guaranteed. Alternatively, the conditions for violation of C are

$$\begin{aligned} \operatorname{Re}(C_i C_j'^*) &\neq 0 \\ \operatorname{Im}(C_i C_j^*) &\neq 0 \qquad\qquad (5.65) \\ \operatorname{Im}(C_i' C_j'^*) &\neq 0 \end{aligned}$$

Recall that T invariance requires that all of the C_i, C_i' be relatively real. If P is violated (both C_i, C_i' nonvanishing), and T is not violated (C_i, C_i' relatively real) then C must be violated (since otherwise C_i and C_i' must be relatively imaginary). This result is also to be expected from the CPT theorem which will be discussed in the next chapter. The observed large asymmetry parameter of the β^--angular distribution from polarized Co^{60} nuclei shows $\operatorname{Re}(C_T C_T'^* - C_A C_A'^*)$ is nonvanishing. So the classical experiment of Wu, *et al.* (1957) already implies that charge conjugation invariance is violated (even though they did not study anti Co^{60} decay) provided that β decay can be properly described by a Hermitian interaction. A similar conclusion can be drawn from the very fact that β^- (β^+) helicity is $-\frac{v}{c}\left(+\frac{v}{c}\right)$.

Suppose we have calculated various correlation coefficients for β^- decay. How can we find the corresponding correlation coefficients for (a) the β^+ decay of anti-matter, (b) the β^+ decay of "ordinary" matter (more realistic case)?

(a) Express H.C. using the charge conjugate spinors $p = \eta_p C\overline{p^c}^T$, $\bar{n} = -\eta_n^* n^{cT} C^{-1}$, etc.,

$$\text{H.C.} = \eta_n^*\eta_p\eta_\nu^*\eta_e \sum_i \{C_i^*(\overline{p^c}\Gamma_i n^c)(\overline{e^c}\Gamma_i \nu^c) - C_i'^*(\overline{p^c}\Gamma_i n^c)(\overline{e^c}\Gamma_i\gamma_5\nu^c)\} \qquad (5.66)$$

where the ω_i cancel again and the common phase is, of course, unobservable. Now compare (5.66) with (5.61). To obtain the correlation coefficients of β^+ decay of anti-matter when they are known for the β^- decay of ordinary matter just let $C_i \to C_i^*$ and $C_i' \to -C_i'^*$. The new formulae are the same as the old only when all of the C_i have the same relative phase and all of the C_i''s are imaginary relative to C_i, which agrees with our earlier conditions for C invariance. For instance, if the asymmetry parameter for Co^{60} is given by

$$\beta = \frac{-2\operatorname{Re}(C_A C_A'^*)}{|C_A|^2 + |C_A'|^2}\frac{v}{c}$$

then the asymmetry parameter for anti-Co^{60} is given by

$$\frac{2\operatorname{Re}(C_A^* C_A')}{|C_A|^2 + |C_A'|^2}\frac{v}{c} = -\beta$$

(b) Express e and $\bar{\nu}$ in terms of the charge conjugate spinors *but leave $\bar{n}$ and p as they are*

$$\text{H.C.} = \eta_\nu^* \eta_e \sum_i \omega_i \{C_i^*(\bar{n}\Gamma_i p)(\overline{e^c}\Gamma_i \nu^c) - C_i'^*(\bar{n}\Gamma_i p)(\overline{e^c}\Gamma_i \gamma_5 \nu^c)\} \tag{5.67}$$

This time the ω_i do not cancel. To obtain the formulae for β^+ decay when those for β^- decay are known, let $C_i \rightarrow \omega_i C_i^*$, $C_i' \rightarrow -\omega_i C_i'$. (In this way one can check the formulae of Jackson, Treiman, and Wyld (1957). Their upper sign is for β^- decay and the lower for β^+ decay.) In particular note

$$\begin{aligned} C_V &\rightarrow -C_V^*, \qquad & C_V' &\rightarrow C_V'^* \\ C_A &\rightarrow C_A^*, \qquad & C_A' &\rightarrow -C_A'^* \end{aligned} \tag{5.68}$$

If the e^- helicity is given by $\dfrac{-2\,\text{Re}\,(C_V C_V'^*)}{|C_V|^2 + |C_V'|^2}\dfrac{v}{c}$, then the e^+ helicity is $+\dfrac{2\,\text{Re}\,(C_V C_V'^*)}{|C_V|^2 + |C_V'|^2}\dfrac{v}{c}$. If a scalar interaction could also contribute to β^- decay, there would be a term proportional to $2\,\text{Re}\,(C_S C_V^* + C_S' C_V'^*)$, known as a Fierz interference term, which would distort a straight line Kurie plot. The sign of such a term for β^+ decay must be opposite since $\omega_i = 1$ for S but -1 for V.

In Table 5.1 we summarize the criteria for testing invariance under C, P, T in β decay.

TABLE 5.1

Criteria for Testing Invariance Principles in β Decay

Coefficient of observable	P	T	C
$\text{Re}\,(C_i C_j^*)$	✓	✓	✓
$\text{Re}\,(C_i' C_j'^*)$	✓	✓	✓
$\text{Im}\,(C_i C_j^*)$	✓	×	×
$\text{Im}\,(C_i' C_j'^*)$	✓	×	×
$\text{Re}\,(C_i C_j'^*)$	×	✓	×
$\text{Im}\,(C_i C_j'^*)$	×	×	✓

The ✓ means compatible with the invariance in question.
The × means violation of that invariance principle.

We can summarize the behavior of the β decay Hamiltonian under P, T, and C as follows (Lee, Oehme, and Yang, 1957).
Under parity

$$PH(C_i, C_i')P^{-1} = H(\eta_P C_i, -\eta_P C_i') \tag{5.69}$$

e.g., $(\bar{p}\Gamma_i n)(\bar{e}(C_i\Gamma_i + C_i'\Gamma_i\gamma_5)\nu] \rightarrow \eta_P(\bar{p}\Gamma_i n)(\bar{e}(C_i\Gamma_i - C_i'\Gamma_i\gamma_5)\nu]$
where $\eta_P = \eta_P^*(p)\eta_P(n)\eta_P^*(e)\eta_P(\nu)$.

Under time reversal

$$TH(C_i, C_i')T^{-1} = H(\eta_T C_i^*, \eta_T C_i'^*) \tag{5.70}$$

(using $B^{-1}\Gamma_i B = \pm\Gamma_i$ and $B^{-1}\Gamma_i^*\gamma_5^* B = \pm\Gamma_i\gamma_5$).

Under charge conjugation

$$CH(C_i, C_i')C^{-1} = H(\eta_C^* C_i^*, -\eta_C^* C_i'^*). \tag{5.71}$$

where the complex conjugate occurs because one compares CHC^{-1} with the Hermitian conjugate part of H. Under CPT, TCP, PTC, etc.

$$H(C_i, C_i') \xrightarrow{CPT} H(C_i, C_i') \tag{5.72}$$

with suitable choices of the phases. To find the condition on the phases consider TCP (read from *right* to *left*)

$$C_i \xrightarrow{TCP} \eta_T(\eta_C^*(\eta_P C_i)^*)^* = \eta_P\eta_C\eta_T C_i$$

while for CTP

$$C_i \xrightarrow{CTP} \eta_P\eta_C^*\eta_T^* C_i$$

These overall phases must be unity. One can show that no other independent condition results from other permutations of CPT. Thus

$$\eta_P = \eta_C\eta_T = \eta_C^*\eta_T^* = \pm 1 \tag{5.73}$$

This means that it is quite difficult to construct a Hamiltonian that violates the product of C, P, and T taken in any order. This makes us think that there is something fundamental about the product.

CHAPTER 6

Strong Reflection and the CPT Theorem

6.1. Strong Reflection

Using specific examples we have seen that it is rather hard to write down Hamiltonians that violate invariance under CPT (or any other permutation of C, P, and T). We may naturally ask: What are the minimal set of assumptions on a field theory of elementary particles in order that the product CPT is "good?" We shall make as few assumptions as possible. We do assume invariance under proper orthochronous Lorentz transformations. As we go along, it becomes necessary for us to make a few more assumptions such as the "normal" spin statistics connection.

Instead of considering the product CPT from the very beginning we first ask: Is there any transformation such that the transformation property depends only on the oddness or evenness of the rank of a tensor density or of a field quantity (rank = number of uncontracted indices)? Specifically we want for a tensor density of rank n

$$\underbrace{O_{\mu\nu\ldots\lambda\sigma}}(x) \rightarrow (-1)^n \underbrace{O_{\mu\nu\ldots\lambda\sigma}}(-x) \tag{6.1}$$
$$n \text{ uncontracted indices}$$

For instance for the charge-current density we want $j_\mu(x) \rightarrow -j_\mu(-x)$, which can be brought about by neither P, C, nor T. If we could find a set of operations that leads to (6.1), then anything we can write down for an interaction density (both for a Lagrangian density which is a "true" scalar density and for a Hamiltonian density which is the 4-4 component of a symmetric tensor density of rank 2) would be automatically invariant under that set of operations. We shall see that there indeed exists the desired set of operations called "strong reflection." Moreover, we will show that the product C, P, T taken in any order is intimately related to strong reflection.

Since we do not assume reflection invariance under parity nor under time reversal, the rank of a tensor is defined solely from its behavior under $\Lambda \varepsilon \mathscr{L}_+^\uparrow$ (proper orthochronous Lorentz transformations)

$$O_{\mu\nu\ldots\lambda\sigma} \rightarrow a_{\mu'\mu} a_{\nu'\nu} \ldots a_{\lambda'\lambda} a_{\sigma'\sigma} O_{\mu\nu\ldots\lambda\sigma} \tag{6.2}$$

where $a_{\mu'\mu}$ stands for a matrix element associated with Λ: $x_\mu \rightarrow a_{\mu'\mu} x_\mu$.

Examples of fields and tensor densities of various ranks are given in Table 6.1.

TABLE 6.1

Fields and Tensor Densities of Various Ranks

Rank	Field (or derivative of field)	Tensor density
0	$\phi(x)$	$\bar{\psi}\psi,\ i\bar{\psi}\gamma_5\psi$
1	$\partial_\mu\phi,\ A_\mu$	$i(\phi^\dagger\partial_\mu\phi - \partial_\mu\phi^\dagger\phi),\ i\bar{\psi}\gamma_5\gamma_\mu\psi$
2	$F_{\mu\nu},\ \varepsilon_{\mu\nu\lambda\sigma}F_{\lambda\sigma}$	$\bar{\psi}\sigma_{\mu\nu}\psi,\ \mathscr{T}_{\mu\nu} = \partial_\mu\phi\dfrac{\partial\mathscr{L}}{\partial(\partial_\nu\phi)} - \mathscr{L}\delta_{\mu\nu}$

The prefix "pseudo" is entirely irrelevant (and in fact has no meaning) in our considerations, e.g., $i\bar{\psi}\gamma_\mu\psi$ and $i\bar{\psi}\gamma_5\gamma_\mu\psi$ are on the same footing.

Now what is the "physical" significance of (6.1)? If space were Euclidean, the answer would be trivial. For 4-dimensional Euclidean space there are only two classes of "Lorentz" transformations, one with $\det(a_{\mu\nu}) = 1$ and the other with $\det(a_{\mu\nu}) = -1$. In particular the transformation

$$x \to -x \qquad a_{\mu\nu} = \begin{pmatrix} -1 & & & 0 \\ & -1 & & \\ & & -1 & \\ 0 & & & -1 \end{pmatrix}$$

is of the first type and can be brought about by continuous rotations in 4-space. (Simply rotate in the 3-4 plane about the 1-2 plane by π and then rotate in the 1-2 plane about the 3-4 plane by π.) States with angular momentum J would transform as

$$|\Psi\rangle \to S|\Psi\rangle$$

$$S = \exp(iJ_{12}\pi)\exp(iJ_{34}\pi)$$

The field operators would transform as

$$\phi(x) \to \phi(-x) \qquad A_\mu(x) \to -A_\mu(-x)$$

$$\psi(x) \to \left(\cos\frac{\pi}{2} + i\sigma_{12}\sin\frac{\pi}{2}\right)\left(\cos\frac{\pi}{2} + i\sigma_{34}\sin\frac{\pi}{2}\right)\psi(-x) = \gamma_5\psi(-x)$$

Now in reality space is Lorentzian. $x \to -x$ belongs to $\mathscr{L}_+^\downarrow$ where $\downarrow$ stands for $a_{44} < 0$ and cannot be brought about continuously from the identity by compounding infinitesimal transformations.

For ordinary Lorentz transformations we have

$$a_{jk}, \text{ real}$$

$$a_{44}, \text{ real}$$

$$a_{4k}, \text{ imaginary}$$

If we give up the above conditions we can consider a complex Lorentz transformation. In particular consider:

$$\begin{cases} x_3 \to x_3 + i\varepsilon x_0 \\ x_0 \to x_0 + i\varepsilon x_3 \end{cases} \qquad \begin{pmatrix} 1 & 0 & 0 & 0 \\ 0 & 1 & 0 & 0 \\ 0 & 0 & 1 & \varepsilon \\ 0 & 0 & -\varepsilon & 1 \end{pmatrix}$$

We can compound such infinitesimal transformations (rotate in the 3-4 plane about the 1-2 plane, etc.) to get:

$$x_3 \to ix_0 \to -x_3$$

$$x_0 \to ix_3 \to -x_0$$

Then apply an ordinary 3-dimensional rotation in the 1-2 plane to get $x_1 \to -x_1$, $x_2 \to -x_2$. Thus, $x \to -x$ can be obtained continuously from the identity if we allow complex Lorentz transformations.

So we suspect that even if space is Lorentzian, the transformation $\psi \to \gamma_5\psi(-x)$ etc., has something to do with $O_{\mu\nu\ldots\lambda\sigma} \to (-1)^n O_{\mu\nu\ldots\lambda\sigma}$. The actual situation is a little less simple, and it turns out that we need something else. At this stage we *define* a set of operations called *strong reflection* (SR) as follows:

$$\begin{cases} x \to -x \\ \phi(x) \to +\phi(-x) \qquad \phi^\dagger(x) \to \phi^\dagger(-x) \\ A_\mu(x) \to -A_\mu(-x) \\ \psi(x) \to \gamma_5\psi(-x) \qquad \bar{\psi}(x) \to -\bar{\psi}(-x)\gamma_5 \\ \text{Reverse the order of factors in operator algebra} \end{cases} \tag{6.3}$$

The importance of the last line will be demonstrated shortly. Some authors (e.g. Pauli) prefer to require ψ and ψ^c to transform in the same way, in which case

$$\psi(x) \to i\gamma_5\psi(-x), \qquad \bar{\psi}(x) \to i\bar{\psi}(-x)\gamma_5$$

In any case the phase is irrelevant. We set the phase factor to be unity as would be the case if space were Euclidean.

In order to see the physical significance of strong reflection consider how $\phi(x)$ (non-Hermitian) transforms:

$$\frac{1}{(2\pi)^{3/2}} \int d^3p \frac{1}{\sqrt{2p_0}} [a(\vec{p}) \exp(+ip\cdot x) + b^\dagger(\vec{p}) \exp(-ip\cdot x)]$$

$$\to \frac{1}{(2\pi)^{3/2}} \int d^3p \frac{1}{\sqrt{2p_0}} [a(\vec{p}) \exp(-ip\cdot x) + b^\dagger(\vec{p}) \exp(ip\cdot x)]$$

So $\left.\begin{aligned} a(\vec{p}) &\to b^\dagger(\vec{p}) \\ a^\dagger(\vec{p}) &\to b(\vec{p}) \end{aligned}\right\}$

The creation operator for a particle goes into the annihilation operator for the corresponding antiparticle *with the same momentum.*

In the Dirac field case we have to be slightly careful with the spin indices

$$\gamma_5 u^{(1)}(\vec{p}) = \begin{pmatrix} 0 & 0 & -1 & 0 \\ 0 & 0 & 0 & -1 \\ -1 & 0 & 0 & 0 \\ 0 & -1 & 0 & 0 \end{pmatrix} \begin{pmatrix} 1 \\ 0 \\ p_3/(E+m) \\ (p_1+ip_2)/(E+m) \end{pmatrix}$$

$$= -\begin{pmatrix} p_3/(E+m) \\ (p_1+ip_2)/(E+m) \\ 1 \\ 0 \end{pmatrix} = -u^{(3)}(-\vec{p}) = -v^{(2)}(\vec{p})$$

so that the right correspondence is:

$$a^{(1)}(\vec{p}) \to -b^{(2)\dagger}(\vec{p}),\ b^{(1)\dagger}(\vec{p}) \to -a^{(2)}(\vec{p}),\ \text{etc.}$$

Thus invariance under *SR* means a symmetry between the creation of a left-handed particle and annihilation of a right-hand antiparticle.

Consider strong reflection followed by Hermitian conjugation (H.C.)

$$a^{(1)}(\vec{p}) \xrightarrow{\text{SR}} -b^{(2)\dagger}(\vec{p}) \xrightarrow{\text{H.C.}} -b^{(2)}(\vec{p})$$

Compare with the effect of *CPT*, ignoring irrelevant phases:

$$a^{(1)}(\vec{p}) \xrightarrow{\text{T}} a^{(2)}(-\vec{p}) \xrightarrow{\text{P}} a^{(2)}(\vec{p}) \xrightarrow{\text{C}} b^{(2)}(\vec{p})$$

Hence SR followed by H.C. gives the same effect on creation and annihilation operators as the product *CPT*.

6.2. Proof of the *CPT* Theorem (Hamiltonian Formalism)

Now we are in a position to prove one of the most far reaching theorems in quantum field theory. We shall prove the *CPT* theorem in two steps. (1) A wide class of field theories are invariant under strong reflection. (2) The product of *C*, *P*, *T* taken in any order is identical to SR followed by Hermitian conjugation.

First check the equation of motion under SR. The boson case is trivial. For the Dirac equation we see that the transformed equation

$$\frac{\partial}{\partial(-x_\mu)} \gamma_\mu \gamma_5 \psi + \gamma_5 m \psi = 0$$

is equivalent to the original Dirac equation (proof: multiply by γ_5 from the left). As for the commutation relations for bosons

$$[\phi(x), \phi^\dagger(x')] = i\Delta(x - x'),$$

the right-hand side changes sign since $\Delta(x - x')$ is odd. This is compensated by the change of order on the left-hand side. The spinor case can be treated similarly. Note that the change of order is absolutely essential. We could have seen this from the Heisenberg equation

$$[P_0, f] = i\frac{\partial f}{\partial x_0} \qquad P_0 = -\int d^3x\, \mathscr{T}_{44}$$

The time derivative changes sign, but the total energy P_0 should remain invariant. Hence we must have $[P_0, f] \to [f, P_0]$ to preserve the original form. (Contrast this case with $i \to -i$ for the Wigner time reversal case.)

For densities constructed out of boson fields and derivatives (of at most finite order) of fields, just note that with each index a minus sign is associated under SR. Change of order is irrelevant since the boson fields commute. (Properly we should have symmetrized all bilinear forms of boson fields according to Bose-Einstein statistics just as we antisymmetrized bilinear covariants made up of spinors according to the Heisenberg rule.) Thus (6.1)

$$O(x)_{\mu\nu\ldots\lambda\sigma} \to (-1)^n O(x)_{\mu\nu\ldots\lambda\sigma}$$

is satisfied. $\mathscr{L}$ and $\mathscr{T}_{44}$ are made up of $O_{\mu\nu\ldots\lambda\sigma}$ and field operators with indices contracted. With each contraction the rank decreases by 2. Hence

$$\mathscr{L}(x) \to \mathscr{L}(-x), \qquad \mathscr{T}_{44}(x) \to \mathscr{T}_{44}(-x)$$

are satisfied.

The behavior of bilinear covariants made up of spinors is much more interesting. The transformation

$$\psi(x) \to \gamma_5\psi(-x), \qquad \bar{\psi}(x) \to -\bar{\psi}(-x)\gamma_5$$

followed by a change in the order of factors is the same as

$$\bar{\psi}_2\Omega\psi_1 \to [(-\bar{\psi}_2\gamma_5)\Omega\gamma_5\psi_1)]^T = -\psi_1^T\gamma_5^T\Omega^T\gamma_5^T\bar{\psi}_2^T \tag{6.4}$$

[Because "change of order" causes:

$$\bar{\psi}_\beta O_{\beta\alpha}\psi_\alpha \to \psi_\alpha O_{\beta\alpha}\bar{\psi}_\beta = \psi_\alpha(O^T)_{\alpha\beta}\bar{\psi}_\beta = (\bar{\psi}O\psi)^T]$$

If the "normal" spin statistics relations hold, we must antisymmetrize bilinear covariants using the Heisenberg rule.

$$:-\psi_1^T(\gamma_5\Omega\gamma_5)^T\bar{\psi}_2^T: \;=\; :\bar{\psi}_2\gamma_5\Omega\gamma_5\psi_1: \tag{6.5}$$

Now

$$\gamma_5\Omega\gamma_5 = (-1)^n\Omega \qquad \text{where } n = \text{even for } S, T, P$$
$$n = \text{odd for } V, A$$

[Note that $C_V(\bar{p}\gamma_\mu n)(\bar{e}\gamma_\mu\nu)$ and $C_V'(\bar{p}\gamma_\mu n)(\bar{e}\gamma_\mu\gamma_5\nu)$ transform exactly in the same way under SR. The bilinear covariants involved are both odd since their rank is one.]

Thus (6.4) and (6.5) lead to

$$\bar{\psi}_2\Omega\psi_1 \xrightarrow{\text{SR}} (-1)^n\bar{\psi}_2\Omega\psi_1 \tag{6.6}$$

Note that *without the normal spin statistics relation the result would be just the opposite of what we want.*

To summarize $\mathscr{L}$ and $\mathscr{T}_{44}$ are invariant under SR regardless of whether they are made up of boson fields, fermion fields, or both, provided that they are properly symmetrized or antisymmetrized according to the "normal" spin-statistics relations.

In the second part of the proof we shall show that strong reflection followed by Hermitian conjugation is equivalent to CPT. If we operate on bosons with CPT the result is

$$\begin{gathered}
\phi(x) \xrightarrow{C} \eta_C\phi^\dagger(x) \xrightarrow{P} \eta_C\eta_P^*\phi^\dagger(-\vec{x}, x_0) \xrightarrow{T} \eta_C^*\eta_P\eta_T^*\phi^\dagger(-x) \\
A_\mu(x) \xrightarrow{C} -A_\mu(x) \xrightarrow{P} \begin{cases} A_k(-\vec{x}, x_0) \\ -A_4(-\vec{x}, x_0) \end{cases} \xrightarrow{T} \begin{cases} -A_k(-x) \\ -(-i)A_0(-x) \end{cases} \\
\partial_\mu \xrightarrow{C} \partial_\mu \xrightarrow{P} \begin{cases} -\partial_k \\ \partial_4 \end{cases} \xrightarrow{T} \begin{cases} -\partial_k \\ (-i)(-\partial_0) \end{cases} \\
c\ (\text{number}) \xrightarrow{C} c \xrightarrow{P} c \xrightarrow{T} c^*
\end{gathered} \tag{6.7}$$

Similarly operating with SR and then taking the Hermitian conjugate, we have

$$\begin{gathered}
\phi(x) \xrightarrow{\text{SR}} \phi(-x) \xrightarrow{\text{H.C.}} \phi^\dagger(-x) \\
A_\mu(x) \xrightarrow{\text{SR}} -A_\mu(-x) \xrightarrow{\text{H.C.}} \begin{cases} -A_k(-x) \\ -(-i)A_0(-x) \end{cases} \\
\partial_\mu \xrightarrow{\text{SR}} -\partial_\mu \xrightarrow{\text{H.C.}} \begin{cases} -\partial_k \\ -(-i)\partial_0 \end{cases} \\
c\ (\text{number}) \xrightarrow{\text{SR}} c \xrightarrow{\text{H.C.}} c^*
\end{gathered} \tag{6.8}$$

Compare the two results. We note that the product TPC is the same as SR followed by Hermitian conjugation if $\eta_C^*\eta_P\eta_T^* = 1$. A slightly different condition on the phase can be obtained if the order of T, P, C is different (cf. beta decay example at the end of Chapter 5.)

For Dirac fields

$$\begin{gathered}
\psi \xrightarrow{C} \eta_C C\bar{\psi}^T \xrightarrow{P} \eta_C\eta_P^*\gamma_4 C\bar{\psi}^T(-\vec{x}, x_4) \xrightarrow{T} \eta_C^*\eta_P\eta_T^* B\gamma_4 C\bar{\psi}^T(-x) \\
= \eta_C^*\eta_P\eta_T^*\gamma_4^* BC\bar{\psi}^T(-x) = \eta_C^*\eta_P\eta_T^*\gamma_4^T\gamma_5^T\bar{\psi}^T(-x) \\
\gamma_\mu \xrightarrow{C} \gamma_\mu \xrightarrow{P} \gamma_\mu \xrightarrow{T} \gamma_\mu^*
\end{gathered} \tag{6.9}$$

where $BC = (-B^T)(-C^T) = (CB)^T = \gamma_5^T$ has been used.
Under SR we have the following situation. Because

$$\bar{\psi}_2\Omega\psi_1 \to \psi_1^T\gamma_5^T\Omega^T(-\gamma_5^T\bar{\psi}_2^T)$$

we have

$$\psi \to \psi^T \gamma_5^T, \qquad \bar{\psi} \to -\gamma_5^T \bar{\psi}^T$$

$$\psi \xrightarrow{\text{SR}} \psi^T \gamma_5^T \xrightarrow{\text{H.C.}} \gamma_5^T \psi^{T\dagger} = \gamma_5^T \gamma_4^T \gamma_4^T \psi^{\dagger T} = -\gamma_4^T \gamma_5^T \bar{\psi}^T \tag{6.10}$$

$$\gamma_\mu \xrightarrow{\text{SR}} \gamma_\mu^T \xrightarrow{\text{H.C.}} \gamma_\mu^*$$

The order is reversed by SR but is restored by Hermitian conjugation. Thus TPC and SR followed by H.C. give the identical results if $\eta_C^* \eta_P \eta_T^* = 1$ (the phase condition is again somewhat modified if T, P, C are applied in a different order.)

This formally completes the proof that a wide class of field theories are invariant with respect to the product of C, P, and T taken in any order. We now summarize exactly what we have assumed.

(1) Invariance under proper orthochronous Lorentz transformations. No assumption about reflection has been made.
(2) Interaction densities are local and constructed out of field operators and derivatives of field operators of at most finite order.
(3) The normal spin statistics relation. (We have assumed that kinematically independent fermion fields anticommute. This condition may be too strong. It is irrelevant in beta decay, for instance, whether the neutrino field commutes or anticommutes with the proton field.)
(4) Interaction densities are properly symmetrized or antisymmetrized.
(5) The Hermiticity of interaction densities. (The Hermiticity was not necessary for proving invariance under SR, but we used it in identifying CPT with SR followed by H.C.)

Brief remarks about the history of the theorem: Schwinger (1951, 1953) seemed to have assumed something like CPT invariance to deduce the spin-statistics connection. Lüders (1954) noted that C invariance and T invariance impose the same restriction (since he did not consider parity violation). Pauli (1955) showed that SR follows from proper Lorentz invariance and the normal spin-statistics connection. He also proved higher spin cases. The proof we have given parallels a more recent work of Lüders (1957).

Pauli (1955) prefers to use Schwinger's approach to time reversal. In Schwinger's approach complex conjugation does not directly appear, but he makes use of transposition (and bra ⇄ ket). For Hermitian operators complex conjugation and transposition amount to the same thing. The two approaches are physically equivalent since all observables are Hermitian. However, for non-Hermitian operators such as the charged boson field the Schwinger-Pauli approach to time reversal and our approach differ by H.C. For example:

$$\text{Schwinger-Pauli} \quad \phi(\vec{x}, x_0) \to \phi^\dagger(\vec{x}, -x_0)$$
$$\text{Ours (Lüders)} \quad \phi(\vec{x}, x_0) \to \phi(\vec{x}, -x_0) \qquad \text{etc.}$$

under time reversal. In terms of creation and annihilation operators

$$\text{Schwinger-Pauli} \quad a(\vec{p}) \to a^\dagger(-\vec{p})$$
$$\text{Ours} \quad a(p) \to a(-\vec{p})$$

i.e., annihilation and creation operators get interchanged in the Schwinger-Pauli approach. In particular, for the creation and annihilation operators of Dirac particles

$$\text{Schwinger-Pauli:} \quad a^{(1)}(\vec{p}) \to a^{(2)\dagger}(-\vec{p})$$
$$\text{Ours:} \quad a^{(1)}(\vec{p}) \to a^{(2)}(-\vec{p})$$

Thus the annihilation operator is interchanged with the creation operator with opposite spin and momentum. In the Pauli article, SR is the product of C, P and what he (or Schwinger) calls time reversal. Pauli also defines Weak Reflection, (WR), which amounts to the product of P and the Schwinger-Pauli time reversal; in our approach WR followed by H.C. is our PT.

6.3. *CPT* and Microcausality

Thus far in everything we have done we have relied upon the existence of Hamiltonians. There is a school of field theory in which Hamiltonians do not explicitly appear. Its program starts with a very general set of assumptions such as Lorentz invariance, microscopic causality, translational invariance, the non-existence of negative energy states, asymptotic conditions, etc. The most central task in such an approach turns out to be to study the vacuum expectation values of field operators in the Heisenberg representation, e.g.

$$\langle[\phi_A(x_1), \phi_B(x_2)]\rangle_0, \qquad \langle T(\phi_1(x_1), \phi_2(x_2)\ldots\phi_n(x_n))\rangle_0$$

where $\langle O\rangle_0$ stands for $\langle\Phi_0|O|\Phi_0\rangle$ where Φ_0 is the vacuum state.

A particular class of products we shall be concerned with is of the form

$$W^{(n)} = \langle\phi(x_1)\phi(x_2)\ldots\phi(x_n)\rangle_0 \tag{6.11}$$

which is called a Wightman product. Wightman (1956) has shown that given a set of $W^{(n)}$ one can construct a theory of neutral fields. Instead of studying the behavior of Hamiltonians under symmetry operations as we have done in the Hamiltonian formalism we may talk about how Wightman products transform under symmetry operations.

Before we discuss the behavior of Wightman products under strong reflection we first familiarize ourselves with some other properties in Wightman products. We take the simplest kind

$$W_{AB}(\xi) = \langle\phi_A(x)\phi_B(x')\rangle_0, \qquad x - x' = \xi \tag{6.12}$$

where both ϕ_A and ϕ_B are scalar fields. The special case where ϕ_A and ϕ_B refer to the same field has been considered by Lehmann (1954). We assume that the total energy-momentum vector P_μ exists with the property

$$[iP_\mu, \phi_A(x)] = -\frac{\partial\phi_A(x)}{\partial x_\mu} \tag{6.13}$$

Assume that the vacuum is the state of lowest energy, and normalize the energy scale so that

$$P_\mu|0\rangle = 0$$

Consider eigenstates of P_μ

$$P_\mu|p\rangle = p_\mu|p\rangle \tag{6.14}$$

Using (6.14) we can integrate the expectation value of the equation of motion (6.13) taken between $\langle 0|$ and $|p\rangle$

$$\begin{aligned}\langle 0|[iP_\mu, \phi_A(x)]|p\rangle &= \langle 0|iP_\mu\phi_A(x)|p\rangle - i\langle 0|\phi_A(x)p_\mu|p\rangle \\ &= -ip_\mu\langle 0|\phi_A(x)|p\rangle \\ &= -\frac{\partial}{\partial x_\mu}\langle 0|\phi_A(x)|p\rangle\end{aligned}$$

because $\langle 0|P_\mu = 0$; so

$$\langle 0|\phi_A(x)|p\rangle = \langle 0|\phi_A(0)|p\rangle \exp(ip\cdot x), \qquad p_0 \geq 0 \text{ only}$$

Therefore

$$\begin{aligned}W_{AB}(\xi) = \langle 0|\phi_A(x)\phi_B(x')|0\rangle &= \sum_{\vec{p},\, p_0 \geq 0} \langle 0|\phi_A(x)|p\rangle\langle p|\phi_B(x')|0\rangle \\ &= \sum_{\vec{p},\, p_0 \geq 0} \exp(ip\cdot x)\exp(-ip\cdot x')\langle 0|\phi_A(0)|p\rangle\langle p|\phi_B(0)|0\rangle\end{aligned}$$

$$\therefore \qquad W_{AB}(\xi) = \sum_{\vec{p},\, p_0 \geq 0} \exp(ip\cdot\xi)\langle 0|\phi_A(0)|p\rangle\langle p|\phi_B(0)|0\rangle \tag{6.15}$$

In integral form we have

$$W_{AB}(\xi) = \frac{i}{(2\pi)^3}\int \theta(p_0)\rho_{AB}(-p^2)\exp(ip\cdot\xi)d^4p : -p^2 > 0 \tag{6.16}$$

where

$$\rho_{AB}\frac{d^4p}{(2\pi)^3} = \sum_{\vec{p},\, p_0 > 0} \langle 0|\phi_A(0)|p\rangle\langle p|\phi_B(0)|0\rangle$$

For $A = B$ and ϕ_A free, W_{AB} becomes the $\Delta^{(+)}$ function.

$$W_{AA}(\xi) \xrightarrow{\text{free}} \Delta^{(+)}(\xi) = \tfrac{1}{2}(\Delta - i\Delta^{(1)})$$

$\rho_{AA} = \rho(-p^2)$ is called a spectral function. Eq. (6.16) is the spectral representation for the Wightman function. For the free field case $\rho(-p^2) = \delta(p^2 + \mu^2)$ where μ is the physical mass. Crudely speaking, $\rho(-p^2)$ represents the probability of finding the physical particle in a virtual state with "mass" $= \sqrt{-p^2}$, e.g. if there exists an interaction that breaks up the particle into two particles, then in addition to the δ-function singularity at $p^2 = -\mu^2$, there is a continuum contribution starting at $p^2 = -(2\mu)^2$. For further reading on this subject see Lehmann (1954). $W_{AA}(\xi)$ is usually denoted by $\Delta^{(+)'}$ (Heisenberg representation in contrast to $\Delta^{(+)}$ that appears in perturbation theory based on the interaction representation.)

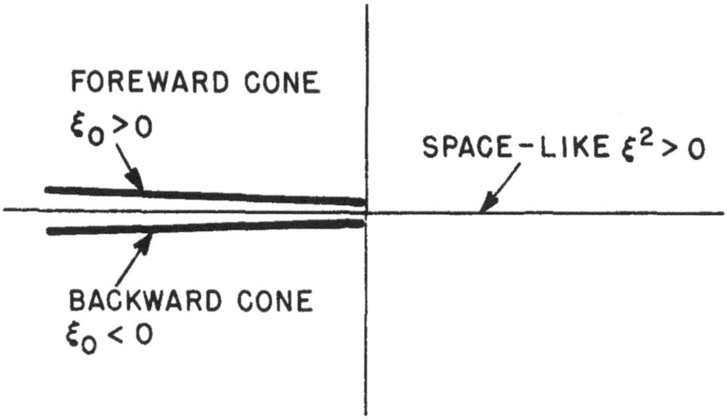

Figure 6.1. ζ^2 plane

$W_{AB}(\xi)$ is not well defined since the integrand oscillates. Let us regard W_{AB} as a function of a complex four vector

$\zeta = \xi - i\eta$, with $\begin{Bmatrix} \xi_k, \xi_0 \\ \eta_k, \eta_0 \end{Bmatrix}$ all real. $\eta^2 < 0$, $\eta_0 > 0$, i.e., η within the forward cone. We define $W_{AB}(\xi)$ so that

$$W_{AB}(\xi) = \lim_{|\vec{\eta}|\to 0,\, \eta_0 \to 0+} \frac{-i}{(2\pi)^3} \int \theta(p_0)\rho_{AB}(-p^2) \exp{(p\cdot\eta)} \exp{(ip\cdot\xi)} d^4p \tag{6.17}$$

which is convergent. ($p\cdot\eta = -p_0\eta_0$ by choosing a frame in which $\vec{\eta} = 0$. Then the oscillations are damped.) Moreover, $W_{AB}(\zeta)$ is analytic as long as η is in the forward cone. $W_{AB}(\xi)$ is the boundary value of an analytic function of a Lorentz invariant scalar product ζ^2. Note that the analyticity is a direct consequence of the assumption that there are no negative energy states. The presence of $\theta(p_0)$ in (6.16) and (6.17) is crucial in our argument. $\zeta^2 = \xi^2 - \eta^2 - 2i\xi\cdot\eta$ with η in the forward cone fills up the entire ζ^2 plane except for the negative real axis and the origin. (See Fig. 6.1.)

Consider a Lorentz transformation $\xi \to -\xi$. If $\xi^2 > 0$, $W_{AB}(\xi) = W_{AB}(-\xi)$. But if ξ is timelike, under $\xi_0 \to -\xi_0$, the W function must jump between the two Riemann sheets. So $W_{AB}(\xi) \neq W_{AB}(-\xi)$. All this can be exhibited explicitly in the free field case.

$$W_{AA}(\xi) \xrightarrow{\text{free}} \Delta^{(+)}(\xi) = -\frac{i}{(2\pi)^3}\int d^4p\ \theta(p_0)\exp(ip\cdot\xi)\delta(p^2+\mu^2)$$

$$= -\frac{i}{2(2\pi)^3}\int_{p_0=(\vec{p}^2+\mu^2)^{1/2}}\frac{d^3p}{p_0}\exp(i\vec{p}\cdot\vec{\xi} - p_0\xi_0)$$

$$= -\frac{\mu}{8\pi is}H_1^{(1)}(i\mu s) \tag{6.17}$$

$$s = \sqrt{|\vec{\xi}|^2 - \xi_0^2} \qquad \text{spacelike}$$

$$s = i\sqrt{\xi_0^2 - |\vec{\xi}|^2} \qquad \text{forward cone}$$

$$s = -i\sqrt{\xi_0^2 - |\vec{\xi}|^2} \qquad \text{backward cone}$$

Clearly $W_{AA}(\xi)$ is not equal to $W_{AA}(-\xi)$ if ξ is timelike. This is true in general since W_{AB} can be regarded as superposition of free field propagators of the form $\Delta^{(+)}(\xi)$ of different masses weighted by ρ_{AB}. (Compare (6.16) with (6.17).)

Now we are in a position to show an interesting and important connection between *CPT* and microcausality first pointed out by Jost (1957). In our special case of a two-fold vacuum expectation value, Jost's theorem reads as follows: Invariance of $\langle\phi_A(x)\phi_B(x')\rangle_0$ under SR and hence under *CPT* is completely equivalent to microcausality in the sense of vacuum expectation value, or more precisely, invariance under SR

$$\langle\phi_A(x)\phi_B(x')\rangle_0 = \langle\phi_B(-x')\phi_A(-x)\rangle_0 \tag{6.18}$$

if and only if

$$\langle[\phi_A(x), \phi_B(x')]\rangle_0 = 0, \text{ for } \xi^2 = (x - x')^2 > 0 \tag{6.19}$$

Note that (6.19) is weaker than the usual microcausality condition

$$[\phi_A(x), \phi_B(x')] = 0, \text{ for } \xi^2 > 0$$

and is often referred to as "weak local commutativity (WLC)." In terms of W_{AB}

$$\text{SR}: W_{AB}(\xi) = W_{BA}(\xi)$$

since $\langle\phi_B(-x')\phi_A(-x)\rangle_0 = W_{BA}(-x' - (-x)) = W_{BA}(\xi)$, and

$$\text{WLC}: W_{AB}(\xi) = W_{BA}(-\xi), \text{ for } \xi^2 > 0$$

To prove this, first consider

$$F_{AB}(\xi) = W_{AB}(\xi) - W_{BA}(\xi)$$

$F_{AB}(\xi)$ vanishes for $\xi^2 > 0$ if WLC holds since $W_{BA}(\xi) = W_{BA}(-\xi)$ for

$\xi^2 > 0$. But both $W_{AB}(\xi)$ and $W_{BA}(\xi)$ are boundary values of analytic functions in the ζ^2 plane. Hence the vanishing of $F_{AB}(\xi)$ for $\xi^2 > 0$ on the real axis implies $F_{AB}(\xi) = 0$ everywhere. Thus WLC implies SR. (WLC $\Rightarrow$ SR). To go backward (SR $\Rightarrow$ WLC) note that for the set of real points the region of analyticity coincides with the region for which WLC holds. (For details see Jost, 1957.)

To prove this theorem for the spinor case we must show

$$-(\gamma_5)_{\alpha\alpha'}\langle\psi_{\alpha'}(-x)\bar{\psi}_{\beta'}(-x')\rangle_0(\gamma_5)_{\beta'\beta} = \langle\bar{\psi}_\beta(x')\psi_\alpha(x)\rangle_0$$

if and only if

$$\langle\{\psi_\alpha(x),\, \bar{\psi}_\beta(x')\}\rangle_0 = 0, \text{ for } \xi^2 > 0$$

If we take out γ matrix dependence there are four invariant scalar functions, whose analytic properties are to be studied. Proof is essentially the same.

In the case of n-fold vacuum expectation values, Jost's theorem for neutral boson fields reads: WLC which implies $\langle\phi_1(x_1)\phi_2(x_2)\ldots\phi_n(x_n)\rangle_0 = \langle\phi_n(x_n)\ldots\phi_1(x_1)\rangle_0$ for every $x_1\ldots x_n$ such that $(\sum_{i=1}^{n-1}\lambda_i\xi_i)^2 > 0$ for any set $\{\lambda_i\}$ satisfying $\sum\lambda_i = 1$, $\lambda_i \geq 0$, where $\xi_i = x_i - x_{i+1}$, is equivalent to SR

$$\langle\phi_1(x_1)\ldots\phi_n(x_n)\rangle_0 = \langle\phi_n(-x_n)\ldots\phi_1(-x_1)\rangle_0$$

Again the essential step in the proof is the observation that the set of real points for which W can be analytically continued under $\xi \to -\xi$ is precisely the set of points for which WLC holds. This is not trivial to show for $n \geq 3$. The general proof makes use of a theorem of Hall and Wightman (1957) concerning the analytic behavior of Lorentz invariant functions under complex Lorentz transformations.

To summarize, if we have (1) invariance under $\Lambda \varepsilon \mathscr{L}_+^\uparrow$ (2) the existence of P_μ with $i[P_\mu, \phi] = \dfrac{\partial\phi}{\partial x_\mu}$ (eigenvalues of $P_0 > 0$; $P_\mu = 0$ for the vacuum state), then invariance under SR, or equivalently under CPT of the vacuum expectation value, is completely equivalent to weak local commutativity (normal spin-statistics relation implicit). A theory in which CPT invariance does not hold has to be very different from present field theories that postulate microcausality.

6.4. Consequences of *CPT* Invariance

First we consider properties of antiparticles. Suppose H, the total Hamiltonian, is invariant under CPT. Consider a single particle state at rest with definite magnetic quantum number J_z.

$$\begin{aligned}\langle a|H|a\rangle &= \langle a^\tau|THT^{-1}|a^\tau\rangle \\ &= \langle\bar{a}|PCTHT^{-1}C^{-1}P^{-1}|\bar{a}\rangle \\ &= \langle\bar{a}|H|\bar{a}\rangle\end{aligned} \tag{6.20}$$

where $|a^\tau\rangle \equiv T|a\rangle$, $|\bar{a}\rangle \equiv PCT|a\rangle$.

$|\bar{a}\rangle$ is antiparticle state with $-J_z$ (except for i^{2m} which is irrelevant). In the absence of external field, H acting on $|a\rangle$ gives the mass of the particle. Since the mass cannot depend on the orientation of the z axis,

$$CPT \Rightarrow \text{mass of a particle} = \text{mass of antiparticle.}$$

How about other "attributes" such as magnetic moment? Let the internal energy of $|a\rangle$ with magnetic quantum number J_z in a static external $\vec{B}$ be E_{mag}. In the CPT reversed state we have $\vec{B}$ in the *same* direction (since $\vec{B}$ changes sign under C and T but not under P). Therefore the internal energy of $|\bar{a}\rangle$ in $\vec{B}$ is also E_{mag}. But $|a\rangle$ and $|\bar{a}\rangle$ differ (apart from phase factor and charge, particle number) by their magnetic quantum numbers

$$|a\rangle : J_z$$
$$|\bar{a}\rangle : -J_z$$

Hence the magnetic moments are opposite in sign and equal in magnitude. This result is applicable to the anomalous moments of p and $\bar{p}$, etc.

It may appear that if C invariance is violated, the lifetime of a particle might not be identical to that of its antiparticle. This is not the case. Consider the decay of a particle state $|a\rangle$ into a final state $|\beta\rangle$ via the decay interaction Hamiltonian H_W, which is assumed to be "weak." Let

$$H_W = H_+ + H_-$$

where $PH_\pm P^{-1} = \pm H_\pm$

$$\begin{aligned}\langle\beta|H_\pm|a\rangle &= \langle a^\tau|TH_\pm T^{-1}|\beta^\tau\rangle = \langle\bar{a}|CPTH_\pm T^{-1}P^{-1}C^{-1}|\bar{\beta}\rangle \\ &= \langle\bar{a}|H_\pm|\bar{\beta}\rangle \end{aligned} \tag{6.21}$$

If there are final state interactions, $|\beta\rangle$ and $|\bar{\beta}\rangle$, in addition to being different by spin direction, charge, particle number, etc. may differ by phase. For example, if the phase of $|\beta\rangle$ is given by $\exp(i\delta_\beta)$, then the phase of $|\bar{\beta}\rangle$ is given by $\exp(-i\delta_\beta)$ so that the relative phase between $|\beta\rangle$ and $|\bar{\beta}\rangle$ is given by $\exp(2i\delta_\beta)$. But for particle lifetime, phase is irrelevant provided that $|\beta\rangle$ is an eigenstate of the *strong* interaction Hamiltonian.

$$\begin{aligned}\text{“Particle lifetime”} &= \frac{1}{2\pi}\Big[\sum_\beta (|\langle\beta|H_+|a\rangle|^2 + |\langle\beta|H_-|a\rangle|^2) \\ &\qquad\qquad \times (\text{phase space})_\beta\Big]^{-1} \\ &= \text{“antiparticle lifetime”}\end{aligned} \tag{6.22}$$

where rotational invariance (independence of J_z) was also assumed. We may mention here that the equality between the *partial* decay rate of Σ^+ into $p + \pi^0$ and that of $\overline{\Sigma^+}$ into $\bar{p} + \pi^0$ cannot be deduced on the basis of the CPT theorem alone; the theorem does require that the transition rate of Σ^+ into the $T = \frac{3}{2}(T = \frac{1}{2})$ state of the πN system be equal to that of $\overline{\Sigma^+}$ into the $T = \frac{3}{2}(T = \frac{1}{2})$ state of the $\pi \bar{N}$ system (Okubo, 1958).

As illustrative examples of the mass and lifetime equalities we may quote the following experimental numbers for the charged pion (Crowe, 1957)

$$\frac{m(\pi^+)}{m(\pi^-)} = 1.0021 \pm 0.0027$$

$$\tau(\pi^+) = (2.53 \pm 0.10) \times 10^{-8} \text{ sec.}$$

$$\tau(\pi^-) = (2.55 \pm 0.19) \times 10^{-8} \text{ sec.}$$

More recently the mass ratio for p and $\bar{p}$ has been determined at CERN (Cocconi *et al.*, 1960).

$$\frac{m(\bar{p})}{m(p)} = 1.008 \pm 0.005.$$

For K^0 and $\bar{K}^\circ$ it is possible to obtain a much more stringent limit on the mass equality as we shall discuss in Chapter IV.

Note added in proof. According to a recent Columbia experiment, the equality of the lifetimes of μ^+ and μ^- is now known to an accuracy of 0.2% (Anderson *et al.*, 1962).

In deriving the lifetime theorem we have used the fact that states of opposite parity cannot interfere if we are measuring the total decay probability which is scalar. Instead, we can sometimes utilize the fact that two states with opposite spin directions cannot interfere. Consider

$$\pi^+ \rightarrow \mu^+ + \nu$$

$$|\alpha\rangle : \pi^+ \text{ at rest}$$

$$|\beta\rangle : \mu^+ \text{ and } \nu \text{ both left-handed}$$

Then the relation $\langle\beta|H_W|\alpha\rangle = \langle\bar{\alpha}|H_W|\bar{\beta}\rangle$ now means $w(\mu^+L) = w(\mu^-R)$ where $w(\mu^+L)$ stands for the transition probability of π^+ into a left-handed μ^+ (and left-handed ν by momentum and angular momentum conservation).

$$\begin{aligned}\pi^+ \text{ lifetime} &= [w(\mu^+R) + w(\mu^+L)]^{-1} \\ &= [w(\mu^-L) + w(\mu^-R)]^{-1} \\ &= \pi^- \text{ lifetime.}\end{aligned}$$

As for the helicity of μ in pion decay

$$\mathscr{H}(\mu^+) = \frac{w(\mu^+R) - w(\mu^+L)}{w(\mu^+R) + w(\mu^+L)} = \frac{w(\mu^-L) - w(\mu^-R)}{w(\mu^-L) + w(\mu^-R)} = -\mathscr{H}(\mu^-)$$

provided that *CPT* invariance holds.

Immediate corollary: If μ^+ in π^+ decay is longitudinally polarized, using *CPT* we can conclude that charge conjugation invariance is violated even if we did not know anything about the polarization of μ^- in π^- decay.

PROOF:

$$C \text{ invariance} \Rightarrow \mathscr{H}(\mu^+) = \mathscr{H}(\mu^-)$$
$$CPT \text{ invariance} \Rightarrow \mathscr{H}(\mu^+) = -\mathscr{H}(\mu^-)$$

The two conditions are compatible only if $\mathscr{H}(\mu^+) = \mathscr{H}(\mu^-) = 0$. Another way to see this is to recall that the observation of a term of the form $\langle\vec{\sigma}\cdot\vec{p}\rangle_\mu$ is compatible with T invariance since the final state interaction between μ and ν is practically null. $\langle\vec{\sigma}\cdot\vec{p}\rangle \neq 0$ violates parity but not time reversal. The CPT theorem then requires that C be violated.

Λ decay can be treated in an analogous way. $\langle\vec{\sigma}_\Lambda\rangle\cdot\vec{p}_\pi$ violates C invariance if we assume that no final state interaction exists between π^- and p. Since the coefficient of the C violating $\langle\vec{\sigma}_\Lambda\rangle\cdot\hat{p}_\pi$ term is $2\,\mathrm{Re}(a_s a_p^*)/(|a_s|^2 + |a_p|^2)$ (cf. Chapter 3, Section 5), C invariance would require that a_s and a_p be relatively imaginary. In reality π^- and p can interact strongly once they have been created. But the effect of this final state interaction can be estimated in the same way as in the photoproduction case from the known phase shifts for πp scattering at 100 Mev./c. Gatto (1957) estimated an upper limit to $|\alpha|$ when C invariance holds.

$$C \text{ invariance} \Rightarrow |\alpha| \leq 0.18 \pm .02$$

Experimentally, $|\alpha| \geq 0.6$. Hence Λ decay violates charge conjugation invariance.

6.5. *CP* Invariance and the Mach Principle

If time reversal invariance turns out to be an exact symmetry principle, we can conclude with the aid of the CPT theorem that CP invariance is also an exact symmetry principle. (Note, however, that it is entirely incorrect to say that the CPT theorem claims $CP = T$). The right-left symmetry can be still retained if we interchange matter and antimatter when we take a mirror image, since for every possible state there exists a corresponding state in "antiworld" such that the mirror image of the original state looks identical to the corresponding state in "antiworld." Right and left are still indiscernible to a "detached" observer (unwilling to take the risk of annihilating himself) who is faced with both a world and its antiworld. Once we define matter and antimatter by convention, the sense of right and left can be uniquely determined from the asymmetric behavior of the weak interactions without recourse to any convention. Conversely, once we decide what is meant by right and left by convention, we can now differentiate matter from antimatter. (The "disturbing possibility" that CP but not P and C separately might turn out to be the only exact symmetry operation was first discussed by Wick, Wightman, and Wigner (1952). The possibility, according to the authors, was considered to be rather "remote at the moment.")

Should time reversal invariance turn out to be not valid, we would have an absolute way of telling right from left. This can be most clearly seen in the following problem.

PROBLEM 8. Suppose $\langle \vec{\sigma}_{\mu^+} \cdot (\vec{p}_{\mu^+} \times \vec{p}_{\pi^0}) \rangle > 0$ in $K^+_{\mu 3}(\rightarrow \mu^+ + \pi^0 + \nu)$ decay and $\langle \vec{\sigma}_{\mu^+} \cdot \vec{p}_{e^+} \rangle > 0$ in μ^+ decay. By comparing the $K^+_{\mu 3} - \mu^+ - e^+$ chain with the $K^-_{\bar{\mu} 3} - \mu^- - e^-$ chain show how we can define the sense of right and left in an absolute manner without recourse to matter and antimatter.

Prior to the discovery of parity nonconservation one of the *a priori* arguments in favor of parity conservation was that the idea of parity conservation is in accordance with the spirit of Mach's principle: The laws of physics should not depend on the particular geometrical coordinate system we happen to choose. If *CP* invariance turns out to be exact, the observed right-left asymmetry in one kind of matter can be made compatible with the Mach principle by considering the exactly opposite right-left asymmetry exhibited by the corresponding antimatter (Lee and Yang, 1957b).

If *CP* invariance should fail, in order to save the Mach principle, we might conjecture the possible existence of two kinds of elementary particles with the same charge, but in different "handedness," say p_+ and p_- etc. At the *present* time in *our* laboratory system there is only one kind, p_+. (Otherwise there would be difficulty with the exclusion principle as *we* know it.) The observed asymmetry is then due to a cosmological preponderance of one kind of elementary particle. Free oscillation time between p_+ and p_- is presumably greater than the age of the universe. (There arises the question of whether or not we have to double bosons too. Consider for instance $\gamma \rightarrow p_+ + \bar{p}_+$. If we do not double photons, $\gamma \rightarrow p_+ + \bar{p}_+$ and $\gamma \rightarrow p_- + \bar{p}_-$, must be equally probable. But $\bar{p}_-$ would not annihilate with p_+, so the antiproton of the $(-)$ type would live forever in *our* laboratory. Of course, we need not worry about all this "ugliness" if *CP* invariance is exact, which seems to be the case experimentally for the weak as well as for the strong interactions.)

CHAPTER 7

$_5$ *Invariance and Weak Interactions*

7.1. Two Component Theory of the Neutrino

Thus far we have been concerned with finding the most general interaction density that may not conserve P, C, and T. One of the most surprising things we have learned in recent years is that when nature violates P and C, she does so in a very specific and elegant manner! We will first talk about the two component neutrino theory and then we will discuss an even more specific theory—universal V-A theory.

Start with the Dirac equation in the Weyl representation (cf., Chapter 4, Section 4 and also Appendix B)

$$\gamma_k = \begin{pmatrix} 0 & i\sigma_k \\ -i\sigma_k & 0 \end{pmatrix}, \qquad \gamma_4 = \begin{pmatrix} 0 & I \\ I & 0 \end{pmatrix}$$

The Dirac equation

$$(\gamma_\mu \partial_\mu + m)\psi = 0$$

becomes

$$\begin{pmatrix} m & i\vec{\sigma}\cdot\vec{\nabla} + \dfrac{\partial}{\partial x_4} \\ -i\vec{\sigma}\cdot\vec{\nabla} + \dfrac{\partial}{\partial x_4} & m \end{pmatrix} \begin{pmatrix} \phi^{(R)} \\ \phi^{(L)} \end{pmatrix} = 0$$

where

$$\psi = \begin{pmatrix} \phi^{(R)} \\ \phi^{(L)} \end{pmatrix}$$

So we have two coupled equations:

$$\left(i\vec{\sigma}\cdot\vec{\nabla} - i\frac{\partial}{\partial x_0}\right)\phi^{(L)} + m\phi^{(R)} = 0$$

$$\left(-i\vec{\sigma}\cdot\vec{\nabla} - i\frac{\partial}{\partial x_0}\right)\phi^{(R)} + m\phi^{(L)} = 0 \qquad (7.1)$$

Note that

$$\gamma_5 = \begin{pmatrix} -I & 0 \\ 0 & I \end{pmatrix}$$

in the Weyl representation. Hence

$$\psi^{(R)} = \tfrac{1}{2}(1 - \gamma_5)\psi = \begin{pmatrix} \phi^{(R)} \\ 0 \end{pmatrix}$$

$$\psi^{(L)} = \tfrac{1}{2}(1 + \gamma_5)\psi = \begin{pmatrix} 0 \\ \phi^{(L)} \end{pmatrix} \tag{7.2}$$

Recall $\frac{1}{2}(1 + \gamma_5)\psi$ and $\frac{1}{2}(1 - \gamma_5)\psi$ project the left-handed and right-handed $p_0 > 0$ solutions as $v/c \to 1$ which justifies the notation (R) and (L). $\psi^{(R)}$ and $\psi^{(L)}$ are sometimes called "chiral spinors." As $m \to 0$, the two equations become *uncoupled*

$$\left(i\vec{\sigma}\cdot\vec{\nabla} + i\frac{\partial}{\partial x_0}\right)\phi^{(R)} = 0 \tag{7.3a}$$

$$\left(-i\vec{\sigma}\cdot\vec{\nabla} + i\frac{\partial}{\partial x_0}\right)\phi^{(L)} = 0 \tag{7.3b}$$

It is perhaps of interest to remark here that the chiral spinors $\psi^{(R)}$ and $\psi^{(L)}$ obey the following equal-time commutation relations *regardless of whether or not the field is massless.*

$$\begin{aligned} \{\psi_\alpha^{(L)}(\vec{x}, x_0), \psi_\beta^{(R)\dagger}(\vec{x}', x_0)\} &= a_{\alpha\beta}^{(R)}\delta^{(3)}(\vec{x} - \vec{x}') \\ \{\psi_\alpha^{(L)}(\vec{x}, x_0), \psi_\beta^{(L)\dagger}(\vec{x}', x_0)\} &= a_{\alpha\beta}^{(L)}\delta^{(3)}(\vec{x} - \vec{x}') \end{aligned} \tag{7.4}$$

where

$$a^{(R)} = \tfrac{1}{2}(1 - \gamma_5)$$

$$a^{(L)} = \tfrac{1}{2}(1 + \gamma_5)$$

All other anticommutators vanish. (To prove (7.4) just apply $a^{(R)}$ from the left and also from the right to the ordinary equal-time commutation relations keeping in mind that $a^{(R)2} = a^{(R)}$.)

So far there has been no physics. Physics enters when we demand that only one of the two two-component equations is actually realized in nature for an $m = 0$ particle, namely for the neutrino. Such a hypothesis was first advanced by Weyl. (It was rejected by Pauli (1933) on the grounds that the equation of motion is not invariant under parity. It was revived by Salam (1957), Landau (1957), and Lee and Yang (1957a).)

From Problem 1 we know that even if we have only one of the two equations it is possible to construct a bilinear 4-vector invariant under proper orthochronous Lorentz transformations. Moreover, such a current is conserved if the equation of motion is given by (7.3a) (if $\vec{j} = \phi^\dagger\vec{\sigma}\phi$, $j_0 = \phi^\dagger\phi$) or (7.3b) (if $\vec{j} = \phi^\dagger\vec{\sigma}\phi$, $j_0 = -\phi^\dagger\phi$).

Assume the neutrino field satisfies (7.3b). As in the Dirac equation there are two energy states. For $p_0 > 0$ solution,

$$\vec{\sigma}\cdot\vec{p}\phi^{(L)} = -p_0\phi^{(L)} = -E\phi^{(L)} \tag{7.5a}$$

For $p_0 = -E < 0$ solution,

$$\vec{\sigma}\cdot\vec{p}\phi^{(L)} = -p_0\phi^{(L)} = +E\phi^{(L)} \tag{7.5b}$$

Thus, in this case, the positive energy neutrino is left-handed and the negative energy neutrino is right-handed.

We can easily find a relationship for the helicities of the neutrino and antineutrino. First note

$$\mathscr{H}(\bar{\nu}; p_0 > 0) = \mathscr{H}(\nu; p_0 < 0) \tag{7.6}$$

where $\mathscr{H}(\bar{\nu}; p_0 > 0)$ is the helicity for a physical antineutrino and $\mathscr{H}(\nu; p_0 < 0)$ is the helicity of the corresponding hole. The helicities are the same because both the spins and the momenta are opposite for hole and antiparticle. From (7.3b) we see that

$$\mathscr{H}(\bar{\nu}; p_0 < 0) = -\mathscr{H}(\nu; p_0 > 0)$$

Thus

$$\mathscr{H}(\bar{\nu}; p_0 > 0) = -\mathscr{H}(\nu; p_0 > 0) = +1 \tag{7.7}$$

and we see that the antineutrino is right-handed if we develop a hole theory in exact correspondence with the Dirac theory. In this approach the neutrino and antineutrino are fundamentally different: the neutrino is always left-handed and the antineutrino always right-handed.

An alternative approach is to regard (7.3) as the $m = 0$ limit of the Majorana theory in which the neutrino and antineutrino are identical. (See Serpe (1957), McLennan (1957), and Case (1957).) Case emphasizes that it is misleading to say parity is not conserved if (7.3b) holds. All we have to do is to let under space inversion

$$\phi^{(L)} \to i\sigma_2\phi^{(L)*}, \qquad \vec{\sigma}\cdot\vec{\nabla} \to -\vec{\sigma}\cdot\vec{\nabla}$$

Contrast this with the Maxwell case, where the Maxwell equations

$$\vec{\nabla}\times\vec{B} = \frac{\partial\vec{E}}{\partial x_0}, \qquad \vec{\nabla}\times\vec{E} = -\frac{\partial\vec{B}}{\partial x_0}, \qquad \vec{\nabla}\cdot\vec{B} = 0, \qquad \vec{\nabla}\cdot\vec{E} = 0$$

may be written as

$$i\frac{\partial\Phi}{\partial x_0} = -i\vec{S}\cdot\vec{\nabla}\Phi$$

$$\Phi = \begin{pmatrix} B_1 - iE_1 \\ B_2 - iE_2 \\ B_3 - iE_3 \end{pmatrix}, \qquad (S_j)_{ik} = i\varepsilon_{ijk}$$

Under parity

$$\Phi_k = B_k - iE_k \to B_k + iE_k = \Phi_k^*$$

In any case, weak interactions violates parity whether or not the transformation property of the free field equation is such that it can be made invariant under P.

What kind of interaction will produce only left-handed neutrinos? Naturally any interaction will do if we require invariance under the transformation

$$\psi_\nu \rightarrow \gamma_5 \psi_\nu \tag{7.8}$$

so that ψ_ν appears only in the combination $\frac{1}{2}(1 + \gamma_5)\psi_\nu$. Invariance under (7.8) is called γ_5 invariance. Note that the Dirac equation with $m = 0$ is invariant under the above transformation. So we can say that the vanishing of the neutrino rest mass admits the possibility of *exact* invariance under $\psi_\nu \rightarrow \gamma_5 \psi_\nu$.

7.2. Lepton Conservation and the πμe Sequence

In order that the two component theory make a definite prediction we need a principle that tells us under what circumstances neutrinos (or antineutrinos) are emitted or absorbed. For this reason we postulate the principle of *lepton conservation* as first stated by Konopinski and Mahmoud (1953):

> Assign lepton numbers that can take values $+1$, 0, -1. The lepton numbers of light fermions are ± 1, and of baryons, pions, photons and K mesons, zero. In any reaction the algebraic sum of lepton numbers is conserved.

Take e^- to be a lepton by definition. Then e^+ must be an antilepton, because any conserved attribute must transform like $\int \psi^\dagger \psi \, d^3x$, which changes sign under charge conjugation. If lepton conservation holds in beta decay, then the neutral particle emitted in β^+ decay, which we may call "neutrino," is a lepton. We shall see that the neutrino in this sense turns out to be left-handed so that (7.3b) is the right equation of motion.

A non-trivial question is: What is the lepton number of μ^+? Assuming lepton conservation, we have two possibilities:

$$\mu^+(-1) \rightarrow e^+(-1) + \nu(+1) + \bar{\nu}(-1) \tag{7.9a}$$

$$\mu^+(+1) \rightarrow e^+(-1) + \nu(+1) + \nu(+1) \tag{7.9b}$$

In the second case, if we further assume the two component neutrino theory, it can be shown that the positron spectrum must vanish at the maximum e^+ energy. Experimentally there are a lot of high energy positrons at the top of the spectrum. Therefore the lepton number of the μ^+ is -1. The situation is summarized in Table 7.1.

Consider now π^+ decay. Because of lepton conservation $\pi^+ \rightarrow \mu^+ + \nu$ and not $\mu^+ + \bar{\nu}$ if our assignment is correct. The neutrino is left-handed, and therefore the μ^+ must also be left-handed because of the conservation of angular momentum. Thus $\mathscr{H}(\mu^+) = -1$ and we see that the π^+ decay at rest is a perfect polarizer of the μ^+ spin. Similarly we expect that the helicity of μ^- in π^- decay is $+1$.

TABLE 7.1

Lepton Number

	e^-, μ^-, ν_L	$e^+, \mu^+, \bar{\nu}_R$
Lepton number	$+1$	-1

Polarized muons slow down and stop before they decay, but depending on the material (graphite, aluminum, etc.) the muon spin direction is still preserved. So we have a source of polarized muons. If parity is not conserved in muon decay either, there will be a forward-backward asymmetry in the positron distribution with respect to the original μ^+ direction. Friedman and Telegdi (1957) and Garwin, Lederman, and Weinrich (1957) observed such an asymmetry. More positrons are emitted backward with respect to the μ^+ direction, which shows that parity is not conserved in both π decay and μ decay. (Actually this conclusion is subject to the assumption that muons do not exist in degenerate states of opposite parity, i.e. no parity doublet for muons.)

We can test the two component neutrino theory with lepton conservation by measuring the e^+ helicity in μ^+ decay (cf. Fig. 7.1) For the favored

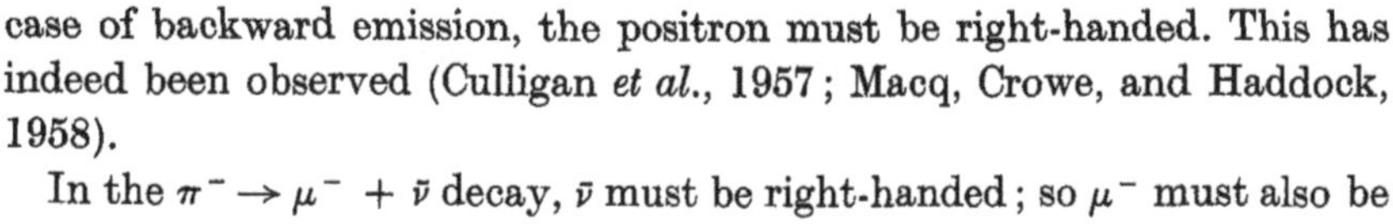

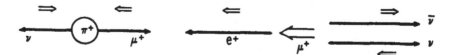

Figure 7.1. $\pi\mu e$ sequence. A double arrow stands for spin, a single arrow for momentum

case of backward emission, the positron must be right-handed. This has indeed been observed (Culligan *et al.*, 1957; Macq, Crowe, and Haddock, 1958).

In the $\pi^- \to \mu^- + \bar{\nu}$ decay, $\bar{\nu}$ must be right-handed; so μ^- must also be right-handed. This agrees with observation (Alikanov *et al.*, 1960); Backenstoss *et al.*, 1961; Bardon, Franzini, and Lee, 1961). From the CPT theorem we know that the asymmetry parameter (the coefficient of $\langle\vec{\sigma}_\mu\rangle\cdot\vec{p}_e$) must change sign as we go from μ^+ decay to μ^- decay. So the *net* effect is that the e^- goes backward with respect to the original μ^- direction. This has been confirmed. Actually the observed asymmetry effect is less marked in the μ^- case, since negative muons get depolarized much more readily. In any case the e^- from μ^- decay has to be left-handed even if the μ^- polarization is zero. This has also been confirmed (Macq, Crowe, and Haddock, 1958).

To make the story more quantitative let us write down the μ decay Hamiltonian using the two component condition with the assumption that ν and $\bar{\nu}$ are emitted.

$$H_{\text{int}} = g_{SP}[\bar{\nu}(1 - \gamma_5)\mu][\bar{e}(1 + \gamma_5)\nu] + g_{VA}[\bar{\nu}(1 - \gamma_5)\gamma_\mu\nu][\bar{e}\gamma_\mu(1 + \gamma_5)\nu] + \text{H.C.} \qquad (7.10)$$

Note that S and P can be grouped together. Similarly for V and A. Tensor vanishes identically, i.e.,

$$\tfrac{1}{2}[\bar{e}\sigma_{\mu\nu}(1 + \gamma_5)\nu][\bar{\nu}(1 - \gamma_5)\sigma_{\mu\nu}\mu] = 0$$

PROOF: Use $\gamma_5\sigma_{4k} = \sigma_{ij}\varepsilon_{ijk}$, $\gamma_5\sigma_{ij} = \sigma_{4k}\varepsilon_{ijk}$ (ijk cyclic). Then the ij term in the sum is

$$[\bar{e}\sigma_{ij}(1 + \gamma_5)\nu][\bar{\nu}(1 - \gamma_5)\sigma_{ij}\mu] = [\bar{e}(\sigma_{ij} + \sigma_{4k})\nu][\bar{\nu}(\sigma_{ij} - \sigma_{4k})\mu]$$

and the $k4$ term is

$$[\bar{e}\sigma_{k4}(1 + \gamma_5)\nu][\bar{\nu}(1 - \gamma_5)\sigma_{k4}\mu] = [\bar{e}(-\sigma_{4k} - \sigma_{ij})\nu][\bar{\nu}(-\sigma_{4k} + \sigma_{ij})\mu]$$

Thus the ij term cancels exactly with the $k4$ term. Similarly the ji term cancels with the $4k$ term.

The energy and angular distribution of electrons as first computed by Landau (1957) and Lee and Yang (1957a) is (assuming $m_e \approx 0$ which can be justified since the mean electron energy is $\approx$40 Mev)

$$f(x)(1 + a(x)\cos\theta)\,dx\,\frac{d\Omega}{4\pi} = 2[(3 - 2x) \mp \xi\cos\theta(1 - 2x)]x^2dx\,\frac{d\Omega}{4\pi} \qquad (7.11)$$

where the upper sign is for μ^- decay and the lower sign for μ^+ decay,

$$x = \frac{p_e}{(p_e)_{\max}} = \frac{E_e}{(E_e)_{\max}}$$

and ξ is the asymmetry parameter given by

$$\xi = \frac{-|g_{SP}|^2 + 4|g_{VA}|^2}{|g_{SP}|^2 + 4|g_{VA}|^2} \qquad (7.12)$$

The integrated asymmetry coefficient is

$$\bar{a} \equiv \frac{\int_0^1 a(x)f(x)dx}{\int_0^1 f(x)dx} = \mp\frac{\xi}{3}$$

which is at most $\frac{1}{3}$ in magnitude. Note that $a(x)$ at low energies is of opposite sign to $\bar{a}$, and is small in magnitude. The energy dependence of $a(x)$ given by the two-component neutrino theory is in good agreement with experiments. (See Plano, 1960).

As for the energy dependence of the isotropic term it is customary to compare it with Michel's formula derived under the assumption of a general four-fermion interaction (Michel, 1950).

$$f(x)dx = 4[3(1 - x) + 2\rho\left(\frac{4x}{3} - 1\right)x^2]\,dx \tag{7.13}$$

The well-known Michel parameter ρ, which is a complicated function of coupling constants, characterizes the end point of the electron distribution. (See Fig. 7.2.) By comparing (7.11) with (7.13) we see that the two component neutrino theory predicts a unique value for ρ, namely $\rho = 0.75$.

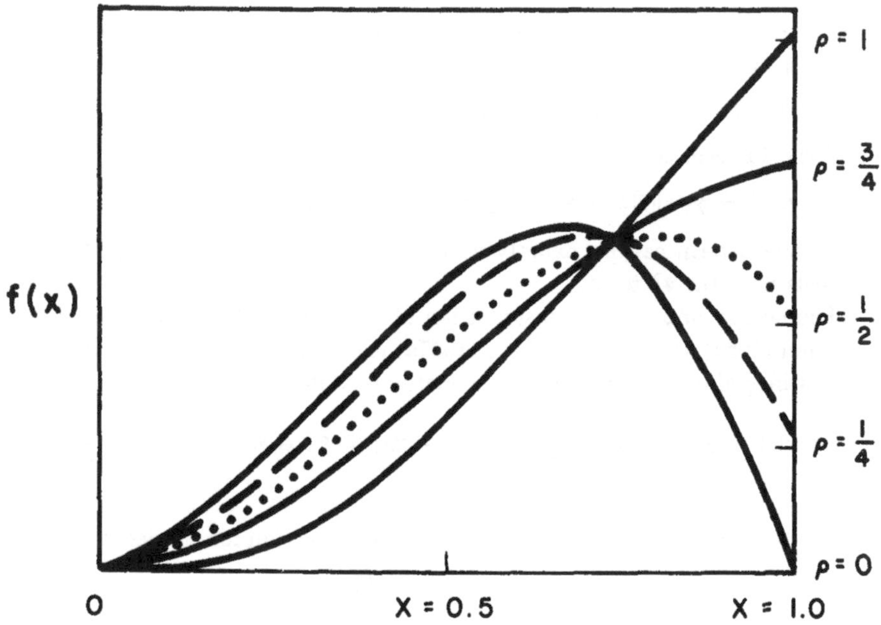

Figure 7.2. The electron energy spectrum in muon decay. $x = E_e/E_e^{max}$

Experimentally $\rho = 0.68$–0.80 depending on which experimentalists you ask. (Plano (1960), for instance, gives $\rho = 0.780 \pm 0.025$. Radiative corrections (calculated by Kinoshita and Sirlin (1959)) have already been taken into account in such experimental numbers; so these numbers should be compared directly with 0.75.) Incidentally if we had $\mu^+ \rightarrow e^+ + 2\nu$ in the sense of the two-component neutrino theory (see (7.9b)), i.e. if the μ^+ rather than the μ^- were a lepton, then ρ would be equal to zero in sharp disagreement with experiments.

In the two component neutrino theory ξ may take any value from -1 to 1 (and can even be zero!). Experimentally, Bardon, Berley, and Lederman (1959) gives $\xi \geq 0.97 \pm 0.05$, which corresponds to $|g_{SP}|^2/|g_{VA}|^2 \leq$ 0 05 $\pm$ 0.09. The experimental fact that ξ is roughly $+1$ corresponding to $g_{VA} \neq 0$, $g_{SP} = 0$ shows that the two component neutrino theory *alone* might not be enough, and suggests a very particular form of coupling. Within the framework of the two component neutrino theory, one can show that the $e^{\pm}$ helicity is related to the asymmetry parameter by

$$\mathscr{H}(e^+) = -\mathscr{H}(e^-) = \xi$$

Since ξ is essentially maximal from angular distribution experiments, the positron helicity must also be maximal. We need some kind of "rule" that requires that the e^+ helicity in μ^+ decay be $+1$. The rule we need is simply the following: All fermions (e^-, ν, etc.) are produced left-handed and all antifermions (e^+, $\bar{\nu}$, etc.) are produced right-handed as $v/c \to 1$.

7.3. The Form of the Four-Fermion Weak Coupling

In looking for a principle that determines the form of the four-fermion weak coupling, perhaps the most natural approach is to extend γ_5 invariance to fermions other than the neutrino. It is true that the free field equation is not invariant under

$$\psi \to \gamma_5 \psi \quad (\text{or } -\gamma_5 \psi) \tag{7.14}$$

when the mass term is present (unless we switch simultaneously the sign of the mass term). In other words, only the kinetic part of the free field Lagrangian is γ_5 invariant. Nevertheless it is of interest to examine the consequences of requiring the weak coupling to be invariant under (7.14) for every fermion field. Initial attempts along this line were made by Stech and Jensen (1955) and Tiomno (1955) even before the discovery of parity nonconservation.

We recall (cf. Chapter 5, Section 5) that if $\psi \to \gamma_5 \psi$ for some fermion field, then $\psi^c \to -\gamma_5 \psi^c$ for the corresponding antifermion field. Without loss of generality we may define a "true fermion" field in such a way that we always have $\psi \to +\gamma_5 \psi$ rather than $\psi \to -\gamma_5 \psi$ under the transformation in question. In this connection it is worth mentioning that for a massless fermion field, the conventional distinction between the fermion and the antifermion becomes ambiguous because of the so-called Pauli transformation which mixes the "particle" state and the "antiparticle" state, as we shall discuss in Chapter 8. However, if we define the sense of particle versus antiparticle via their "chiral" properties, the particle state and the antiparticle state can never be mixed up.

The requirement that the weak coupling be invariant under an individual transformation of the form (7.14) for every fermion field leads to the following interesting possibility: The fermion field ψ appears only in the form $\frac{1}{2}(1 + \gamma_5)\psi$. In other words, nature somehow picks out only one of the two chiral spinors, namely the lower two components of the full Dirac spinor in the Weyl representation (representation in which γ_5 is diagonal). (Cf. Equation (7.2).) Because $[(1 + \gamma_5)\psi]^\dagger = \bar{\psi}(1 - \gamma_5)$, $\bar{\psi}$ must appear only in the form $\frac{1}{2}\bar{\psi}(1 - \gamma_5)$. Since $(1 - \gamma_5)(1 + \gamma_5) = 0$, the only non-vanishing bilinear form constructed out of two different spinors turns out to be

$$J_\mu^{(ij)} = \bar{\psi}_i \gamma_\mu (1 + \gamma_5)\psi_j \tag{7.15}$$

where it is understood that ψ_i and ψ_j are "true fermion" fields. All other bilinear forms vanish

$$\bar{\psi}_i(1 - \gamma_5)\begin{Bmatrix} 1 \\ \gamma_5 \\ \sigma_{\mu\nu} \end{Bmatrix}(1 + \gamma_5)\psi_j = 0$$

This means that the four-fermion interaction (presumably in its "bare" form) is of the form

$$[\bar{\psi}_1\gamma_\mu(1 + \gamma_5)\psi_2][\bar{\psi}_3\gamma_\mu(1 + \gamma_5)\psi_4] + \text{H.C.} \tag{7.16}$$

as written down by Sudarshan and Marshak (1958). Feynman and Gell-Mann (1958), Sakurai (1958a) in advance of the confirming experiments.

Of course, it is quite possible (and even plausible from some point of view) that (7.16) is a phenomenological effective interaction which is approximately valid only in the limit of low momentum transfer. This may be the case, if the fundamental weak coupling turns out to be the coupling of $J_\mu^{(ij)}$ in (7.15) with a massive vector field. In such a case the interaction between $J_\mu^{(12)}$ and $J_\mu^{(34)}$ is transmitted via the exchange of a vector boson, W, but if the boson is massive enough (say $\mu_W > \mu_K$), the effective interaction would still appear to be essentially a point interaction of the form (7.16).

There are a few aspects of (7.16) worth listing. First of all, if one (or two) of the fermion fields involved turns out to be the neutrino field, we have the two-component neutrino theory. The left-handed neutrino is, in our language, a true fermion. If lepton conservation in the sense of the previous section holds, e^- and μ^- must also be true fermions. The interaction necessarily violates P and C, but CP invariance still holds. In discussing muon decay we noted that we need some kind of rule that guarantees that all fermions (ν, e^-, μ^-) are produced left-handed and all antifermions ($\bar{\nu}$, e^+, μ^+) are produced right-handed as $v/c \rightarrow 1$. We see that (7.16) precisely fulfills this requirement. (Recall that $\frac{1}{2}(1 + \gamma_5)\psi$ creates right-handed antifermions and $\frac{1}{2}\bar{\psi}(1 - \gamma_5)$ creates left-handed fermions as $v/c \rightarrow 1$.)

If we examine the implications of (7.16) in the beta decay case, we see why (7.16) is called the $V-A$ coupling. Assume that n and p as well as e^- and ν are true fermions. We then have

$$[\bar{p}\gamma_\mu(1 + \gamma_5)n][\bar{e}\gamma_\mu(1 + \gamma_5)\nu] = [\bar{p}\gamma_\mu n][\bar{e}\gamma_\mu(1 + \gamma_5)\nu] - [\bar{p}i\gamma_5\gamma_\mu n][\bar{e}i\gamma_5\gamma_\mu(1 + \gamma_5)\nu]$$

Compare this with

$$[\bar{p}\gamma_\mu n][\bar{e}\gamma_\mu(C_V + C_V'\gamma_5)\nu] + [\bar{p}i\gamma_5\gamma_\mu n][\bar{e}i\gamma_5\gamma_\mu(C_A + C_A'\gamma_5)\nu]$$

We see that

$$C_V = C'_V, \quad C_A = C'_A \quad \text{(two-component neutrino condition)}$$

$$C_V = -C_A \qquad (V-A \text{ condition})$$

Note that all other coefficients vanish

$$C_i = C'_i = 0 \text{ for S, T, and P}$$

If we had regarded $\bar{n}$ and $\bar{p}$ as true fermions, then we would have had $C_V = C_A$ corresponding to $V + A$ since

$$[\bar{p}^c\gamma_\mu(1 + \gamma_5)n^c] = -\eta_C(p)\eta_C(n)[\bar{p}\gamma_\mu(1 - \gamma_5)n]$$

Experimentally $C_V \approx -C_A$. Hence n and p are true fermions just as are e^- and ν.

One rather general remark, which might turn out to be a "deep" one in the future theory of elementary particles—the coupling (7.16) is reminiscent of the electromagnetic coupling in the sense that the notion of a "current" plays a vital role. Now, in electromagnetism, the appearance of the charge-current four-vector in the fundamental interaction is intimately tied in with the law of conservation of electric charge, as we shall see in Chapter 8. Similarly we may be tempted to speculate that the conservation law of "fermionic charge" might have something to do with the origin of the so-called weak interactions. (See e.g. Bludman, 1958.) Of course, this analogy is not perfect since the quantity $J_\mu = \bar{\psi}_i\gamma_\mu(1 + \gamma_5)\psi_j$ is not a conserved quantity unless ψ_i and ψ_j are massless fields. Yet the analogy might be of some value especially if the masses of the fermions could be attributed solely to the strong and electromagnetic interactions. (The muon, of course, is a deep mystery.)

7.4. Nuclear Beta Decay

We now discuss nuclear beta decay in a more systematic manner than before. In particular we show that all beta decay experiments are in full accord with the predictions of the $V-A$ theory.

We start with the Hamiltonian

$$H_{\text{int}} = \sum_i (\bar{p}\Gamma_i n)[\bar{e}(C_i\Gamma_i + C'_i\Gamma_i\gamma_5)\nu] + \text{H.C.};$$

$$\Gamma_i = 1, \gamma_\mu, \frac{1}{\sqrt{2}}\sigma_{\mu\nu}, i\gamma_5\gamma_\mu, \gamma_5$$

This Hamiltonian contains 10 arbitrary constants. If we relax the condition of time reversal invariance, there are 19 real parameters (not 20, because an arbitrary phase factor does not make any difference). Pauli (1957) has pointed out that further doubling of coupling constants is possible if we relax lepton conservation, i.e. if we admit the possibility of antineutrinos as well as neutrinos being emitted in β^+ decay. Then there will be 35 real parameters (rather than 40, for reasons which will be discussed in Chapter 8). At present we assume lepton conservation since in discussing single beta decay the question of lepton conservation can be shown to be irrelevant.

If we assume that nucleons are slow—that is, they may be treated nonrelativistically—then

$$\bar{p}\Gamma_i n \xrightarrow{\text{NR}} \begin{cases} \psi_p^{(P)\dagger}\psi_n^{(P)} & \text{for } \Gamma_i = 1, \gamma_4 \\ \psi_p^{(P)\dagger}\sigma_k\psi_n^{(P)} & \text{for } \Gamma_i = \sigma_{ij}, i\gamma_5\gamma_k \\ 0 & \text{for } \Gamma_i = \gamma_5, \gamma_k, \sigma_{i4} = -\sigma_{4i}, i\gamma_5\gamma_4 \end{cases}$$

(where $\psi^{(P)}$ is a Pauli two-component wave function). Note that P (pseudoscalar) "escapes" in this approximation. Even if we analyze "forbidden" transitions, no evidence for P is known. If we believe in the Universal Fermi Interaction (UFI) hypothesis, then the presence of P in beta decay is incompatible with the $(\pi \rightarrow e + \nu)/(\pi \rightarrow \mu + \nu)$ ratio as we shall show later.

In nuclear beta decay the nucleon that decays is not free, but is bound to the rest of the nucleus both in the initial and in the final state. Thus the transition matrix elements will look like

$$\int (\psi_f^\dagger\psi_i)(\bar{\psi}_e C_V \gamma_4 \psi_\nu)\, d^3x, \text{ etc.}$$

where ψ_f and ψ_i are nuclear wave functions.

If the neutron and proton are free as in elementary beta decay we can immediately take out their plane wave dependence and integrate to get

$$(u_p^{(P)\dagger}u_n^{(P)})(\bar{u}_e C_V \gamma_4 u_\nu)$$

which we had before in Chapter 3, Section 8. But in nuclear beta decay we must first approximate $\exp[(i\vec{p}_e + i\vec{p}_\nu)\cdot\vec{x}]$ by its leading term 1, which is justified by the fact that the de Broglie wave length of the leptons $\sim 10^{-11}$ cm is large compared to the radius of the interaction $\lesssim 10^{-13}$ cm. So we get

$$\begin{aligned} &\int (\psi_f^\dagger\psi_i)\, d^3x\, [\bar{u}_e(C_i\Gamma_i + C_i'\Gamma_i\gamma_5)u_\nu] \text{ for } \Gamma_i = 1, \gamma_4 \\ &\int (\psi_f^\dagger\sigma_k\psi_i)\, d^3x\, [\bar{u}_e(C_i\Gamma_i + C_i'\Gamma_i\gamma_5)u_\nu] \text{ for } \Gamma_i = \frac{1}{\sqrt{2}}\sigma_{ij}, i\gamma_5\gamma_k \end{aligned} \tag{7.17}$$

where it is customary to define

$$M_F = \int \psi_f^\dagger\psi_i\, d^3x$$

$$M_{GT} = \int \psi_f^\dagger\vec{\sigma}\psi_i\, d^3x$$

Note we have made two approximations

(1) The nucleons are slow ($v/c \approx 0$ for n and p).

(2) The de Broglie wave lengths of the leptons are long.

For "allowed" transitions M_F and/or M_{GT} are non-vanishing.

If the symmetry of the nuclear states is such that both M_F and M_{GT} are zero, the transition is called a forbidden transition. In such a case we would have to consider the second term in the expansion of $\exp[i(\vec{p}_e + \vec{p}_\nu)\cdot\vec{x}]$. So we have an integral of the form:

$$i \int (\psi_f^\dagger \vec{x} \psi_i)\, d^3x \cdot (\vec{p}_e + \vec{p}_\nu)[\bar{u}_e(C_i\Gamma_i + C_i'\Gamma_i\gamma_5)u_\nu].$$

This gives rise to the so-called first forbidden spectrum. The comparative lifetime (which is the reciprocal of the transition probability divided by phase space) is much greater in this case. From now on we will consider only allowed transitions.

We are now in a position to find the selection rules for allowed transitions in nuclear beta decay. If the relative parity of the initial and final states is odd, then both M_F and M_{GT} vanish, because the operators which act on the initial state in these integrals are 1 and $\vec{\sigma}$, both of which commute with the parity operator. Also M_F vanishes unless there is no change in angular momentum, because the operator 1 cannot change angular momentum. On the other hand $\vec{\sigma}$ can change angular momentum by 1 unit, so M_{GT} vanishes unless $\Delta J = 0, 1$; except that $0 \to 0$ transitions are forbidden, because $\vec{\sigma}$ vanishes between states of zero angular momentum.

To sum up:

$M_F \neq 0$ Fermi transition	$\Delta J = 0$	no parity change
$M_{GT} \neq 0$ Gamow-Teller transition	$\Delta J = 0, 1$ no $0 \to 0$	no parity change

The actual evaluation of M_F and M_{GT} belongs to the theory of nuclear structure. For an elementary discussion on this subject, consult e.g. Chapter 10, Appendix 3 of Mayer and Jensen (1955).

The energy spectrum and angular distribution for beta decay is found to be (cf. Chapter 3, Section 8)

$$\begin{aligned} &N(E_e, \Omega_e, \Omega_\nu)\, dE_e\, d\Omega_e\, d\Omega_\nu \\ &= \frac{1}{(2\pi)^5} p_e E_e (E_e^{(\max)} - E_e)^2\, dE_e\, d\Omega_e\, d\Omega_\nu\, \xi\left(1 + a\,\frac{\vec{p}_e \vec{p}_\nu}{E_e \cdot E_\nu} + b\,\frac{m_e}{E_e}\right) \end{aligned} \tag{7.18}$$

where

$$\begin{aligned} \xi &= |M_F|^2(|C_S|^2 + |C_S'|^2 + |C_V|^2 + |C_V'|^2) \\ &\qquad + |M_{GT}|^2(|C_T|^2 + |C_T'|^2 + |C_A|^2 + |C_A'|^2) \\ a\xi &= |M_F|^2(-|C_S|^2 - |C_S'|^2 + |C_V|^2 + C_V'|^2) \\ &\qquad + (1/3)\,|M_{GT}|^2(|C_T|^2 + |C_T'|^2 - |C_A|^2 - |C_A'|^2) \\ b\xi &= \pm |M_F|^2 2\,\mathrm{Re}\,(C_S C_V^* + C_S' C_V'^*) \pm |M_{GT}|^2 2\,\mathrm{Re}\,(C_T C_A^* + C_T' C_A'^*) \end{aligned}$$

The following facts were known prior to the discovery of parity non-conservation.

(1) Allowed $0 \rightarrow 0$ transitions, which are possible by the Fermi rule but not by the Gamow-Teller rule, definitely exist. For example, C^{14}, O^{14}, N^{14*} form a $J = 0$ isotriplet, and $O^{14} \rightarrow N^{14*}$ is observed. Therefore S and/or V is present in beta decay.

(2) Allowed $0 \rightarrow 1$ transitions, which are possible by the Gamow-Teller, but not by the Fermi rule, definitely exist. For example $He^6 \xrightarrow{\beta^-} Li^6$. Now He^6 consists of an alpha particle, plus a "dineutron" in a relative s state. The alpha spin is zero, and by the exclusion principle, the spin of the "dineutron" must also be zero. Thus the spin of He^6 is zero. On the other, Li^6 consists of an alpha particle and a deuteron, which has spin 1, so Li^6 has spin 1. Therefore T and/or A must be present.

(3) From the lifetimes of various nuclei for which $|M_F|^2$ and $|M_{GT}|^2$ can be estimated, it was found that

$$\frac{|C_T|^2 + |C_T'|^2 + |C_A|^2 + |C_A'|^2}{|C_S|^2 + |C_S'|^2 + |C_V|^2 + |C_V'|^2} = 1.3 \pm 0.1$$

(4) The Fierz interference term (b term) is absent.

$$b = 0 \pm 0.10 \text{ for Fermi transitions}$$
$$b = 0 \pm 0.04 \text{ for Gamow-Teller transitions}$$

(5) Studies on the energy and/or the angular distribution of the recoil nucleus were made in order to get information on the electron-neutrino angular correlation term ($\vec{p}_e \cdot \vec{p}_\nu$ term in (7.18)). Conclusions drawn from such recoil experiments which were generally considered to be trustworthy prior to 1957 have since been shown to have been completely wrong.

From parity experiments we learn the following:

(1) Beta asymmetry experiments, e.g., $Co^{60} \rightarrow Ni^{60} + e^- + \nu$ (which is a Gamow-Teller transition with $J = 5 \rightarrow J = 4$), shows that electrons are emitted preferentially backward relative to the direction of the nuclear spin, with an asymmetry parameter of v/c. This can be shown to imply that

$$\frac{2\,\mathrm{Re}\,(C_T C_T'^* - C_A C_A'^*)}{|C_T|^2 + |C_T'|^2 + |C_A|^2 + |C_A'|^2} \approx -1$$

Thus either $C_T = -C_T'$ and/or $C_A = C_A'$. The first possibility implies a right-handed neutrino and a left-handed antineutrino, because the tensor term in the interaction Hamiltonian will then contain

$$C_T(\bar{e}\sigma_{\mu\nu}(1 - \gamma_5)\nu)$$

On the other hand $C_A = C_A'$ implies a left-handed neutrino and a right-handed antineutrino, because in this case the interaction Hamiltonian contains

$$C_A[\bar{e}\gamma_\mu\gamma_5(1 + \gamma_5)\nu]$$

Pictorially the situation is shown in Fig. 7.3.

Figure 7.3. Co^{60} decay

(2) The longitudinal polarization of beta-particles is experimentally always found to be v/c for positrons and $-v/c$ for negatrons to an accuracy of 5–10%. (See e.g. Frauenfelder *et al.*, 1957; Benczer-Koller *et al.*, 1958.) Thus in the interaction Hamiltonian, $\bar{e}$ must always be found in the combination $\bar{e}(1 - \gamma_5)$. So the lepton fields must appear in the form

$$[\bar{e}(1 - \gamma_5)\Gamma_i\nu] = [\bar{e}\Gamma_i(1 \pm \gamma_5)\nu] \begin{cases} \text{upper sign for } V, A \\ \text{lower sign for } S, T \end{cases}$$

Thus, for V and A, $C_V = C'_V$, $C_A = C'_A$, which implies ν_L, ν_R, whereas for S and T, $C_S = -C'_S$, $C_T = -C'_T$ which implies ν_R, $\bar{\nu}_L$.

(3) Beta-gamma circular polarization experiments give information on the amount of interference of Gamow-Teller transitions with Fermi transitions. After a beta particle is emitted, the recoil nucleus is in general polarized. In the subsequent gamma decay, which conserves parity, the photon spin can be correlated with the nucleon spin. A measurement of the circular polarization of the photon in coincidence with the beta direction gives information on

$$2 \operatorname{Re} (C_S C_T'^* + C_S' C_T^* - C_V C_A'^* - C_V' C_A^*)$$

The Fermi channel can interfere with the Gamow-Teller channel only if the neutrino emitted in the Fermi channel has the same helicity as that emitted in the Gamow-Teller channel. The experimental conclusion is that ST and/or VA interference is maximal so that pure SA or pure VT are excluded. Similar conclusions may be reached by studying electron asymmetry in the decay of polarized neutrons.

From these experiments the following three possibilities are seen:

(1) S and T with ν_R, $\bar{\nu}_L$;

(2) V and A with ν_L, $\bar{\nu}_R$;

(3) S and T with ν_R, $\bar{\nu}_L$, and V and A with ν_L, $\bar{\nu}_R$.

The first two possibilities are compatible with the two-component neutrino theory.

To decide among these three possibilities, we look at recoil experiments. According to an old He^6 experiment carried out before the discovery of parity nonconservation, T was dominant in this Gamow-Teller transition, with practically no A. Experiments on Ne^{19} and n showed that the beta interaction is either ST and/or VA. Therefore it was believed for a long time the beta decay goes via S and T. But in May, 1957, this time honored assignment was challenged by J. S. Allen's group at Illinois (Hermannsfeldt *et al.*, 1957). In the decay of A^{35}, which is almost a pure Fermi transition, recoil experiments showed that the positron and neutrino tend to go preferentially in the same direction with an angular correlation coefficient, $a \approx 0.95$. This behavior is characteristic of V. (See (7.18).) This experiment caused a state of confusion that lasted for half a year!

Actually, it is not necessary to perform hard recoil experiments in order to decide between S and T with ν_R and V and A with ν_L. All that is necessary is to measure the helicity of the neutrino. This sounds even harder, but it is actually not so hard, if one is sufficiently clever, as Goldhaber, Grodzins, and Sunyar (1958) were. They considered K-capture by Eu^{152} (see Fig. 7.4) $e^- + E_u^{152} \rightarrow \nu + S_m^{152*}$. We notice that the helicity of the

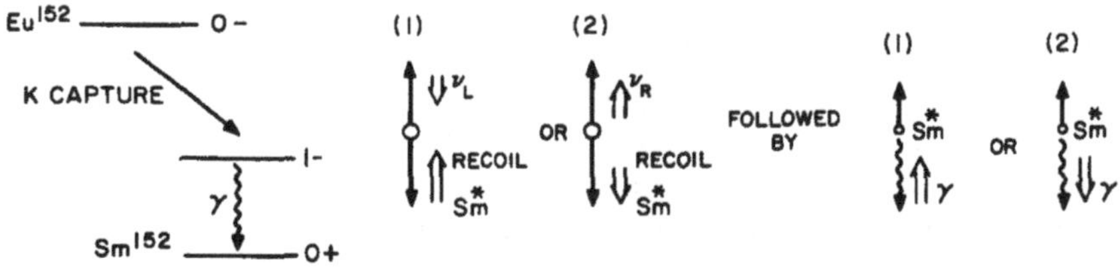

Figure 7.4. Experiment to determine the helicity of the neutrino

recoil Sm^{152*} is the same as the helicity of the neutrino because of the laws of conservation of linear and angular momentum. (Note that $J_{\text{total}} = \frac{1}{2}$ initially.) Similarly the photon emitted *in the direction of motion of the excited recoil nucleus* will have the same helicity as the nucleus, and hence as the neutrino. The experiment of Goldhaber, Grodzins, and Sunyar was to measure the polarization of such photons—the "correct" photons were chosen by having them resonance-scatter from a target of Sm^{152}. This experiment showed very clearly that the neutrino must be left-handed. Hence, V and A must be dominant in beta decay.

Neutron decay experiments at Argonne not only confirmed this conclusion, but also showed that the interaction is $V - A$ (i.e., $C_V \approx -C_A$) and not $V + A (C_V \approx C_A)$ (Burgy *et al.*, 1958a). The results are summed up in Table 7.2.

TABLE 7.2

Neutron Asymmetry Experiments

	S + T	S − T	V + A	V − A	Experiments
$\mathscr{H}(\bar{\nu})$	L	L	R	R	R (deduced)
β^- asymmetry	$-1.00\,\frac{v}{c}$	$-0.07\,\frac{v}{c}$	$1.00\,\frac{v}{c}$	$-0.07\,\frac{v}{c}$	$(-0.09 \pm 0.03)\,\frac{v}{c}$
$\bar{\nu}$ asymmetry	-0.07	-1.00	-0.07	1.00	0.88 ± 0.15

Moreover, a new recoil experiment with He^6 was performed at Illinois. The experimental data with $a = -0.39 \pm 0.05$ (where we expect $a = -0.33$ for pure A, $+0.33$ for pure T) conclusively established the dominance of A. (Allen *et al.*, 1959.)

To sum up, we have found that the interaction Hamiltonian is

$$[\bar{p}\gamma_\mu n][\bar{e}\gamma_\mu(C_V + C'_V\gamma_5)\nu] + [\bar{p}i\gamma_5\gamma_\mu n][\bar{e}i\gamma_5\gamma_\mu(C_A + C'_A\gamma_5)\nu] + \text{H.C.}$$
$$= [\bar{p}\gamma_\mu(C_V - C_A\gamma_5)n][\bar{e}\gamma_\mu(1 + \gamma_5)\nu] + \text{H.C.} \quad (7.19)$$

with

$$\frac{|C_A|^2}{|C_V|^2} \approx 1.3\text{–}1.4$$

The ratio 1.3 was estimated from comparative lifetimes. The ratio 1.4 was estimated from the decay rate of O^{14} ($|C_V|^2$) and the neutron decay rate ($|C_V|^2 + 3|C_A|^2$). It was also found that the phase angle between C_V and C_A is $180° \pm 8°$ (cf. discussion of the $\langle\vec{\sigma}_n\rangle \cdot (\vec{p}_e \times \vec{p}_\nu)$ experiment in Chapter 4, Section 4).

If we ignore, for the moment, the difference between 1 and $\sqrt{1.3}$ (we will return to this point later), we find that the interaction can be written as

$$C_V[\bar{p}\gamma_\mu(1 + \gamma_5)n][\bar{e}\gamma_\mu(1 + \gamma_5)\nu] + \text{H.C.}$$

which is of the form (7.16). It is noteworthy that Fermi's first beta interaction was of the vector type. It took 25 years to discover $(1 + \gamma_5)$.

7.5. Universal Fermi Interaction

One of the most widely discussed problems in the weak interactions has centered around the question of their universality. That is: Are the various weak processes "universal" in the sense that they are different manifestations of a single fundamental interaction?

Shortly after the first systematic study of decay and capture interactions of muons, it became apparent that the coupling constants which characterize muon decay and muon capture ($\mu^- + p \to \nu + n$) are of the same order of magnitude as those which characterize beta decay. The beta decay constant in front of the four-fermion covariants has the dimension of (energy $\times$ L^3) since each bilinear product $\bar{\psi}\Gamma\psi$ has the dimension L^{-3}, and

$$\int C_i(\bar{\psi}_1\Gamma_i\psi_2)(\bar{\psi}_3\Gamma_i\psi_4)\, d^3x$$

must have the dimension of energy. The four-fermion coupling constants for beta decay, muon decay and muon capture are all of the order 10^{-49} erg-cm^3. This striking fact led a number of theoreticians to formulate

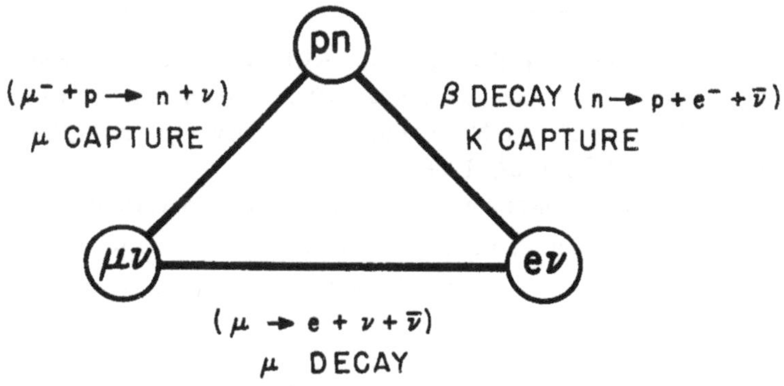

Figure 7.5. Tiomno-Wheeler triangle

what is now known as the Universal Fermi Interaction (UFI) hypothesis. (See Klein (1948), Puppi (1948), Tiomno and Wheeler (1949), and Lee, Rosenbluth, and Yang (1949).) Of course, the form of the interaction was not precisely known. The theoreticians made crude calculations, and showed that regardless of the types of interactions assumed, the coupling constants are of the same order of magnitude. They postulated: There exists a "universal" four-fermion interaction between any pair of the three vertices ($\bar{n}p$), ($\bar{\nu}\mu$), and ($\bar{\nu}e$) of the Tiomno-Wheeler triangle (see Fig. 7.5) leading to

$$n \to p + e^- + \bar{\nu}$$

$$\mu^\pm \to e^\pm + \nu + \bar{\nu},\ \mu^- + p \to n + \nu$$

The early evidence of the UFI hypothesis was summarized by Fermi (1951).

This hypothesis was soon applied to virtual processes such as the decay interaction of the charged pion via a nucleon-antinucleon pair:

$$\pi^+ \xrightarrow{\text{strong}} p + \bar{n} \xrightarrow{\text{weak}} \mu^+ + \nu$$

and was extended to the various decay interactions of newly discovered "strange" particles (Dallaporta, 1953; Gell-Mann, 1956b). The latter extension to strange particle decays is accomplished by adding a fermion pair containing hyperons to the original Tiomno-Wheeler triangle, thus changing it to a tetrahedron. For instance, if we include $(\bar{\Lambda}p)$ we can qualitatively understand decay processes such as

$$\Lambda \xrightarrow{\text{weak}} p + \bar{p} + n \xrightarrow{\text{strong}} p + \pi^-$$

$$K^+ \xrightarrow{\text{strong}} \bar{\Lambda} + p \xrightarrow{\text{weak}} \mu^+ + \nu$$

To appreciate the order of magnitude of $G \approx 10^{-49}$ erg-cm^3, we express G in terms of a dimensionless constant $\mathscr{G}^2$ under the assumption that the radius of the interaction is of the order of the Compton wave length of the K particle

$$\mathscr{G}^2 = \frac{G^2}{(\hbar c)^2}\left(\frac{\mu_K c}{\hbar}\right)^4 \approx 10^{-12}$$

Note that this number is smaller by several orders of magnitude than the coupling constants which characterize the strong interaction ($\approx$1 to 15) and the electromagnetic interaction (1/137).

Prior to the summer of 1957, numerous attempts had been made to understand beta decay, muon decay, the $\pi\mu$-πe ratio and, the $\pi\mu$-$\pi e\gamma$ ratio as different manifestations of a single fundamental interaction with a common *form* (e.g., S, T, P, etc.). Although the UFI hypothesis sounded attractive, the more sophisticated calculations one made, the more evident it became that this hypothesis was not compatible with the experimental data. This is not surprising in view of the fact that in those days people believed in wrong experiments. It was not until the advent of the $V-A$ coupling that attempts along this line became successful.

We now combine the notion of universality with that of $V-A$. Write J_μ as

$$J_\mu = \bar{e}\gamma_\mu(1 + \gamma_5)\nu + \bar{\mu}\gamma_\mu(1 + \gamma_5)\nu + \bar{n}\gamma_\mu(1 + \gamma_5)p + \dots \quad (7.20)$$

The weak interaction can be regarded as an interaction of the overall current J_μ with itself,

$$H_{\text{weak}} = GJ_\mu^\dagger J_\mu + \text{H.C.} \quad (7.21)$$

where $G = C_V$, and C_V is the vector coupling constant in beta decay. We have made the assumption that the current J_μ always lowers the electric charge by one unit, so that a bilinear current is formed out of one neutral and one charged fermion field. This assumption is somewhat *ad hoc* but unless this assumption is made we have a lot of undesirable processes that have never been observed; e.g.,

$$\mu^- + p \rightarrow e^- + p$$
$$K^+ \rightarrow \pi^+ + \nu + \bar{\nu}$$
$$K^+ \rightarrow \pi^+ + \mu^+ + e^-$$
$$K^0 \rightarrow \mu^+ + \mu^-$$

are allowed if we start including "neutral currents" such as $\bar{\nu}\gamma_\mu(1 + \gamma_5)\nu$, $\bar{\mu}\gamma_\mu(1 + \gamma_5)e$. However, it has to be admitted that there is no experimental evidence that forces us to exclude neutral currents constructed out of baryon fields such as $\bar{p}\gamma_\mu(1 + \gamma_5)p$, $\bar{n}\gamma_\mu(1 + \gamma_5)\Lambda$.

To obtain the exact numerical value of G, we integrate the differential transition rate for beta processes given by (7.18). The mean life is the reciprocal of the total transition rate.

$$\frac{1}{\tau} = \frac{(4\pi)^2}{(2\pi)^5}\,\xi \int p_e^2\,(\sqrt{(p_e^{(\max)})^2 + m_e^2} - \sqrt{p_e^2 + m_e^2})^2\,dp_e$$

where $p_e dp_e = E_e dE_e$ was used.
Define the Fermi function $f(\eta_0)$

$$f(\eta_0) = \int_0^{\eta_0} \eta^2(\sqrt{\eta_0^2 + 1} - \sqrt{\eta^2 + 1})^2\,d\eta$$

We get

$$\frac{1}{\tau} = \frac{m_e^5}{2\pi^3} f(\eta_0), \qquad \eta_0 = \frac{p_e^{(\max)}}{m_e}$$

It is more customary to work with the half-life $t_{1/2}$

$$\tau = t_{1/2}/0.693$$

Restoring $\hbar$ and c in the appropriate places and remembering that ξ has the dimension (erg-cm.^{3})^2, we find

$$\frac{1}{f(\eta_0)t_{1/2}} = \frac{m_e^5 \xi}{2\pi^3(0.693)} \left(\frac{c^4}{\hbar^7}\right)$$

If we know the end point energy and lifetime we can compute ξ. (Actually our expression for $f(\eta_0)$ must be modified to account for Coulomb corrections.) Usually $f(\eta_0)t_{1/2}$ is called *ft*. We list values of *ft* for three $T = 1$ nuclei which decay via $0 \rightarrow 0$ transitions

O^{14}	3057 ± 20
A^{26}	3100 ± 53
Cl^{34}	3188 ± 110

For these transitions $|M_F|^2 = 2$, $|M_{G\cdot T}|^2 = 0$ and we get from the O^{14} data

$$C_V = \frac{1}{\sqrt{2}}\sqrt{|C_V|^2 + |C_V'|^2} = 1.00 \times 10^{-49} \text{ erg-cm.}^3 = 6.25 \times 10^{-44} \text{ Mev.-cm.}^3$$

$$= \frac{10^{-5}}{\sqrt{2}} M_p c^2 \left(\frac{\hbar}{M_p c}\right)^3 \tag{7.22}$$

For muon decay we have already shown

$$H_{\text{int}} = g_{VA}[\bar{\nu}\gamma_\mu(1 + \gamma_5)\mu][\bar{e}\gamma_\mu(1 + \gamma_5)\nu]$$

correctly describes the asymmetry and polarizations properties. In particular the fact that e^+ is right-handed in both μ^+ decay and in β^+ decay is a direct consequence of the $(1 + \gamma_5)$ which appears in front of every fermion field. As for the strength of the coupling constant, g_{VA} can be shown to be equal to C_V in beta decay within a few percent as first pointed out by Feynman and Gell-Mann (1958). More recent measurements on the muon lifetime seem to indicate that there may exist about a 4% discrepancy between the observed muon lifetime and the muon lifetime calculated from C_V on the basis of the Universal Fermi Interaction hypothesis, after the radiative corrections of Berman (1958) and Kinoshita and Sirlin (1959) are taken into account (see Butler and Bondelid, 1961). Setting this discrepancy aside, the reader may still wonder why we are comparing g_{VA} with C_V and not with C_A. We will try to answer this question when we discuss the effect of the strong interactions on weak processes in Chapter 9.

Note added in proof. For a more recent discussion of the β decay-μ decay universality see Bardin *et al.* (1962).

Problem 9. Given:

$$H_{\text{int}} = g_{VA}[\bar{\nu}\gamma_\mu(1 + \gamma_5)\mu][\bar{e}\gamma_\mu(1 + \gamma_5)\nu) + \text{H.C.}$$

(1) Prove

$$H_{\text{int}} = g_{VA}(\bar{\nu}\gamma_\mu(1 + \gamma_5)\nu][\bar{e}\gamma_\mu(1 + \gamma_5)\mu] + \text{H.C.}$$

(The $V - A$ coupling is permutation symmetric.)

(2) Prove

$$H_{\text{int}} = 2g_{VA}(\bar{\mu}^c(1 + \gamma_5)\nu)(\bar{e}(1 - \gamma_5)\nu^c) + \text{H.C.}$$

except for a phase factor.

(3) Derive (for $m_e \approx 0$)

$$f(x)(1 + a(x)\cos\theta)\,dx\,\frac{d\Omega}{4\pi} = 2[(3 - 2x) \mp \cos\theta(1 - 2x)]x^2\,dx\,\frac{d\Omega}{4\pi}$$

(4) Given (cf., Reiter *et al.*, 1960; Lundy 1962)

$$\tau_\mu = (2.20 - 2.21) \times 10^{-6}\,\text{sec}$$

evaluate g_{VA} in erg-cm^3 and compare it with C_V in beta decay.

As for the third side of the Tiomno-Wheeler triangle, muon capture, $\mu^- + p \rightarrow n + \nu$, we have the beta decay Hamiltonian with μ replacing e. Negative muons are captured from the ground state of mu-mesic atoms. For $Z = 1$, assuming that the four hyperfine states are equally populated (this assumption cannot be justified for the realistic case of μ^- capture in liquid hydrogen), we have

$$\frac{1}{\tau_{\text{capture}}} = 2\pi\frac{4\pi p_\nu^2}{(2\pi^3)}\frac{dp_\nu}{dE}|\psi(0)|^2(|C_S + C_V|^2 + |C_S' + C_V'|^2 + 3|C_T + C_A|^2 + 3|C_T' + C_A'|^2) \qquad (7.23)$$

The capture probability should be proportional to the probability of finding the muon at the nucleus, which explains the appearance of $|\psi(0)|^2$ where $\psi(\vec{x})$ is the 1 S state atomic wave function

$$\psi(\vec{x}) = \frac{1}{(4\pi)^{1/2}}\left(\frac{1}{a}\right)^{3/2} 2 \exp\left(-\frac{|\vec{x}|}{a}\right)$$

$$a = 2.55 \times 10^{-11} \text{ cm } (\mu \text{ mesic Bohr radius})$$

Noting

$$\frac{dp_\nu}{dE} = \left(\frac{p_\nu}{E_n} + \frac{p_\nu}{E_\nu}\right)^{-1} \approx 1 \quad (\text{since } E = \sqrt{p_\nu^2 + M_n^2} + p_\nu)$$

we obtain

$$\frac{1}{\tau_{\text{capture}}} = \frac{p_\nu^2}{\pi^2 a^3}\left(|C_S + C_V|^2 + |C_S' + C_V'|^2 + 3|C_T + C_A|^2 + 3|C_T' + C_A'|^2\right) \tag{7.24}$$

Inserting numerical values, one finds, on the basis of the universal VA hypothesis,

$$\frac{1}{\tau_{\text{capture}}} \approx \frac{1}{3000}\frac{1}{\tau_{\text{decay}}}$$

where τ_{decay} stands for the mean lifetime of the free muon. So in practice pure hydrogen may not seem too suitable for studying muon capture. The feasibility of "pure" muon capture experiments is discussed by Weinberg (1960b).

Note added in proof. Recently long-awaited muon capture experiments in hydrogen have been performed at three different places (Chicago, Columbia, and CERN). In μ^- capture in liquid hydrogen, what one actually measures can be shown to be the capture rate of the muon in a $p\mu^-p$ molecule (ion) in a $J^{(\text{tot.})} = \frac{1}{2}$, $1s$ ortho state with the two protons in $L = 1, S = 1$. The capture rate predicted by the universal $V - A$ theory (induced pseudoscalar and weak magnetism included) is about 560 sec^{-1}, which is to be compared with the experimental values (400 ± 120) sec^{-1} (Hildebrand, 1962), (515 ± 81) sec^{-1} (Bleser *et al.*, 1962), and (420 ± 75) sec^{-1} (Bertolini *et al.*, 1962). It is to be pointed out that the theoretically expected rate on the basis of the $V + A$ combination is < 200 sec^{-1}.

For $Z > 1$, note that the Bohr radius shrinks by a factor of Z. There are Z protons that can capture μ^-, so we expect that the capture rate increases as Z^4. The actual story is not so simple, because first of all, the radius of the nucleus is not small compared to the Bohr radius, and then, also, the rate is suppressed because of the exclusion principle (Wheeler, 1949). Furthermore, the fact that the proton is not free reduces the rate.

Primakoff (1959) took these complications into account and found

$$\frac{1}{\tau_{\text{cap}}(Z, A)} = \frac{1}{\tau_{\text{cap}}(1, 1)} Z_{\text{eff}}^4 \gamma \left[1 - \frac{A - Z}{ZA} \delta\right] \tag{7.25}$$

$$\gamma \approx .73$$

$$\delta \approx 3$$

From more fundamental points of view, there are two refinements that have to be made—induced pseudoscalar and weak magnetism, both of which will be discussed later. After all these corrections we can evaluate the following combination of coupling constants (assuming V and A)

$$|C_V|^2 + |C_V'|^2 + 3|C_A|^2 + 3|C_A'|^2$$

from capture rates in various complex nuclei measured extensively at Chicago and Liverpool. (See e.g., Sens, 1959.) The results agree with the value expected from beta decay within 20%.

Experiments which we have described so far just measure the capture rates. More information on the nature of the μ^- capture interaction can be obtained by studying neutron asymmetry, circular polarization of the gamma-rays emitted following muon capture, relative difference in capture rates between two hyperfine states, etc. Of these the hyperfine effect first discussed by Bernstein *et al.* (1958) is of particular interest. If we ignore induced pseudoscalar, the capture rates in the $I = 1$ state ($\vec{\sigma}_\mu$ and $\vec{\sigma}_p$ parallel) and the $I = 0$ state ($\vec{\sigma}_\mu$ and $\vec{\sigma}_p$ antiparallel) are proportional to $|C_V + C_A|^2$ and $|C_V - 3C_A|^2$ respectively since, for the axial vector contribution, we have $\vec{\sigma}_\mu, \vec{\sigma}_p$ whose expectation value is 1 (triplet) or -3 (singlet). Telegdi (1959) has pointed out that the relative sign between C_V and C_A may be determined in case there is an appreciable amount of atomic conversion between the two hyperfine states due to the ejection of Auger electrons.

Note added in proof. Using Telegdi's idea, Culligan *et al.* (1961) were able to show that the capture rate in F^{19} is sufficiently large for a configuration in which $\vec{\sigma}_\mu$ and σ_p are antiparallel to rule out the $V + A$ combination.

There is another process in which muon capture may be compared to beta decay even though in this case muon capture and beta decay occur only virtually. The process is the charged pion decay $\pi^- \to e^- + \bar{\nu}$, $\mu^- + \bar{\nu}$. In the lowest order this may be visualized as

$$\pi^- \xrightarrow{\text{strong}} n + \bar{p} \xrightarrow{\text{weak}} \begin{cases} e^- + \bar{\nu} \\ \mu^- + \bar{\nu} \end{cases}$$

as shown in Fig. 7.6(a). Needless to say, we cannot just consider the lowest order diagram, nor can we overlook the fact that there may be other baryon pairs in the intermediate state, e.g., Λ and $\overline{\Sigma^+}$, or Ξ^- and $\overline{\Xi^0}$. Fortunately all these complications due to strong interactions are irrelevant if we are just interested in the *πe–$\pi\mu$ ratio*. We just have a "black box" common to both πe and $\pi\mu$ as shown in Fig. 7.6(b). Also note that μ and ν (or e and ν) do not interact in the final state.

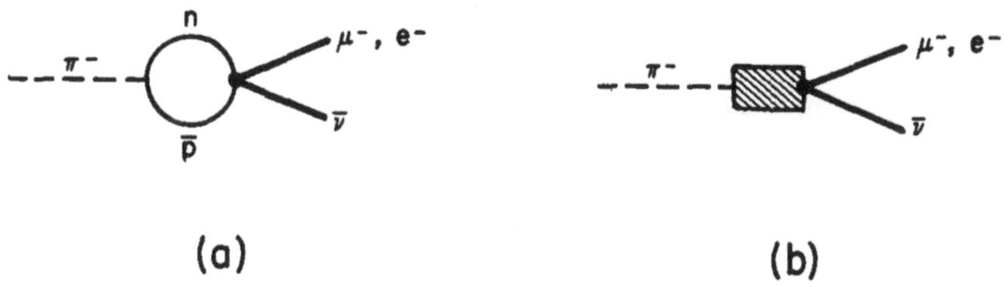

Figure 7.6. Mechanism for π^- decay

Now the general form of the matrix element describing these transitions must be

$$M = L_\mu \bar{u}_l \gamma_\mu (1 + \gamma_5) u_\nu \qquad l = \mu, e$$

where L_μ describes the black-box portion of the matrix element. The only four-vector available in the problem, upon which L_μ can depend is

$$p_\pi = p_l + p_\nu$$

Therefore

$$L_\mu = i f_A (p_\pi)_\mu$$

where f_A is a constant common to πe and $\pi\mu$ decay. But

$$\bar{u}_l (i\gamma \cdot p - m_l) = 0$$

so

$$M = f_A m_l \bar{u}_l (1 + \gamma_5) u_\nu$$

Thus, the transition probability is proportional to

$$\sum_{\text{spin}} MM^\dagger \frac{m_l m_\nu}{E_l E_\nu} p^2 \frac{dp}{dE} = m_l^2 f_A^2 \,\mathrm{Tr}\,\{(1 - \gamma_5)(i\gamma \cdot p_l + m_l)(1 + \gamma_5)(-i\gamma \cdot p_\nu)\} \times \frac{m_l m_\nu p^2}{E_l E_\nu \left(\dfrac{p}{E_l} + 1\right)}$$

$$= 2m_l^2 f_A^2 (E_l E_\nu + p^2) p^2 / (E_l + p) E_\nu$$

$$= \frac{2m_l^2 f_A^2 (m_\pi^2 - m_l^2)^2}{4m_\pi^2}$$

where we have used the kinematical relationship $p \equiv |\vec{p}_l| = (m_\pi^2 - m_l^2)/2m_\pi$. Thus the ratio of the πe decay to $\pi\mu$ decay rates is

$$\frac{W(\pi \to e + \nu)}{W(\pi \to \mu + \nu)} = \frac{(m_\pi^2 - m_e^2)^2}{(m_\pi^2 - m_\mu^2)^2} \left(\frac{m_e}{m_\mu}\right)^2 = 1.3 \times 10^{-4} \tag{7.26}$$

as first derived by Ruderman and Finkelstein (1949). There is a simpler way to look at this effect. The $V-A$ theory says that if the e^- is relativistic, it must be left-handed, whereas the associated $\bar{\nu}$ must be right-handed. If the mass of the electron were strictly zero, the electron would be completely relativistic, and the πe decay would be absolutely forbidden because of angular momentum and momentum conservation. The amplitude for the occurrence of the decay should be proportional to the parameter that characterizes the "non-relativistic nature" of the electron, namely its mass. The $\pi\mu$ decay is fully allowed, since the muon is completely non-relativistic (i.e., $v/c \ll 1$).

For a long time experimentalists did not find $\pi^+ \to e^+ + \nu$, and the absence of this mode had been considered to be evidence against the universal $V-A$ theory (at least in its original form) until September, 1958. Various theoreticians tried to invent clever tricks to "explain" why this mode should be forbidden. Then a group at CERN (Fazzini *et al.*, 1958) conclusively established the existence of this mode. Subsequent reports from various laboratories (Columbia, Chicago, Berkeley) show that even the branching ratio is correctly given by the universal $V - A$ theory. With radiative corrections taken into account, the experimental branching ratio $(1.21 \pm 0.07) \times 10^{-4}$ is in excellent agreement with the theoretically expected value 1.23×10^{-4} (Anderson *et al.*, 1960).

We can now use the "clever tricks" in reverse. For instance, it has been shown that a slight amount of the pseudoscalar interaction in the *fundamental* beta decay interaction (e.g., $C_P \approx C_A/1800$) could considerably modify the πe–$\pi\mu$ ratio (Huang and Low, 1957). The fact that the experiments and the theory agree so well shows that we cannot admit even a tiny amount of the pseudoscalar interaction. This means that when there is no complication due to the strong interactions, γ_5 invariance holds to an amazing degree of accuracy.

The K meson decay modes $K_{e2}^- \to e^- + \bar{\nu}$, $K_{\mu 2}^- \to \mu^- + \bar{\nu}$ can be treated in a similar way. In the lowest order $K^- \to \bar{p} + \Lambda \to \mu^-(e^-) + \nu$, but as in the pion case, the argument is unchanged even if we consider other more complicated virtual processes. The theoretically expected ratio we obtain is

$$\frac{W(K \to e + \nu)}{W(K \to \mu + \nu)} = 2.5 \times 10^{-5}$$

It is worth noting that $K_{e3} \to e + \pi + \nu$ is fully allowed because for the three body decay the electron and the neutrino need not be emitted back to back. Experimentally

$$\frac{W(K \to e + \nu)}{W(K \to \mu + \nu)} < \frac{1}{100}$$

and

$$\frac{W(K \to e + \pi + \nu)}{W(K \to \mu + \pi + \nu)} \approx 1$$

We also know that the helicity of the μ^+ in $K^+ \to \mu^+ + \nu$ is the same as the helicity of the μ^+ in $\pi^+ \to \mu^+ + \nu$, as expected from the two-component theory (Coombes *et al.*, 1957).

Note added in proof. The energy and the angular distribution of π and e in K_{e3} decay ($K_2^0 \to e^\pm + \pi^\mp + \nu$, $K^+ \to e^+ + \pi^0 + \nu$) have recently been studied by Leurs *et al.* (1961) and by Brown *et al.* (1961). When the observed distributions are compared with the theoretically expected distributions derived by Pais and Treiman (1957), Furuichi *et al.* (1957), and MacDowell (1957), rather strong evidence for a vector type coupling is obtained. A pure tensor-type coupling can definitely be ruled out, and a pure scalar type coupling can be made consistent with observation only when an unreasonably large variation in the $K\pi$ form factor is assumed.

Figure 7.7. Non-leptonic decay of strange particles

In contrast, a pure vector type coupling with an approximately constant form factor reproduces the observed data. Further, Brown *et al.* (1962) show that if the K_{l3} decay matrix element can be written as

$$[f_V^{(l)}(p^K + p^\pi)_\rho + g_V^{(l)}(p^K - p^\pi)_\rho]\bar{u}_\nu\gamma_\rho(1 + \gamma_5)u_l,\ l = \mu, e$$

then $f_V^{(\mu)}/f_V^{(e)} = 1.09 \pm 0.15$, in striking agreement with the universality hypothesis.

If the notion of universality is to be extended to strange particles, and if the term $\bar{\Lambda}\gamma_\mu(1 + \gamma_5)p$ appears in J_μ, we can immediately calculate the decay rate for $\Lambda \to p + e^- + \bar{\nu}$. Experimentally there are not many events of this type. The observed decay rate (based on 120 events) seem to be smaller than the calculated rate based on the simple-minded "universal" hypothesis by a factor of 10. (Crawford *et al.*, 1958b; Nordin *et al.*, 1958; Humphrey *et al.*, 1961; Ely *et al.*, 1962b.) This is very puzzling.

It would be nice if the $V-A$ theory could say something quantitative about weak processes not involving leptons. This turns out to be very difficult since we do not know how to treat properly the strong interactions. We simply mention that processes such as Λ decay and $K_{\pi2}(\theta)$ and $K_{\pi3}(\tau)$ decay can be understood *qualitatively* by diagrams shown in Fig. 7.7. In general, parity is not conserved since the diagram contains one four-fermion vertex at which parity is not conserved.

CHAPTER 8

Generalized Gauge Transformations and "Number Laws"

8.1. Gauge Invariance and Charge Conservation in Electrodynamics

From now on we shall be concerned with the conservation laws of "internal" attributes, such as electric charge, baryon number, lepton number, isospin, and strangeness. In field theory such conservation laws can be regarded as consequences of invariance under gauge transformations of one kind or another and stem from the fact that the phase of a non-Hermitian field is not measurable.

Before we present a formal discussion of electric charge conservation, we first discuss available evidence for the conservation of electric charge. If charge conservation were violated, the lightest electrically charged particle, namely the electron, would decay into lighter neutral particles such as neutrino and photon. The possibility of the disappearance of an inner shell atomic electron was investigated by de Mateosian and Goldhaber (unpublished). If the electron disappeared, the subsequent emission of a K X ray would be expected. Their experiment shows that the lifetime of the electron is greater than 10^{17} years. The experiment also indicates that, if the particular decay mode $e \rightarrow \nu + \gamma$ were assumed, the partial lifetime for this mode would be greater than 10^{19} years. From this Feinberg and Goldhaber (1959) estimate that if there were a charge violating interaction of the form

$$\frac{\lambda}{m_e} \bar{\nu}\sigma_{\mu\nu}eF_{\mu\nu}$$

then the dimensionless coupling constant λ^2 must be smaller than 2×10^{-47}.

We see from the foregoing discussion that the existing evidence for electric charge conservation is quite strong. In the quantum mechanical language this implies that there exists no transition element that connects a state with $Q = 0$ with a state with $Q = 1$, and that the relative phase of the $Q = 0$ state and the $Q = 1$ state is nonmeasurable and devoid of physical significance. In the framework of field theory, the invariance under a set of simultaneous phase transformations applied to all fields that bear electric charges indeed leads to the conservation of electric charge. Considerations along this line were originated by London (1927) and Weyl (1929).

A space-time independent gauge transformation of the first kind for a non-Hermitian scalar field is defined by the transformation

$$\phi(x) \to \exp(i\lambda)\phi(x) \tag{8.1}$$

where λ is some arbitrary real parameter. For an infinitesimal parameter we have

$$\phi(x) \to \phi(x) + i\lambda\phi(x) \tag{8.2}$$

Let the unitary operator U be such that

$$U\phi U^{-1} = (1 + i\lambda)\phi \tag{8.3}$$

Then in terms of the Hermitian generator Q of the infinitesimal transformation we have

$$\begin{gathered} U = 1 + iQ\lambda \\ (1 + iQ\lambda)\phi(1 - iQ\lambda) = (1 + i\lambda)\phi \end{gathered} \tag{8.4}$$

Then

$$[Q, \phi] = \phi \tag{8.5}$$

A formal solution for Q can be easily obtained

$$Q = i \int \left(\frac{\partial \phi^\dagger}{\partial x_0} \phi - \phi^\dagger \frac{\partial \phi}{\partial x_0} \right) d^3x \tag{8.6}$$

In verifying this remember that

$$\left[\frac{\partial \phi^\dagger}{\partial x_0} (\vec{x}', x_0), \phi(\vec{x}, x_0) \right] = -i\delta^{(3)}(\vec{x} - \vec{x}')$$

$$[\phi(\vec{x}', x_0), \phi(x, x_0)] = 0 \text{ etc.}$$

In exact analogy to the case of momentum conservation, one can argue that, if the system is unchanged by U, then Q is a constant of motion. Q, which generates an infinitesimal gauge transformation, is to be identified with the total charge operator. The case of Dirac spinors can be treated similarly except that we have to be a little careful with the ordering of field operators.

We now derive the continuity equation $\frac{\partial}{\partial x_\mu} j_\mu = 0$ from the requirement that the Lagrangian density be invariant under the infinitesimal gauge transformation (8.2). We here present a slightly generalized formalism of gauge invariance developed by Gell-Mann and Lévy (1960) which can be readily extended to the case where the conservation law in question is only approximate, since the notion of "partially conserved current" has become increasingly important in recent discussions of weak interaction. (See Chapter 11.)

We assume that a theory is derivable from the Lagrangian density $\mathscr{L}$ containing ψ_i and $\partial_\mu \psi_i \equiv \frac{\partial \psi_i}{\partial x_\mu}$. We subject each field ψ_i to an infinitesimal transformation of the following kind

$$\psi_i(x) \to \psi_i(x) + \Lambda(x) F_i[\psi_1(x), \psi_2(x), \ldots] \tag{8.7}$$

where $\Lambda(x)$, called a gauge function, is taken to be space-time *dependent*. We examine the variation of $\mathscr{L}$ due to (8.7) where $\Lambda(x)$ is understood to be infinitesimal.

$$\delta\mathscr{L} = \frac{\delta\mathscr{L}}{\delta\Lambda}\Lambda + \frac{\delta\mathscr{L}}{\delta(\partial_\mu \Lambda)}\partial_\mu \Lambda \tag{8.8}$$

Using (8.7) and

$$\partial_\mu \psi_i \to \partial_\mu \psi_i + \partial_\mu \Lambda F_i + \Lambda \partial_\mu F_i \tag{8.9}$$

we have

$$\frac{\delta\mathscr{L}}{\delta\Lambda} = \frac{\delta\mathscr{L}}{\delta\psi_i} F_i + \frac{\delta\mathscr{L}}{\delta(\partial_\mu \psi_i)}\partial_\mu F_i, \qquad \frac{\delta\mathscr{L}}{\delta(\partial_\mu \Lambda)} = \frac{\delta\mathscr{L}}{\delta(\partial_\mu \psi_i)} F_i$$

Now consider

$$\partial_\mu\left(\frac{\delta\mathscr{L}}{\delta(\partial_\mu \Lambda)}\right) = \partial_\mu\left(\frac{\delta\mathscr{L}}{\delta(\partial_\mu \psi_i)}\right) F_i + \frac{\delta\mathscr{L}}{\delta(\partial_\mu \psi_i)}\partial_\mu F_i$$

Because of the Lagrange's equation

$$\partial_\mu\left(\frac{\delta\mathscr{L}}{\delta(\partial_\mu \psi_i)}\right) = \frac{\delta\mathscr{L}}{\delta\psi_i}$$

we obtain

$$\partial_\mu\left(\frac{\delta\mathscr{L}}{\delta(\partial_\mu \Lambda)}\right) = \frac{\delta\mathscr{L}}{\delta\Lambda} \tag{8.10}$$

Suppose that $\mathscr{L}$ is invariant under (8.7) with an infinitesimal gauge function Λ *taken independently of space-time coordinates.* Then $\delta\mathscr{L}/\delta\Lambda$ vanishes so that

$$\partial_\mu\left(\frac{\delta\mathscr{L}}{\delta(\partial_\mu \Lambda)}\right) = 0 \tag{8.11}$$

In such a case we identify $\delta\mathscr{L}/\delta(\partial_\mu \Lambda)$ as the current which is conserved as a result of the invariance of the theory under the gauge transformation with constant (space-time independent) Λ.

Let us apply this formalism to the case of a charged (non-Hermitian) scalar field. To generate a conserved vector current we simply set $F = i\phi$. So we have

$$\phi \to \phi + i\Lambda(x)\phi \tag{8.12}$$

which can be regarded as a generalization of (8.2) where the phase is now local (i.e. space-time dependent) in character. The free-field Lagrangian is

$$\mathscr{L} = -\left(\frac{\partial\phi^\dagger}{\partial x_\mu}\frac{\partial\phi}{\partial x_\mu} + \mu^2\phi^\dagger\phi\right) \tag{8.13}$$

Under (8.12) we have

$$\mathscr{L} \to -(\partial_\mu\phi^\dagger - i\partial_\mu\Lambda\phi^\dagger)(\partial_\mu\phi + i\partial_\mu\Lambda\phi) - \mu^2(\phi^\dagger - i\Lambda\phi^\dagger)(\phi + i\Lambda\phi)$$

$$\frac{\delta\mathscr{L}}{\delta(\partial_\mu\Lambda)} = i(\phi^\dagger\partial_\mu\phi - \partial_\mu\phi^\dagger\phi) \tag{8.14}$$

The Gell-Mann-Lévy principle tells us that to the extent that $\delta\mathscr{L}/\delta\Lambda$ is zero for constant Λ, which is certainly satisfied in our case, the continuity equation

$$\frac{\partial}{\partial x_\mu} j_\mu = 0$$

holds, where

$$j_\mu \equiv \frac{\delta\mathscr{L}}{\delta(\partial_\mu\Lambda)}$$

The zero-th component of j_0 integrated over all space is indeed the total charge obtained earlier in (8.6). For the spinor case we obtain

$$\partial_\mu(i\bar\psi\gamma_\mu\psi) = 0$$

as a result of the invariance of the Lagrangian

$$\mathscr{L} = -\bar\psi(\gamma_\mu\partial_\mu\psi + m\psi)$$

under

$$\psi \to \psi + i\Lambda\psi$$

In a more general case where we have several fields, some charged, some neutral, we consider

$$\psi_i \to \psi_i + i\Lambda(x)[Q, \psi_i] \tag{8.15}$$

where $[Q, \psi_i]$ is $\pm\psi_i$ if the electric charge of ψ_i is ± 1 and zero if ψ_i is neutral. The gauge function $\Lambda(x)$ is assumed to commute with all fields. We note that, if we had the charge violating interaction $(\lambda/m_e)(\bar e\sigma_{\mu\nu}\nu F_{\mu\nu})$ + H.C. in the Lagrangian, the current $i\bar e\gamma_\mu e$ would not satisfy the continuity equation since

$$\frac{\delta\mathscr{L}}{\delta\Lambda} = \left(\frac{i\lambda}{m_e}\right)\bar e\sigma_{\mu\nu}\nu F_{\mu\nu} + \text{H.C.} \neq 0$$

PROBLEM 10. Given:

$$\mathscr{L} = -\bar e(\gamma_\mu\partial_\mu + m_e)e - \bar\nu\gamma_\mu\partial_\mu\nu - f\bar e\gamma_\mu(1+\gamma_5)\nu W_\mu - f\bar\nu\gamma_\mu(1+\gamma_5)eW_\mu^\dagger,$$

find the current generated by

$$e \to [1 + i\Lambda(x)(1+\gamma_5)]e$$
$$\nu \to [1 + i\Lambda(x)(1+\gamma_5)]\nu$$
$$W_\mu \to W_\mu$$

Which terms in $\mathscr{L}$ prevent the resulting current from being exactly conserved?

To see the relation between the gauge transformations we have considered and the gauge transformations in classical electrodynamics, recall that in electrodynamics observable effects depend only on the field strength $F_{\mu\nu} = \frac{\partial A_\nu}{\partial x_\mu} - \frac{\partial A_\mu}{\partial x_\nu}$. This implies an arbitrariness in the definition of A_μ since $F_{\mu\nu}$ is unchanged by

$$A_\mu \rightarrow A_\mu + \frac{\partial \Lambda}{\partial x_\mu} \tag{8.16}$$

This is the usual gauge transformation in classical electrodynamics which, in field theory, is called the gauge transformation of the second kind. Because of the Lorentz condition

$$\frac{\partial}{\partial x_\mu} A_\mu = 0 \tag{8.17}$$

Λ must satisfy

$$\Box^2 \Lambda = 0 \tag{8.18}$$

In particular $\Lambda =$ constant times t is admissible. In electrostatics this means that the absolute scale of the potential is arbitrary.

Even in the classical theory of electromagnetism there is a deep connection between charge conservation and the arbitrariness of the absolute gauge of the potential. The following enlightening example is attributed to Wigner (1949). One can show that gauge invariance and nonconservation of charge would lead to a contradiction. Consider a shielded cage whose potential is maintained at V. Suppose that there is a process in the cage in which charge is created. Let W be the work necessary to create charge Q. Consider the following cycle assuming that no physical processes depend on the absolute scale of the electrostatic potential. Step (1): Create Q by performing work, W, on the inside of the cage. Step (2): Remove the charge creating apparatus including the charge, Q, far away, to infinity, where the potential is zero. In doing this the energy gained is QV. Step (3): Destroy the charge at infinity. The W lost in step (1) is regained since by assumption the work required to create or destroy charge does not depend on the absolute scale of the electrostatic potential. Step (4): Return the charge-creating apparatus to the cage. The net result is a gain of QV; thus energy conservation is violated. Hence nonconservation of charge and gauge invariance are incompatible.

In classical mechanics it is well known that in the presence of the electromagnetic field the canonical momentum of a charged particle conjugate to the coordinate $\vec{x}$ is not $m\dot{\vec{x}}$ but $m\dot{\vec{x}} + Q\vec{A}$. Likewise in quantum field theory the proper electromagnetic interaction can be obtained simply by replacing every $\partial_\mu\psi$ that appears in the Lagrangian in the absence of the electromagnetic interaction by $(\partial_\mu - ieA_\mu)\psi$. It is assumed that only those electromagnetic interactions which can be generated in this manner exist. This postulate is called the principle of minimal electromagnetic coupling. Crudely speaking, this principle implies that the electromagnetic field possesses no interaction except the usual one with the electric charges and currents of the various particles. One might argue that the existence of the "anomalous" moment of the electrically neutral neutron indicates that there are other interactions than the usual ones; but it is believed instead that the anomalous moment emerges as a result of virtual processes $n \to p + \pi^-$, etc., and that the interaction between the photon and the physical neutron in various dissociated states appears phenomenologically as the anomalous moment interaction (Wick, 1935). The point is that once we allow Yukawa processes, we do not need any special electromagnetic interaction to account for the anomaly.

The electromagnetic interaction obtained in conformity with the principle of minimal electromagnetic coupling is automatically invariant under the set of simultaneous transformations

$$\begin{aligned} \psi &\to \exp\,(ie\Lambda(x))\psi \\ A_\mu &\to A_\mu + \frac{\partial \Lambda}{\partial x_\mu} \end{aligned} \tag{8.19}$$

where e is the electric charge associated with ψ. To verify this note

$$\begin{aligned} \left(\frac{\partial}{\partial x_\mu} - ieA'_\mu\right)\psi' &= \left[\frac{\partial}{\partial x_\mu} - ie\left(A_\mu + \frac{\partial \Lambda}{\partial x_\mu}\right)\right] \exp\,(ie\Lambda)\psi \\ &= \exp\,(ie\Lambda)\left(\frac{\partial}{\partial x_\mu} + ie\frac{\partial \Lambda}{\partial x_\mu} - ieA_\mu - ie\frac{\partial \Lambda}{\partial x_\mu}\right)\psi \\ &= \exp\,(ie\Lambda)\left(\frac{\partial}{\partial x_\mu} - ieA_\mu\right)\psi \end{aligned} \tag{8.20}$$

Thus $\exp\,(ie\Lambda(x))$ behaves as if it were a constant (i.e., space-time independent) phase factor. Hence any Lagrangian constructed out of ψ and $\partial_\mu\psi$ that is invariant under space-time independent gauge transformations is automatically invariant under the wider set (8.19) provided that the electromagnetic coupling is minimal.

Note that the space-time dependent phase transformation on the charged field must be counterbalanced by the gauge transformation of the second kind on the electromagnetic field if the invariance is to be preserved. We might be tempted to argue that the invariance under space-time dependent gauge transformation necessitates the existence of A_μ coupled linearly to the conserved electric charge-current four vector density. At first, it might appear that this is purely a matter of taste. But if such an argument is of any value, it should work for other conservation laws—baryon conservation, isospin conservation, etc. We will return to this point later.

The reader should carefully note that, while all interactions constructed in accord with the principle of minimal electromagnetic coupling are necessarily gauge invariant in the sense of (8.19), all gauge invariant interactions do not satisfy the principle of minimal electromagnetic coupling. The Pauli moment interaction $\bar{e}\sigma_{\mu\nu}eF_{\mu\nu}$ is certainly invariant under space-time dependent gauge transformations, but it does not satisfy the principle of minimal electromagnetic coupling.

Note added in proof. Wentzel (unpublished) has emphasized that it is difficult to formulate the principle of minimal electromagnetic coupling in a mathematically unambiguous way. For instance, suppose we add to the usual free-field Lagrangian of the Dirac field an additional 4-divergence term

$$\frac{\lambda}{e}\frac{\partial}{\partial x_\mu}\left(\bar{\psi}\sigma_{\mu\nu}\frac{\partial}{\partial x_\nu}\psi\right)$$

The equation of motion of the free field is still unchanged. If we make the substitution $\partial_\mu \rightarrow \partial_\mu - ieA_\mu$ in our new Lagrangian, the resulting interaction Lagrangian now contains an anomalous moment term. More generally speaking, the electromagnetic interaction generated by the substitution $\partial_\mu \rightarrow \partial_\mu - ieA_\mu$ is ambiguous to the extent that there is arbitrariness in choosing the form of the free field Lagrangian. It appears that the electromagnetic coupling generated by $\partial_\mu \rightarrow \partial_\mu - ieA_\mu$ is "minimal" only when the free field Lagrangian is chosen to be "minimal."

Let us consider the magnetic moment of the electron. Because the electron does not interact strongly, we expect that the magnetic moment can be calculable from purely quantum electrodynamical considerations. The observed value $(g/2 - 1) = 0.0011609 \pm 0.0000024$ where g is the gyromagnetic ratio is in excellent agreement with $(g/2 - 1) = 0.0011596$ calculated on the basis of the assumption that the only electromagnetic interaction is of the form $i\bar{e}\gamma_\mu eA_\mu$. (Schupp, Pidd, and Crane, 1961; Sommerfield 1957; Petermann, 1957.)

Note added in proof. Recently, a precise measurement of the muon magnetic moment has been carried out at CERN (Charpak *et al.*, 1962). The observed value $(g/2 - 1) = 0.001162 \pm 0.000005$ is to be compared with the theoretical value 0.001165.

We now turn to more practical problems. Specifically, what kind of conditions are imposed by gauge invariance on matrix elements? To answer this question we first look at the wave function for a free photon ε_μ exp $(ik\cdot x)$. The Lorentz condition applied to the one-photon wave function reads $\varepsilon\cdot k = 0$. (In quantum electrodynamics the Lorentz condition applied to the electromagnetic field operator itself is too stringent. If we require that the Lorentz condition be satisfied for the annihilation part of the operator, the wave function for the photon does obey the Lorentz condition. See Gupta (1950) and Bleuler (1950).) Under the gauge transformation $A_\mu \rightarrow A_\mu + \partial_\mu\Lambda$, Λ must have the same plane wave dependence if the new wave function is to describe the same photon with the same energy and momentum. So $\Lambda = C \exp(ik \cdot x)$ where C is constant. The polarization vector ε_μ must transform as $\varepsilon_\mu \rightarrow \varepsilon_\mu + iCk_\mu$.

Let us consider transitions in which a single photon is absorbed or emitted; e.g., photopion production, radiative decay of the π^+ etc. The matrix element

$$M = M(k, \varepsilon; p_1, s_1, \ldots)$$

(where p_1 and s_1 etc. denote the momentum and spin of the other particles) for such a process must be linear in ε. Gauge invariance implies

$$M(k, \varepsilon; p_1, s_1, \ldots) = M(k, \varepsilon + iCk; p_1, s_1, \ldots)$$

since ε and $\varepsilon + iCk$ represent the polarization vector of the same photon. But M is linear in ε, so we must have

$$M(k, k; p, S \ldots) = 0 \tag{8.21}$$

i.e., the matrix element must vanish if ε is replaced by k. For instance, in photopion production, the matrix element in the 4×4 Dirac representation must consist of terms like

$$\gamma_5(\gamma \cdot \varepsilon q \cdot k - \gamma \cdot k q \cdot \varepsilon), \quad q \equiv \text{pion four-momentum}$$

and not just any linear combination of $\gamma_5 \gamma \cdot \varepsilon q \cdot k$ and $\gamma_5 \gamma \cdot k q \cdot \varepsilon$.

For a real photon we can always choose a gauge in which the transversality condition

$$\vec{k} \cdot \vec{\varepsilon} = 0$$

is satisfied. The transversality condition together with the Lorentz condition implies that ε_4 vanishes in this gauge. For, if ε_4 is nonvanishing, we can always choose

$$\varepsilon'_\mu = \varepsilon_\mu + iCk_\mu \text{ with } C = i\varepsilon_4/k_4$$

so that in this new gauge ε_4 vanishes.

It is worth emphasizing the connection between the vanishing rest mass of the photon and the transversality condition. We have already shown in Chapter 2 that the relation $\vec{k} \cdot \vec{\varepsilon} = 0$ implies that the photon spin is either parallel or antiparallel to the propagation direction. Thus, although the photon spin is unity, the photon has only two linearly independent spin states. This situation is possible only when the photon mass is strictly zero; otherwise we could find a Lorentz frame in which the photon is at rest, and the photon spin is "parallel to nothing." (For a group-theoretic discussion on this point see Wigner (1939.)) Generally speaking, gauge invariance and the Lorentz condition can be simultaneously satisfied only for a massless vector field. If we are to formulate a gauge invariant theory with a finite-mass vector field, then it is essential that we employ a different subsidiary condition (Stueckelberg, 1938a; Glauber, 1953).

8.2. Baryon Conservation

In the old days the proton and the neutron were the only members of the family of heavy fermions. The family has been enlarged by a factor of four since the 1940's and we have Λ^0, $\Sigma^{\pm,0}$, $\Xi^{-,0}$, which are all heavier than n and p. These heavy particles, together with n and p, are referred to as "baryons" (a terminology originally due to Pais). Unlike leptons they are all capable of interacting "strongly."

We assign baryon number 1 to n, p, Λ^0, $\Sigma^{\pm,0}$, and $\Xi^{-,0}$. The baryon numbers of photon, mesons (π, K), and leptons (ν, e, μ) are taken to be zero. Antibaryons have baryon number -1. The principle of baryon conservation requires that the sum of baryon numbers be conserved in all reactions. The conservation of baryons holds amazingly well; that is why we are here. Although free neutrons ultimately decay into protons and leptons, protons themselves are stable; neutrons that are bound in nuclei are also stable. The law of baryon conservation directly accounts for the stability of matter. Antibaryons can never be created singly. To create an antiproton, for instance, we must also create a proton or a neutron (or $\Lambda + K$ etc.) in association with the antiproton, e.g.

$$p + p \rightarrow p + p + \bar{p} + p$$
$$\pi^- + p \rightarrow p + \bar{p} + \Lambda^0 + K^0$$

More quantitatively, what is the present lower limit to the lifetime of the proton or of the neutron in nuclei? This question was investigated by a number of groups. Most recently Backenstoss *et al.*, (1960) searched for relativistic charged particles emitted by the decay of nucleons. No such events were found, and depending on the decay mode we assume, we obtain as lower limits on the nucleon lifetime $(1.5–2.8) \times 10^{26}$ years for nucleons in many-particle nuclei, and $(2.2–4.7) \times 10^{24}$ years for protons in hydrogen. These limits are considerably larger than the estimated age of the universe ($\approx 2 \times 10^{10}$ years).

Note added in proof. A new measurement has been made of the nucleon lifetime by Giamati and Reines (1962). Lifetime limits obtained range from 1×10^{26} years to 7×10^{27} years, depending on the decay mode assumed.

How do we know that hyperons cannot decay into non-baryons? After all, we know so little about these "strange" fermions. Suppose the reaction $p + \Lambda^0 \rightarrow p + \pi^0$ were allowed. Since the neutron can dissociate virtually into $\Lambda^0 + K^0$, the reaction $p + n \rightarrow p + \Lambda^0 + K^0 \rightarrow p + \pi^0 + K^0$ would take place. Hence "nonstrange" nuclei would not be stable if hyperon reactions violated the principle of baryon conservation.

The conservation of baryons can be treated in a formal theory in the same way as the conservation of electric charge. Indeed, many years before the discovery of hyperons, Stueckelberg (1938) discussed the conservation law of "heavy charge" (Erhaltungssatz der schweren Ladung) in analogy with the conservation law of electric charge. Only fields with non-vanishing baryon numbers suffer phase transformation $\psi \rightarrow \exp(i\lambda)\psi$.

We may naturally ask: Is there any analog of space-time dependent gauge transformation; namely

$$p \rightarrow \exp(i\eta\Lambda(x))p$$
$$n \rightarrow \exp(i\eta\Lambda(x))n$$
$$\Sigma^{\pm,0} \rightarrow \exp(i\eta\Lambda(x))\Sigma^{\pm,0}, \text{ etc.}$$

under which the total Lagrangian is invariant even with $\Lambda \neq$ constant? In the electromagnetic case, a gauge transformation with a space-time dependent phase had to be counterbalanced by a simultaneous change in the electromagnetic field coupled universally to all charged particles with the same strength. Similarly, the space-time dependent phase transformation in the baryon conservation case has to be counterbalanced by a transformation on some vector field coupled universally to all baryon fields with the same strength.

Do we know anything about the nature of the interaction between baryon fields and such a hypothetical vector field? This question was examined by Lee and Yang (1955). If the hypothetical vector field were massless just as the electromagnetic field, there would be a Coulomb-like repulsion between two baryons and attraction between a baryon and an antibaryon. Let η be "baryonic charge" analogous to e. Between substance 1 and substance 2 there would be a long-range force

$$\text{Force} = -\frac{GM_1M_2}{r^2} + \frac{\eta^2}{4\pi}\frac{N_1N_2}{r^2} \tag{8.22}$$

where the first term is the usual gravitational force, and the second term is due to the hypothetical vector field. N_1 and N_2 are the baryon numbers (nucleon numbers for "non-strange" substances) of substance 1 and substance 2. Note that for M_1 and M_2, binding energies of nuclei and the neutron-proton mass difference are counted as "masses,," whereas, for the second term only the total number of nucleons is relevant. Now the ratio M_{nucleus}/NM_p where N stands for the number of nucleons in the nucleus in question varies from substance to substance because the packing fraction varies. The fractional variation in M_{nucleus}/NM_p is of the order of 10^{-3}. Then the ratio of "observed" gravitational mass to inertial mass would vary by $10^{-3}(\eta^2/4\pi)/GM_p^2$. There is an experiment carried out by Eötvös, Pekar, and Fekete (1922) who showed that the ratio of "observed" gravitational mass to inertial mass is constant within one part in 10^8. (A much more accurate Eötvös-type experiment is currently in progress by Dicke and collaborators at Princeton.) Hence

$$(\eta^2/4\pi)/GM_p^2 < 10^{-5}$$

In other words, such a massless vector field, if it existed, would have to interact even more weakly than the gravitational field, which is already too weakly interacting to be of any significance in elementary particle physics.

Note added in proof. Recently, Schwinger (1962a, b) advanced the remarkable argument that a gauge invariant coupling of a zero bare-mass vector field to a fermion field does not necessarily require that the physical mass of the spin 1 particle associated with the vector field be also zero. Crudely speaking, this possibility arises because the polarization of the vacuum due to the presence of a charge can be so strong that the net dynamical manifestation of the original charge plus the induced charge is short-ranged.

8.3. Superselection Rules; the Concept of Charge

The conservation of baryon number and conservation of electric charge are, for all practical purposes, "absolute" selection rules. This means that there are no matrix elements that connect states with different baryon numbers or with different electric charges. Now it is well known that the common phase of state vectors is non-measurable. But if there are several state vectors, all belonging to *different* sub-spaces such that there are no matrix elements between any pair of them, then not only the common phase but also the phase of *each* state vector is non-measurable. In such cases we say that there is a superselection rule that forbids us to compare, in a physically meaningful way, phases of various state vectors. The conservation law of baryon number and the conservation law of electric charge are examples of such superselection rules. Note also that the angular momentum conservation forbids transition between a $1/2$ integral J system and an integral J system since we cannot add integral J's such as boson spins and orbital angular momentum to make up a $1/2$ integral J. Then it is meaningless to compare the phase of a $1/2$ integral J system with the phase of an integral J system.

As an example, let us consider a neutron state. We may take the absolute phase of the vacuum to be zero by definition. The phase of the neutron state is devoid of physical meaning since the $n \rightarrow$ vacuum transition is forbidden by baryon conservation and also by angular momentum conservation. The relative phase between p and n is also non-measurable since they have different electric charges.

This is not the first time we talked about phase factors. Under discrete symmetry operations such as parity, we have seen that there often is an up-to-phase ambiguity. In discussing the reaction $\pi^- + d \rightarrow n + n$, recall that we chose n and p to have the same parity by convention. Then the intrinsic parity of π^-, as we saw, is a meaningful concept, and in fact, the capture experiment told us that it is odd. But we could have started equally well by defining the relative n-p parity to be odd, in which case the same experiment tells us that the π^- parity is even. This is the case because electric charge conservation, which is a superselection rule, does not allow us to compare the phase between p and n in an absolute way. Or even more generally, given a set of η_P's for various fields, we may multiply η_P by $\exp(iQ\alpha)$ where α is an arbitrary real constant and Q is the electric charge of the field. The transformation $\eta_P \to \eta'_P = \eta_P \exp(iQ\alpha)$ for every charged field leads to non-measurable physical effects. Similarly, we can set $\eta_P \to \eta''_P = \eta_P \exp(iB\beta)$ where B stands for baryon number, and no change results. This means, for instance, that the relative parity of p and e^+ is a matter of definition. The concept of a superselection rule and its relation to the concept of intrinsic parity were first discussed by Wick, Wightman, and Wigner (1952).

We now consider the problem of determining the electric charge of an elementary particle keeping in mind limitations arising from superselection rules. It was emphasized by Wigner (1952) that there are, in general, two ways of determining the electric charge of a given particle. First, we may regard electric charge as a simple additive number that is conserved in any reaction. For instance, consider the process $\mu^+ \to e^+ + \nu + \bar{\nu}$. To determine the electric charge of μ^+ when the charges of e^+, ν, $\bar{\nu}$ are known to be 1, 0, 0 respectively, we simply argue that the μ^+ charge is $+1$ because the total charge of the right-hand side is $+1$. Feinberg and Goldhaber (1959) note an arbitrary and ambiguous element in this way of determining electric charge. Because what we usually call baryon number, denoted by B, and electric charge number Q, are *separately* conserved, any linear combination of Q and B,

$$Q' = aQ + bB$$

is also conserved. In fact, a and b need not be integers, so Q' can be non-integral. We have just as much right to regard Q' as "new electric charge" if the concept of electric charge is merely that of an additive number conserved in any reaction. To put it somewhat differently, because reactions such as $p \to e^+ + \pi^0$ are forbidden by baryon conservation, the relative charge of p and e^+ is completely undetermined just as the relative parity of p and e^+ can only be a matter of definition.

It is crucial to note here that there is what Wigner calls a "second way" of determining the electric charge of a particle. Instead of appealing to the conservation law of additive numbers, we let particles interact with an external electromagnetic field. For instance, we may measure the electric charge of the proton by observing how much a beam of protons is deflected in a homogeneous electric field. It is from this type of experiment that we can conclude that the proton charge is equal to the electron charge in magnitude to an accuracy of 4 parts in 10^{19} (Zorn, Chamberlain, and Hughes, 1960). This equality is of particular interest for the following two reasons. First, if the electric charges of the baryons were all slightly displaced from their accepted values, say by the common amount ε, then the conservation of baryons would follow from the conservation of electric charge, instead of being an independent physical principle (Feinberg and Goldhaber, 1959). Second, a slight difference in magnitude (of the order of 2 parts in 10^{18}) between the proton charge and the electron charge would produce powerful electrostatic forces on the cosmological scale that might account for the observed expansion of the universe (Lyttleton and Bondi, 1959). In spite of all these speculative possibilities nature seems to have set the two charges equal to an amazing degree of accuracy.

One of the deepest mysteries in elementary particle physics centers around the question: Why is electric charge quantized? It is true that the present field theoretic formalism can explain why the positron charge is equal to the charge of the *physical* proton (in spite of the fact that the proton has a pion cloud) provided that the bare proton charge in the absence of the strong couplings is equal to the e^+ charge. However, we have as yet no compelling principle to start with that requires that the bare p charge be equal to the bare e^+ charge. Neither the principle of gauge invariance nor the principle of minimal electromagnetic coupling seems to throw light on this important point. It appears that our understanding of electric charge (to speak nothing of "baryonic charge") is deficient in some fundamental respect.

Turning now to baryon conservation, we may naturally ask whether there exists Wigner's second way of determining charge where "charge" now means "baryonic charge." If there exists a field that is coupled universally to all baryons, we may measure the "baryonic charge" of a baryon by letting it interact with the field in question.

It was conjectured by Wigner (1952) that the pion field is the baryonic analog of the electromagnetic field. This point of view was further advocated by Schwinger (1957) and Gell-Mann (1957) in the so-called "global symmetry" model where the pion is coupled universally to all baryons. However, in such a model, the analogy with electromagnetism is superficial; what appears in the global symmetry model is a pseudoscalar density, not a conserved vector density, and because of this there are several difficulties in identifying "baryonic charge" with "π-mesonic charge." We will come back to the possible existence of a vector-type coupling between the conserved baryon current and a hypothetical vector field in Chapter 11.

Another feature of baryon conservation worth emphasizing is the following. In electromagnetism the continuity equation

$$-\vec{\nabla}\cdot\vec{j} = \frac{\partial\rho}{\partial t}$$

follows from the Maxwell equations themselves

$$\vec{\nabla}\times\vec{B} - \frac{\partial\vec{E}}{\partial t} = \vec{j}, \qquad \vec{\nabla}\cdot\vec{E} = \rho$$

However, in the baryon conservation case, the continuity equation stands by itself, so to speak. This is somewhat disturbing.

8.4. Lepton Conservation

Previously we discussed lepton conservation within the framework of the two component neutrino theory. Actually the concept of lepton conservation is quite independent of the two component theory. If we just restrict ourselves to nuclear beta decay, the question of whether or not lepton number is conserved is essentially the old question of whether the neutral particle emitted in β^- decay is different from, or the same as, the neutral particle emitted in β^+ decay. The principle of lepton conservation as we know it today, including muon phenomena was first formulated by Konopinski and Mahmoud (1953).

There are processes that are forbidden by lepton conservation; e.g. neutrinoless double beta decay

$$\begin{array}{r} n \rightarrow p + e^- + \nu \\ +)\nu + n \rightarrow p + e^- \\ \hline \text{net result}: 2n \rightarrow 2p + 2e^- \end{array} \tag{8.23}$$

or equivalently

$$(Z, A) \rightarrow (Z + 2, A) + 2e^- \tag{8.24}$$

Experimentally double beta decay with no neutrino emission has been studied extensively by a number of groups. (See e.g. a review article by Primakoff and Rosen, 1959.) For example, the half-life for

$$0^+ \text{ ground state of Ca}^{48}\ (Z = 20) \rightarrow 0^+ \text{ ground state of Ti}^{48}\ (Z = 22) \tag{8.25}$$

is measured to be $>1 \times 10^{18}$ years (Doborochotov, Lazarenko, and Lukyanov, 1956). Similarly the reaction

$$\text{"pile neutrino"} + \text{Cl}^{37} \rightarrow \text{A}^{37} + e^-$$

is forbidden by lepton conservation since the pile is neutron-rich, and the pile neutrino emitted together with e^- has lepton number opposite to that of e^-. Experimentally, the rate is at most $\frac{1}{4}$ of that expected if the "pile neutrino" were the opposite kind (as shown by Davis). On the other hand,

$$\text{"pile neutrino"} + p \rightarrow n + e^+$$

is fully allowed by lepton conservation and indeed has been observed with the right cross section (Reines and Cowan, 1959).

Let us recall that for finite mass the Dirac equation in the absence of electromagnetic interactions

$$(\gamma_\mu \partial_\mu + m)\psi = 0$$

can be satisfied by $\psi^c = C\bar{\psi}^T$ (choose η_C to be 1). But for $m = 0$

$$\gamma_5 \psi^c = \gamma_5 C \bar{\psi}^T$$

also satisfies the Dirac equation in exactly the same way as $\gamma_5\psi$ satisfies the Dirac equation. Also for $m = 0$, note

$$\exp(ia\gamma_5)\psi = \cos a\psi + i \sin a\gamma_5\psi \tag{8.26}$$

satisfies the same Dirac equation.

In summary, the neutrino field equation is invariant under the transformations

$$\text{(I)} \qquad \begin{aligned} \nu &\rightarrow a\nu + b\gamma_5\nu^c \\ \nu^c &\rightarrow -b^*\gamma_5\nu + a^*\nu^c \end{aligned} \qquad |a|^2 + |b|^2 = 1 \tag{8.27}$$

$$\text{(II)} \qquad \begin{aligned} \nu &\rightarrow \exp(ia\gamma_5)\nu \\ \nu^c &\rightarrow \exp(-ia\gamma_5)\nu^c \end{aligned}$$

provided that $m_\nu = 0$. These are called the Pauli transformations (Pauli, 1957). See also Pursey (1957) and Touschek (1957).

Suppose lepton conservation does not hold in beta decay. Then we must admit the possibility that in β^- decay, ν^c as well as ν is emitted. Instead of 10 coupling constants, we have 20 coupling constants.

$$H_{\text{int.}} = \sum (\bar{p}\Gamma n)\{\bar{e}[\Gamma_i(C_i + C_i'\gamma_5)\nu + \Gamma_i(D_i + D_i'\gamma_5)\nu^c] + \text{H.C.} \tag{8.28}$$

The physical significance of C_i and D_i can be more clearly seen if we define

$$\begin{aligned} \nu_R &= \tfrac{1}{2}(1 - \gamma_5)\nu & C_i^R &= C_i - C_i' \\ \nu_L &= \tfrac{1}{2}(1 + \gamma_5)\nu & C_i^L &= C_i + C_i' \\ (\nu^c)_L &= \tfrac{1}{2}(1 + \gamma_5)\nu^c = (\nu_R)^c & D_i^R &= D_i + D_i' \\ (\nu^c)_R &= \tfrac{1}{2}(1 - \gamma_5)\nu^c = (\nu_L)^c & D_i^L &= D_i - D_i' \end{aligned} \tag{8.29}$$

For a single β^+ decay which proceeds via H.C. we have the following four possibilities for each channel, i,

$$\begin{aligned} A &\rightarrow B + e^+ + C_i^{R*}\nu & \mathscr{H}(\nu) &= +1 \\ A &\rightarrow B + e^+ + C_i^{L*}\nu & \mathscr{H}(\nu) &= -1 \\ A &\rightarrow B + e^+ + D_i^{R*}\bar{\nu} & \mathscr{H}(\bar{\nu}) &= -1 \\ A &\rightarrow B + e^+ + D_i^{L*}\bar{\nu} & \mathscr{H}(\bar{\nu}) &= +1 \end{aligned} \tag{8.30}$$

The Pauli transformations can also be rewritten as

$$\begin{aligned} &\text{(I)} & \nu_R &\rightarrow a\nu_R - b(\nu^c)_R = a\nu_R - b(\nu_L)^c \\ & & \nu_L &\rightarrow a\nu_L + b(\nu^c)_L = a\nu_L + b(\nu_R)^c \\ &\text{(II)} & \nu_R &\rightarrow \exp(-ia)\nu_R \\ & & \nu_L &\rightarrow \exp(ia)\nu_L \end{aligned} \tag{8.31}$$

The crucial point is that ν, $a\nu + b\gamma_5\nu^c$ and $\exp(ia\gamma_5)\nu$ (regarded as c-number wave functions) all represent neutral particles with the same "physical" momentum and spin states. Although lepton numbers may change, helicities do not change under the Pauli transformation. In beta decay experiments the most we can measure is the transition probability into a certain spin and momentum state of the emitted particle. Then Hamiltonians that can be transformed into each other under (8.31) (I) and/or (II) are physically equivalent in the sense that they lead to the same observable effects.

The result of applying (I) and (II) to the field operators in H_{int} is identical to the result of applying the following corresponding transformation (I)′ and (II)′ to the coupling constants *leaving the field operators unchanged.*

$$\text{(I)}' \quad \begin{cases} \begin{pmatrix} C_i^R \\ D_i^L \end{pmatrix} \to \begin{pmatrix} a & -b^* \\ -b & a^* \end{pmatrix} \begin{pmatrix} C_i^R \\ D_i^L \end{pmatrix} \\ \begin{pmatrix} C_i^L \\ D_i^R \end{pmatrix} \to \begin{pmatrix} a & -b^* \\ +b & a^* \end{pmatrix} \begin{pmatrix} C_i^L \\ D_i^R \end{pmatrix} \end{cases}$$

$$\text{(II)}' \quad \begin{cases} \begin{pmatrix} C_i^R \\ D_i^L \end{pmatrix} \to \exp(-ia) \begin{pmatrix} C_i^R \\ D_i^L \end{pmatrix} \\ \begin{pmatrix} C_i^L \\ D_i^R \end{pmatrix} \to \exp(+ia) \begin{pmatrix} C_i^L \\ D_i^R \end{pmatrix} \end{cases} \tag{8.32}$$

Observable effects can depend only on certain combinations of coupling constants if equivalent Hamiltonians are to lead to the same physical effects. For instance, observable effects cannot depend on $C_i^{R*} C_i^L$ since

$$C_i^{R*} C_i^L \xrightarrow{\text{(II)}'} \exp(2ia) C_i^{R*} C_i^L \tag{8.33}$$

This is not surprising in view of the fact that R and L states cannot interfere. One can show that the observable quantities depend only on the following

$$\begin{aligned} K_{ij} &= C_i^{R*} C_j^R + D_i^{L*} D_j^L + C_i^{L*} C_j^L + D_i^{R*} D_j^R \\ L_{ij} &= C_i^{R*} C_j^R + D_i^{L*} D_j^L - C_i^{L*} C_j^L - D_i^{R*} D_j^R \\ I_{ij} &= C_i^R D_j^R + D_i^L C_j^L + C_i^L D_j^L + D_i^R D_j^R \\ J_{ij} &= C_i^R D_j^R + D_i^L C_j^L - C_i^L D_j^L - D_i^R D_j^R \end{aligned} \tag{8.34}$$

Single beta decay depends only on K_{ij} and L_{ij}. Formulae we wrote down previously for single beta decay experiments work just as well if we redefine the coupling constants as

$$\begin{aligned} C_i^{R*} C_j^R &\to C_i^{R*} C_j^R + D_i^{L*} D_j^L \\ C_i^{L*} C_j^L &\to C_i^{L*} C_j^L + D_i^{R*} D_j^R \end{aligned}$$

On the other hand, double beta decay processes and Davis-type capture experiments depend on I_{ij} and J_{ij}. A sufficient condition for lepton conservation is all D_i^R, $D_i^L = 0$ for *some* equivalent Hamiltonian.

It may appear that no information on lepton conservation can be obtained from single beta decay experiments since K_{ij} and L_{ij} are not sensitive to the vanishing or nonvanishing of D_i^R and D_i^L. However, this is not the case for the following special situation, which, according to recent experiments may well correspond to reality. We now show that if

(a) Only V and A contribute in beta decay

(b) The $e^{\mp}$ helicity is $\mp\dfrac{v}{c}$

(c) The real part of the VA interference is maximal

then lepton conservation must be satisfied. From (a) we have $C_i^R = C_i^L = D_i^R = D_i^L = 0$ for $i = S, T, P$. From (b) we have $C_i^R = D_i^L = 0$ for $i = V, A$ since the neutral particle emitted in $\beta^+(\beta^-)$ decay must be left-handed (right-handed). Then (c) imposes the relation

$$|Re(C_V^{L*}C_A^L + D_V^{R*}D_A^R)| = \sqrt{(|C_V^L|^2 + |D_V^R|^2)(|C_A^L|^2 + |D_A^R|^2)} \quad (8.35)$$

Now the invariance of observable quantities under

$$\begin{pmatrix} C_i^L \\ D_i^R \end{pmatrix} \to \begin{pmatrix} a & -b^* \\ b & a^* \end{pmatrix} \begin{pmatrix} C_i^L \\ D_i^R \end{pmatrix} \quad (8.36)$$

implies that we can set one of the D_i^R, say D_A^R, to be zero. (If D_A^R does not vanish, choose $b/a^* = D_A^R/C_A^L$.) Physically this arbitrariness means that the differentiation between left-handed ν and left-handed $\bar{\nu}$ can be given only through their interactions. Then the condition for maximal VA interference (8.35) implies that D_V^R also vanishes in the representation in which D_A^R is zero. (Just use a Schwartz-inequality type argument.) Hence

$$\text{All } D_i^\alpha = 0, \qquad \alpha = R, L \quad (8.37)$$

This is a nontrivial result. It tells us that if we establish from single beta decay experiments that (a)–(c) are satisfied, then neutrinoless double beta decay is forbidden, and so is the Davis reaction. In fact, the rate of double beta decay for Ca^{47} is proportional to

$$\frac{|C_S^R|^2 + |D_S^L|^2}{|C_V^L|^2 + |D_V^R|^2}$$

if the choice $C_S^L = D_S^R = C_V^R = D_V^L = 0$ (which follows from (b)) is made. Primakoff and Rosen (1959) estimate that, if the above S- to -V ratio is 1/3, which corresponds to just about the maximum amount of S we can tolerate from the A^{35} recoil experiment, then the half-life for Ca^{48} double beta decay is $10^{15\pm2}$ years, which is to be compared with the previously quoted experimental value $>1 \times 10^{18}$ years.

In discussing the π-μ-e sequence, we assumed the two-component theory of the neutrino. If we assume the two-component condition, the available experimental data imply lepton conservation in the sense of Chapter 7.

If we relax the two-component condition, however, we cannot learn anything about lepton conservation from the available data. Suppose lepton number is not conserved. We have both

$$\pi^- \to \mu^- + (\bar{\nu})_R \quad \text{(as in the normal theory)}$$

and

$$\pi^- \to \mu^- + \nu_R$$

(Note that the definitions of $(\nu)_R$ and ν_R can be fixed from beta decay experiments so that there is no Pauli-type arbitrariness left.) Consider the following ν capture experiment

$$\text{"}\mu^- \text{ neutrino"} + p \to e^+ + n \tag{8.38}$$

(where "μ^- neutrino" stands for the neutral particle emitted in π^- decay together with μ^-) which may become feasible in the near future with high intensity machines (see e.g., Schwartz, 1960). If this is fully allowed, the "μ^- neutrino" is $(\bar{\nu})_R$ itself, and lepton conservation holds. If this proceeds at a reduced rate, the "μ^- neutrinos" contain both ν and $\bar{\nu}$, and lepton number is not conserved.

Finally we wish to emphasize that the discussion of neutrino capture experiments is complicated by another effect, which may well be realized in nature. If the reaction (8.38) does not go at all, nor does

$$\text{"}\mu^+ \text{ neutrino"} + p \to e^+ + n \tag{8.39}$$

then we are forced to believe that there are two kinds of neutrinos, say ν and ν', such that ν appears only with e^+, and ν' appears only with μ^+. Even in such a case, the physical predictions of the two-component neutrino theory with lepton conservation discussed in Chapter 7, Section 2 may hold exactly in the same way. One simply replaces $\bar{\mu}\gamma_\mu(1 + \gamma_5)\nu$ by $\bar{\mu}\gamma_\mu(1 + \gamma_5)\nu'$. Note that both $\nu(\bar{\nu})$ and $\nu'(\bar{\nu}')$ are left-handed (right-handed). The possible non-identity of the "muon neutrino" to the "electron neutrino" has been discussed by several authors (e.g., Pontecorvo, 1959.).

Note added in proof. Recent work of a Columbia-Brookhaven group (*Danby et al.*, 1962) indicates that

$$\text{"}\mu \text{ neutrino"} + N \to \mu^\pm + N$$

(where N stands for p or n) is fully allowed, whereas

$$\text{"}\mu \text{ neutrino"} + N \to e^\pm + N$$

seems forbidden. This strongly suggests that the "muon neutrino" ν' (emitted in $\pi^+ \to \mu^+$ decay) is different from the "electron neutrino" (that participates in β^+ decay). It is, however, possible to *describe* ν and ν' using a *single four*-component field instead of using *two distinct two*-component fields (Schwinger, 1957; Nishijima, 1957; Kawakami, 1957). The two descriptions lead to completely identical physical consequences as long as the masses of both ν and ν' are strictly zero.

CHAPTER 9

Isospin and Related Topics (S = 0)

9.1. History

The notion of isospin (or isotopic spin, isobaric spin, *i*-spin) was first introduced by Heisenberg (1932) as a convenient mathematical parameter to characterize the two charge states of what is now known as the nucleon. The nucleon is assumed to take isospin "up" ($T_3 = \frac{1}{2}$, proton) and isospin "down" ($T_3 = -\frac{1}{2}$, neutron) in analogy with the electron which can take ordinary spin "up" and ordinary spin "down." This formalism is convenient because we know the algebra of spin $\frac{1}{2}$ systems from atomic physics.

The isospin formalism assumes an important role if we make the hypothesis that the so-called strong interactions are charge independent. In atomic physics we know that in the absence of external magnetic fields observable effects cannot depend on the direction of the spin quantization axis so that there is no distinction between $S_z = \frac{1}{2}$ and $S_z = -\frac{1}{2}$. This arbitrariness is a consequence of invariance under rotations in ordinary space. Similarly we argue that in the absence of electromagnetic interactions which are clearly charge-dependent, there is no distinction between $T_3 = \frac{1}{2}$ (proton) and $T_3 = -\frac{1}{2}$ (neutron), and that this arbitrariness stems from invariance under rotations in some internal space, which is called isospin space. Just as the conservation of angular momentum follows from rotational invariance in ordinary space, the conservation of isospin holds as long as the strong interactions are invariant under rotations in isospin space. The usefulness of isospin as a good quantum number was first discussed by Cassen and Condon (1936) and Wigner (1937).

We should not forget that full rotational invariance in isospin space is only approximate since there exist the electromagnetic couplings which introduce a preferred direction in isospin space. Fortunately the electromagnetic couplings are weaker than the strong couplings by a factor of ~ 100; hence isospin conservation is expected to be valid to an accuracy of a few per cent. In this connection we may note that the neutron-proton mass difference is 1.3 Mev. out of 938 Mev. This mass difference is usually attributed to a weak charge-dependent perturbation brought about by the electromagnetic couplings.

The very analysis of the binding energies of various mirror nuclei already reveals the approximate equality of *p-p* and *n-n* forces. For instance, the difference in binding energy between H^3 and He^3 is 0.7 Mev. out of 8 Mev.

What is perhaps more striking is the equality of *p-p*, *n-n*, and *p-n* forces as revealed in charge triplets ($T = 1$), e.g. O^{14}, N^{14*} (excitation energy 2.3 Mev) and C^{14}. The implications of the charge independence hypothesis in nuclear physics are discussed in many books (e.g., Sachs, 1953).

The first evidence for charge independence in elementary particle physics was presented by Breit, Condon, and Present (1936) who made a comparison between *n-p* scattering data obtained by Fermi and Amaldi (1936) in Rome and *p-p* scattering data obtained by Tuve, Heydenburg, and Hafstad (1936) at the Department of Terrestrial Magnetism of the Carnegie Institution in Washington. It was demonstrated that the interactions responsible for scattering in the $^1S_0$ state are the same both for *n-p* and for *p-p* after the Coulomb effects are taken into account. (Note that the principle of charge independence does not claim that the total *p-p* and *n-p* scattering cross sections are the same. Both the $^1S_0$ and $^3S_1$ states are accessible to *n-p* whereas only the $^1S_0$ state is accessible to *p-p* because of the Pauli principle.) Recent evidence for the equality between the *p-p* interactions and the *n-p* interactions in the $^1S_0$ state is discussed by Noyes (1960b).

It is of interest to examine the role played by the charge independence hypothesis in the development of meson physics. By the time Yukawa (1935) postulated the existence of a particle with mass $\sim 200m_e$ to explain nuclear forces the exchange character of nuclear forces had already been discussed in connection with the saturation property. It was therefore natural for Yukawa to have let his hypothetical particle bear electric charge. However, if we just had charged particles, it would be hard to understand the equality between *p-p* and *n-p* forces demonstrated by Breit and collaborators (Yukawa and Sakata, 1937; Fröhlich, Heitler, and Kemmer, 1938). Let us consider the *n-p* case. Some form of exchange force is expected to arise when the Yukawa particle of negative charge, namely the π^- in the modern language, is emitted by the neutron and subsequently absorbed by the proton. But in the *p-p* case, because the proton can emit only π^+ and absorb only π^-, the two protons *cannot* exchange a *single* charged meson by the same mechanism. This means that, if there were just charged mesons, the range of nuclear forces in the *n-p* case would be of the order of $1/\mu_\pi$ whereas in the *p-p* case, of the order of $1/2\mu_\pi$. Thus the *n-p* interactions would be very different from the *p-p* interactions. To save the situation it is essential that there be a neutral meson of approximately the same mass. In other words, the Yukawa particle must be a triplet in isospin space as emphasized particularly by Kemmer (1938).

We may mention parenthetically that all these arguments were advanced many years before the discovery of the pion. To be sure, the discovery of a charged cosmic-ray particle with mass $\sim 200m_e$ (to be identified with $\mu^\pm$) by Neddermeyer and Anderson (1937) and Street and Stevenson (1937) stimulated the development of meson theory. But as our knowledge of the "experimentalists' mesons" increased, it became more and more evident that the observed meson was incapable of interacting strongly with nucleons. The discrepancy between the observed behavior of the "experimentalists' mesons" and the expected behavior of the "theoreticians' mesons" became more and more serious until Conversi, Pancini, and Piccioni (1947) conclusively demonstrated that the capture rate of the observed cosmic-ray mesons by carbon nuclei is at least 10^{10} times slower than the capture rate expected of the Yukawa particle. Meanwhile it was conjectured that there are actually two mesons, and that the observed meson is a decay product of the Yukawa particle (Sakata and Inoue, 1946; Marshak and Bethe, 1947). This two-meson hypothesis was confirmed when Lattes, Occhialini, and Powell (1947) observed for the first time an example of what has later been known as the π-μ-e sequence. Subsequent experiments showed that the parent π meson can indeed interact strongly with nucleons. The existence of a neutral pion was established a few years afterwards (Bjorklund *et al.*, 1950; Steinberger, Panofsky, and Stellar, 1950) in accelerator experiments at Berkeley.

9.2. Charge Independence in Pion Physics

In order to deduce the consequences of the charge independence hypothesis, one need not know the detailed form of the interaction, nor need one carry out dynamical calculations. This is so because the S matrix is invariant under isospin rotations as long as the basic interaction, whatever it may be, is invariant under isospin rotations, and the S matrix connects initial and final states with the same isospin provided that charge independence holds. (We are again using the fact that the rotation operation in isospin space is unitary.)

The isospin of a system composed of several particles is the vectorial sum of the isospins of the individual constituent particles. All we need to know is the rule for adding angular momenta. As an example, consider a system composed of two nucleons. We have

$$T = 1: \begin{cases} |p\rangle|p\rangle, & T_3 = 1 \\ \dfrac{1}{\sqrt{2}}(|p\rangle|n\rangle + |n\rangle|p\rangle), & T_3 = 0 \\ |n\rangle|n\rangle, & T_3 = -1 \end{cases} \tag{9.1}$$

$$T = 0 : \frac{1}{\sqrt{2}}(|p\rangle|n\rangle - |n\rangle|p\rangle), \qquad T_3 = 0$$

where $|p\rangle|n\rangle$ stands for the state in which the nucleon 1 is in the $T_3 = \frac{1}{2}$ state and the nucleon 2 is in the $T_3 = -\frac{1}{2}$ state. Note that the $T = 1$ (charge triplet) functions are symmetric under the interchange of the particle labels while the $T = 0$ (charge singlet) function is antisymmetric. Since n and p are to be regarded as two different states of the same particle, the following generalized principle holds as a result of the charge independence hypothesis: The total wave function which is the product of the spatial coordinate wave function, the spin wave function and the isospin wave function must change sign as we interchange the coordinate labels, the spin labels, and the isospin labels. Then for even (odd) L states the $T = 1$ state has to be combined with the $S = 0$ ($S = 1$) state, and the $T = 0$ state has to be combined with the $S = 1$ ($S = 0$) state, where L and S denote the orbital and spin angular momenta of the two-nucleon system. In other words

$$
\begin{aligned}
T &= 1\ (p\text{-}p,\ p\text{-}n,\ n\text{-}n):\ {}^1S_0,\ {}^3P_{0,1,2},\ {}^1D_2,\ {}^3F_{2,3,4}\cdots \\
T &= 0\ (p\text{-}n \text{ only})\qquad\ :\ {}^3S_1,\ {}^1P_1,\ {}^3D_{1,2,3},\ {}^1F_3\cdots
\end{aligned}
\tag{9.2}
$$

Consider the deuteron, which is known to be in ${}^3S_1$ with a few per cent mixture of ${}^3D_1$. We see from (9.2) that the deuteron must be in the $T = 0$ state. This is in conformity with the empirical fact that there is no other bound system of two nucleons.

In pion physics π^+, π^0, and π^- form a triplet in isospin space. Since π^+ and π^- are antiparticles of each other, they must have exactly the same mass from *CPT* invariance. On the other hand, since isospin conservation is only approximate, the π^0 mass need not be exactly equal to the $\pi^\pm$ mass. The mass difference turns out to be $\mu(\pi^\pm) - \mu(\pi^0) = 4.6$ Mev. out of 140 Mev.

What kind of experimental evidence do we have in favor of charge independence in pion physics? In particular, is there any *direct* evidence that supports the $T = 1$ assignment for the pion? Consider

$$n + p \rightarrow \pi^0 + d \tag{9.3a}$$

$$p + p \rightarrow \pi^+ + d \tag{9.3b}$$

Since the d isospin is zero, the final π-d system has isospin 1 for both π^0-d and π^+-d. If isospin is a good quantum number, the initial system must have isospin 1. From (9.1) we note that the n-p system is in the $T = 1$ state only half of the time whereas the p-p system is in the $T = 1$ state all the time. So we expect

$$\frac{\sigma(\pi^0 + d)}{\sigma(\pi^+ + d)} = \frac{1}{2}$$

as first noted by Yang. Reaction (9.3a) was first measured by Hildebrand (1953), who compared his results with the inverse of (9.3b) previously measured by Durbin, Loar, and Steinberger (1951) at the same center of mass energy. The angular distributions were the same in both cases, $0.2 + \cos^2\theta$, and the ratio was 1:2 as required by charge independence to an accuracy of about 10%.

There is a similar set of reactions (Messiah, 1952; Ruderman, 1952)

$$\begin{aligned} p + d &\rightarrow \pi^+ + H^3 \\ p + d &\rightarrow \pi^0 + He^3 \end{aligned} \tag{9.4}$$

The initial state is accessible to $T = \frac{1}{2}$ only. From nuclear physics we know that H^3 and He^3 form an isodoublet. So for the final states we add a half unit of isospin and one unit of isospin to obtain a half unit of isospin. We have (see Appendix A)

$$|T = \tfrac{1}{2}, T_3 = \tfrac{1}{2}\rangle = \sqrt{\tfrac{2}{3}}|\pi^+ H^3\rangle - \sqrt{\tfrac{1}{3}}|\pi^0 He^3\rangle$$

so that

$$R \equiv \frac{\sigma(\pi^+ + H^3)}{\sigma(\pi^0 + He^3)} = 2$$

Reactions (9.4) have been studied by many groups at various energies (e.g., Crewe *et al.*, 1959). Most recently Harting *et al.* (1960) obtained

$$R_{\text{exp.}} = 2.13 \pm 0.06$$

with 590 Mev. protons. Köhler (1960) claims that due to charge-dependent corrections caused by the mass-difference and Coulomb effects, $R_{\text{exp.}}$ should be compared with

$$R_{\text{theor.}} = 2.20 \pm 0.07$$

Thus the agreement is excellent.

Isospin conservation suppresses the rates of certain reactions which would otherwise be fully allowed. Consider

$$d + d \rightarrow He + \pi^0$$

Since the isospins of d and He are both zero, the reaction could occur only if the isospin changed by one unit. Experimentally the cross section for this reaction at 400 Mev. lab. is estimated to be smaller than 2×10^{-32} cm^2 (Akimov, Savchenko, and Soroko, 1960). It is expected that the cross section would be of the order of 10^{-30} cm^2 if there were no selection rule. *Note added in proof.* Poirer *et al.* (1962) have shown that the 90° differential cross section for $d + d \rightarrow He^4 + \pi^0$ at 460 Mev is less than 1.5×10^{-34} cm^2. If there were no isospin selection rule, we could estimate the cross section for this process from the known cross section for $p(n) + d \rightarrow He^3$ $(H^3) + \pi^0$ using the impulse approximation method; the expected cross section turns out to be $\sim 4 \times 10^{-32}$ cm^2 (Greider 1961). Thus the isospin selection rule is at least 99.6% "good."

We now turn to π-N scattering. A system composed of one pion and one nucleon can be either in $T = \frac{1}{2}$ or in $T = \frac{3}{2}$. We have (see Appendix A)

$$\begin{aligned} |T = \tfrac{3}{2}, T_3 = \tfrac{3}{2}\rangle &= |p\pi^+\rangle \\ |T = \tfrac{3}{2}, T_3 = -\tfrac{1}{2}\rangle &= \sqrt{\tfrac{1}{3}}|p\pi^-\rangle + \sqrt{\tfrac{2}{3}}|n\pi^0\rangle \\ |T = \tfrac{1}{2}, T_3 = -\tfrac{1}{2}\rangle &= -\sqrt{\tfrac{2}{3}}|p\pi^-\rangle + \sqrt{\tfrac{1}{3}}|n\pi^0\rangle, \text{ etc.} \end{aligned} \tag{9.5}$$

In the π^+-p case we just have a pure $T = \frac{3}{2}$ state. For π^--p we start with

$$|p\pi^-\rangle = \sqrt{\tfrac{1}{3}}|T = \tfrac{3}{2}, T_3 = -\tfrac{1}{2}\rangle - \sqrt{\tfrac{2}{3}}|T = \tfrac{1}{2}, T_3 = -\tfrac{1}{2}\rangle$$

The hypothesis of charge independence states that the orientation of the third axis in isospin space is devoid of physical significance. This means that scattering depends only on T and not on T_3. Thus the two amplitudes $f_{1/2}$ and $f_{3/2}$ should be sufficient to characterize the scattering as first pointed out by Heitler (1946). The final state is given by

$$\begin{aligned}|\psi_f\rangle &= f_{3/2}\sqrt{\tfrac{1}{3}}|T = \tfrac{3}{2}, T_3 = -\tfrac{1}{2}\rangle - \sqrt{\tfrac{2}{3}}f_{1/2}|T = \tfrac{1}{2}, T_3 = -\tfrac{1}{2}\rangle \\ &= \tfrac{1}{3}(f_{3/2} + 2f_{1/2})|p\pi^-\rangle + \frac{\sqrt{2}}{3}(f_{3/2} - f_{1/2})|n\pi^0\rangle\end{aligned}$$

Charge-exchange scattering ($\pi^- + p \to \pi^0 + n$) is then given by $\frac{2}{9}|f_{3/2} - f_{1/2}|^2$ while charge-elastic scattering ($\pi^- + p \to \pi^- + p$) is given by $\frac{1}{9}|f_{3/2} + 2f_{1/2}|^2$. Note that if we sum the two, the $f_{1/2}$-$f_{3/2}$ interference terms drop out, and we simply get $\frac{1}{3}(|f_{3/2}|^2 + 2|f_{1/2}|^2)$.

We can now consider the partial wave expansion in s and p states (cf., Chapter 3, Section 2) taking into account the various charge states. We have

$$\frac{d\sigma}{d\Omega}(\pi^+p \to \pi^+p) = \lambda\!\!\!^-{}^2(|a_3 + (2a_{33} + a_{31})\cos\theta|^2 + |a_{33} - a_{31}|^2 \sin^2\theta)$$

$$\begin{aligned}\frac{d\sigma}{d\Omega}(\pi^-p \to \pi^-p) &= \frac{\lambda\!\!\!^-{}^2}{9}[|a_3 + 2a_1 + (2a_{33} + a_{31} + 4a_{13} + 2a_{11})\cos\theta|^2 \\ &\quad + |a_{33} - a_{31} + 2a_{13} - 2a_{11}|^2 \sin^2\theta] \qquad (9.6) \\ \frac{d\sigma}{d\Omega}(\pi^-p \to \pi^0n) &= \frac{2\lambda\!\!\!^-{}^2}{9}[|a_3 - a_1 + (2a_{33} + a_{31} - 2a_{13} - a_{11})\cos\theta|^2 \\ &\quad + |a_{33} - a_{31} - a_{13} + a_{11}|^2 \sin^2\theta]\end{aligned}$$

where the meaning of the subscripts is explained in Table 9.1.

TABLE 9.1

Scattering Amplitudes

	T	Angular momentum
a_1	$\frac{1}{2}$	$S_{1/2}$
a_3	$\frac{3}{2}$	$S_{1/2}$
a_{11}	$\frac{1}{2}$	$P_{1/2}$
a_{13}	$\frac{1}{2}$	$P_{3/2}$
a_{31}	$\frac{3}{2}$	$P_{1/2}$
a_{33}	$\frac{3}{2}$	$P_{3/2}$

The amplitude a is defined as

$$a \equiv \frac{(\exp(2i\delta) - 1)}{2i}$$

It is customary to denote an s state phase shift by δ_{2T} and a p state phase shift by $\delta_{2T,\,2J}$ just as for a's given in Table 9.1. If we fit the observed angular distribution for the three processes, $\pi^+p \to \pi^+p$, $\pi^-p \to \pi^-p$, $\pi^-p \to \pi^0 n$, with $A + B\cos\theta + C\cos^2\theta$, there are 9 numbers available and 6 phase shifts to be determined. A graphical method has been used to determine these phase shifts. (See Ashkin and Vosko, 1953, and Section 32 of Bethe and de Hoffman, 1955.)

We have already seen that there is a resonance in π^+-p scattering at 190 Mev. Let us suppose

$$|f_{3/2}| \gg |f_{1/2}| \quad \text{around 190 Mev.}$$

Then we must have

$$\frac{\sigma(p\pi^+)}{\sigma(p\pi^- \to p\pi^- \text{ and } n\pi^0)} = 3$$

This is satisfied experimentally within 5% (see Fig. 9.1). We also must have

$$\frac{\sigma(p\pi^- \to p\pi^-)}{\sigma(p\pi^- \to n\pi^0)} = \frac{1}{2}$$

This has been confirmed within 15%. The angular distribution for π^--p elastic and also π^--p charge exchange scattering at this energy are of the form $1 + 3\cos^2\theta$ just as in the π^+-p case. Low energy pion scattering up to 300 Mev. takes place predominantly in $T = \frac{3}{2}$. There is a salient peak at 190 Mev., and this corresponds to a p wave resonance in $T = 3/2$, $J = \frac{3}{2}^+$ (+ parity because the intrinsic negative parity of the π meson compensates the negative orbital parity); δ_{33} passes through 90° at 190 Mev. This resonance is called the 3-3 resonance (cf., Fig. 9.1).

Photoproduction is clearly charge *dependent*, so it may seem that no useful conclusion can be obtained from isospin considerations. This is not necessarily true. Let us suppose that the following resonance picture holds. The incident γ ray excites the target proton to the 3-3 isobar. This isobar does not live long but decays into a nucleon and a pion. Although the first step, namely the formation of the 3-3 isobar by γ is charge dependent, the second step, namely the disintegration of the 3-3 isobar into a nucleon and a pion, is a non-electromagnetic, strong interaction, and is thus charge independent. Hence if this resonance picture holds

$$\frac{\sigma(\gamma + p \to p + \pi^0)}{\sigma(\gamma + p \to n + \pi^+)} = 2$$

where we have used

$$|T = \tfrac{3}{2}, T_3 = \tfrac{1}{2}\rangle = \sqrt{\tfrac{2}{3}}|p\pi^0\rangle + \sqrt{\tfrac{1}{3}}|n\pi^+\rangle$$

This is not the whole story since there are contributions from states other than the 3-3 resonance; particularly important is the "direct" *s* wave production of π^+ which has no counterpart for π^0 photoproduction. The available data crudely support the 2:1 ratio after the non-resonant contributions are subtracted out.

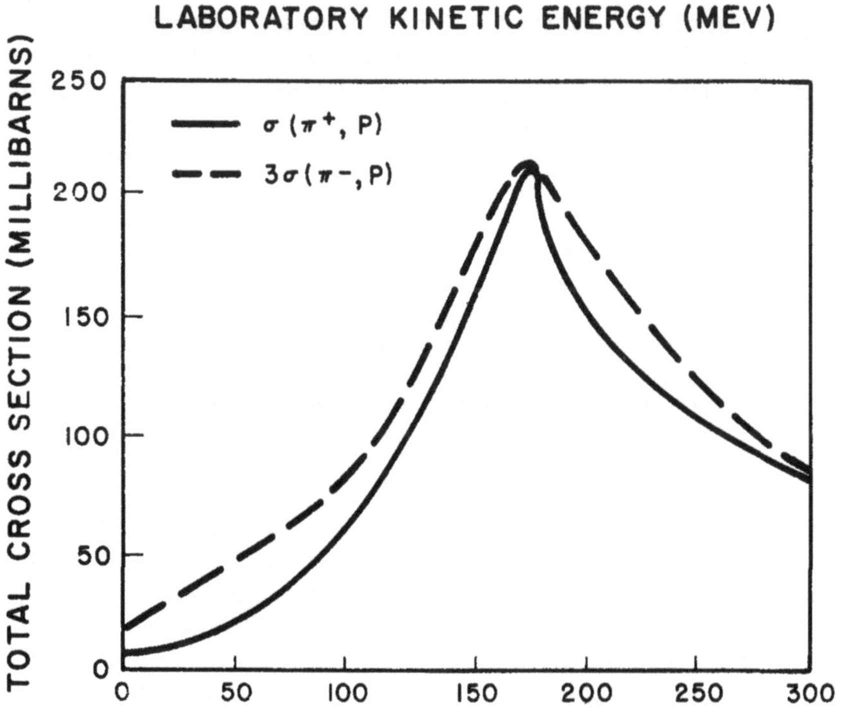

Figure 9.1. Charge independence and the 3-3 resonance

Let us look at the pion-nucleon interaction at higher energies. We note that, even in the presence of inelastic processes, the following relations must hold

$$\sigma_{\text{total}}(\pi^+p) = \sum_i |f^{(i)}_{3/2}|^2$$

$$\sigma_{\text{total}}(\pi^-p) = \tfrac{1}{3}\sum_i |f^{(i)}_{3/2}|^2 + \tfrac{2}{3}\sum_i |f^{(i)}_{1/2}|^2 \tag{9.7}$$

Thus

$$\sigma_{\text{total}}(T = \tfrac{1}{2}) = \tfrac{1}{2}(3\sigma_{\text{total}}(\pi^-p) - \sigma_{\text{total}}(\pi^+p))$$

$$\sigma_{\text{total}}(T = \tfrac{3}{2}) = \sigma_{\text{total}}(\pi^+p) \tag{9.8}$$

The summation in (9.7) is to be taken over spin, angle, and channel, e.g., $\pi + N \to 2\pi + N$ as well as $\pi + N \to \pi + N$.

For a long time it was believed that there was a broad peak in $\sigma_{\text{total}}(\pi^- p)$ from 600 to 1000 Mev. In 1958 this broad peak was resolved by Burrowes *et al.* (1959) into two narrow peaks, one at 600 Mev. and the other at 900 Mev. For recent data, see Brisson *et al.* (1960). (Historically the two-peak picture was first suggested by Wilson (1958) on the basis of photo-production experiments prior to the $\sigma_{\text{total}}(\pi^- p)$ measurements of Burrowes *et al.*) Since these peaks are present only in $\sigma(\pi^- p)$ and not in $\sigma(\pi^+ p)$, they must be $T = \frac{1}{2}$ peaks. Assuming that each peak is a "resonance" due to a single state with definite J and parity, we may make tentative angular

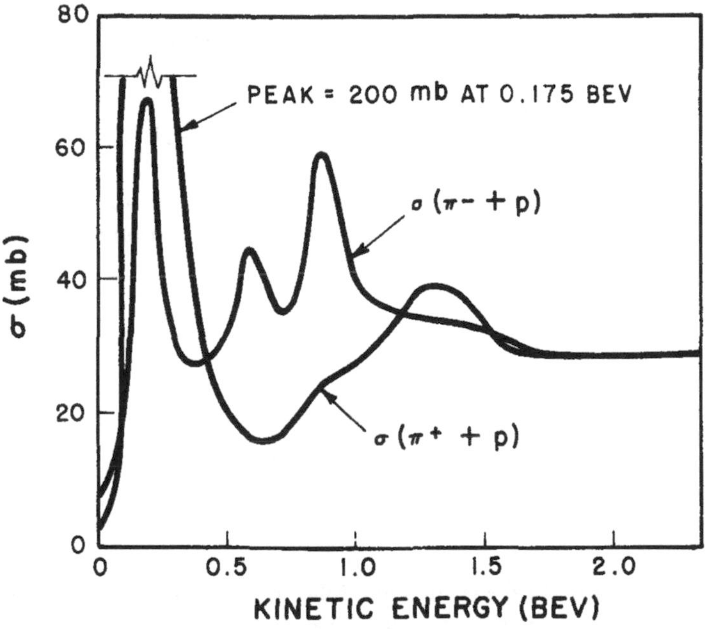

Figure 9.2. The total cross sections for $\pi^{\pm} p$ scattering

momentum and parity assignments. There are some indications that the 600 Mev. resonance is in $T = \frac{1}{2}$, $J = \frac{3}{2}$, odd parity, i.e. behaves as a $d\frac{3}{2}\pi$-N state (Peierls, 1958; Sakurai, 1958b; Stein, 1959; Maloy *et al.*, 1961; Querzoli, Salvini, and Silverman, 1961). For the 900 Mev. resonance a $J = \frac{5}{2}$ assignment seems likely (Wood *et al.*, 1961). There is, in addition, another peak at 1350 Mev. in $\sigma(\pi^+ p)$, but very little is known about this $T = \frac{3}{2}$ peak. Above 2 Bev. the total cross sections (both $\pi^- p$ and $\pi^- p$) seem to stay constant at ≈ 30 mb all the way up to ≈ 10 Bev. The whole situation is summarized in Fig. 9.2.

Note added in proof. The total π^-p and π^+p cross sections have been measured up to 20 Bev/c (Lindenbaum *et al.*, 1961, von Dardel *et al.*, 1962). The π^-p cross section seems higher than the π^+p cross section by about 1.5 mb in the 10–20 Bev range.

9.3. Formal Theory of Isospin Rotations

Isospin rotations in a field theoretic formalism take the form of gauge transformations discussed in the previous chapter. Before discussing full 3-dimensional rotations, we formulate the gauge transformation associated with electric charge conservation in a slightly different manner. Consider a charged scalar (or pseudoscalar) field ϕ. It is necessarily non-Hermitian. Define ϕ_1 and ϕ_2 so that

$$\begin{aligned} \phi_1 &= \frac{1}{\sqrt{2}}(\phi + \phi^\dagger) \qquad & \phi &= \frac{1}{\sqrt{2}}(\phi_1 + i\phi_2) \\ \phi_2 &= \frac{1}{i\sqrt{2}}(\phi - \phi^\dagger) \qquad & \phi^\dagger &= \frac{1}{\sqrt{2}}(\phi_1 - i\phi_2) \end{aligned} \tag{9.9}$$

Then ϕ_1 and ϕ_2 are Hermitian. The relation between ϕ, $\phi^\dagger$ and ϕ_1, ϕ_2 is entirely analogous to the relation between circular polarization and linear polarization discussed in Chapter 2.

The gauge transformation

$$\phi \rightarrow \exp(i\lambda)\phi$$

now becomes

$$\begin{aligned} \phi_1 + i\phi_2 &\rightarrow (\cos\lambda + i\sin\lambda)(\phi_1 + i\phi_2) \\ \phi_1 &\rightarrow \phi_1 \cos\lambda - \phi_2 \sin\lambda \\ \phi_2 &\rightarrow \phi_1 \sin\lambda + \phi_2 \cos\lambda \end{aligned} \tag{9.10}$$

Thus the gauge transformation associated with electric charge conservation is a rotation about the third axis in some kind of internal space. To make the analogy more vivid, consider the generator Q of an infinitesimal gauge transformation.

$$\begin{aligned} Q &= i\int\left(\frac{\partial\phi^\dagger}{\partial x_0}\phi - \phi^\dagger\frac{\partial\phi}{\partial x_0}\right)d^3x \\ &= i\int\left[\frac{1}{\sqrt{2}}\left(\frac{\partial\phi_1}{\partial x_0} - i\frac{\partial\phi_2}{\partial x_0}\right)\frac{1}{\sqrt{2}}(\phi_1 + i\phi_2) - \frac{1}{\sqrt{2}}(\phi_1 - i\phi_2)\right. \\ &\qquad \left.\times\frac{1}{\sqrt{2}}\left(\frac{\partial\phi_1}{\partial x_0} + i\frac{\partial\phi_2}{\partial x_0}\right)\right]d^3x \\ &= -\int\left(\frac{\partial\phi_1}{\partial x_0}\phi_2 - \frac{\partial\phi_2}{\partial x_0}\phi_1\right)d^3x \end{aligned}$$

Then it is extremely tempting to introduce the third Hermitian field ϕ_3 in such a way that

$$\boldsymbol{\phi} = (\phi_1, \phi_2, \phi_3)$$

$$T_3 = Q = -\int \left(\frac{\partial \boldsymbol{\phi}}{\partial x_0} \times \boldsymbol{\phi}\right)_3 d^3x \tag{9.11}$$

Note that the electric charge of ϕ_3 is zero. Charge conservation can now be formulated in terms of invariance under rotations about the third axis in some internal space. Then it is natural to ask whether rotational invariance about any axis other than the 3 axis also holds. More precisely, does invariance under the unitary transformation

$$U = 1 + i\mathbf{T}\cdot\boldsymbol{\epsilon} \tag{9.12}$$

also hold where

$$\mathbf{T} = -\int \left(\frac{\partial \boldsymbol{\phi}}{\partial x_0} \times \boldsymbol{\phi}\right) d^3x \tag{9.13}$$

and $\boldsymbol{\epsilon}$ is an infinitesimal vector in an *arbitrary* direction in this internal space? If such invariance holds, then the orientation of isospin is devoid of physical significance, which is another way of saying that charge independence is valid.

What are the commutation relations among the various components of $\mathbf{T}$? By direct computation

$$\begin{aligned}
[T_1, T_2] &= \left[\int \left(\frac{\partial \phi_2(x)}{\partial x_0}\,\phi_3(x) - \frac{\partial \phi_3(x)}{\partial x_0}\,\phi_2(x)\right) d^3x,\right. \\
&\qquad \left.\int \left(\frac{\partial \phi_3(x')}{\partial x_0'}\,\phi_1(x') - \frac{\partial \phi_1(x')}{\partial x_0}\,\phi_3(x')\right) d^3x'\right] \\
&= \int\!\!\int \left(\left[\phi_3(x), \frac{\partial \phi_3(x')}{\partial x_0'}\right] \frac{\partial \phi_2(x)}{\partial x_0}\,\phi_1(x') + \right. \\
&\qquad \left. + \left[\frac{\partial \phi_3(x')}{\partial x_0'}, \phi_3(x)\right] \phi_2(x)\,\frac{\partial \phi_1(x')}{\partial x_0}\right) d^3x\, d^3x' \\
&= -i \int \left(\frac{\partial \phi_1}{\partial x_0}\,\phi_2 - \frac{\partial \phi_2}{\partial x_0}\,\phi_1\right) d^3x \\
&= iT_3
\end{aligned} \tag{9.14}$$

We could have anticipated this result from the geometrical meaning of $\mathbf{T}$. Since $\mathbf{T}$ generates rotation, it must satisfy such commutation relations.

So far our formalism can adequately describe a charge triplet spin zero field. We know one such example in nature. The pion can assume three charge states $+1$, 0, and -1. Our $(\phi_1 - i\phi_2)/\sqrt{2}$, ϕ_3 and $(\phi_1 + i\phi_2)/\sqrt{2}$ correspond to π^+, π^0, and π^-.

The reader may wonder why $(\phi_1 - i\phi_2)/\sqrt{2}$ rather than $(\phi_1 + i\phi_2)/\sqrt{2}$

corresponds to π^+. This "paradox" can be resolved by noting that the isovector $\boldsymbol{\phi}$ actually stands for

$$\begin{aligned}\boldsymbol{\phi} &= \phi_1 \boldsymbol{e}_1 + \phi_2 \boldsymbol{e}_2 + \phi_3 \boldsymbol{e}_3 \\ &= \left(\frac{\varphi_1 - i\phi_2}{\sqrt{2}}\right)\left(\frac{\boldsymbol{e}_1 + i\boldsymbol{e}_2}{\sqrt{2}}\right) + \left(\frac{\phi_1 + \phi_2}{\sqrt{2}}\right)\left(\frac{\boldsymbol{e}_1 - i\boldsymbol{e}_2}{\sqrt{2}}\right) + \phi_3 \boldsymbol{e}_3\end{aligned}$$

where $\boldsymbol{e}_1$, $\boldsymbol{e}_2$ and $\boldsymbol{e}_3$ form a set of mutually orthogonal, unit base vectors in isospin space.

In extending our formalism to the nucleon, we note that the nucleon has only two charge states $+1$ and zero so that the relation $T_3 = Q$ no longer holds. We define the eight-component spinor

$$N = \begin{pmatrix} p \\ n \end{pmatrix} \tag{9.15}$$

where p and n are the usual four-component spinors in Lorentz space. Then baryon conservation implies

$$B = \int \bar{N}\gamma_4 N d^3x = \int N^\dagger N d^3x = N(p) + N(n) - N(\bar{p}) - N(\bar{n}) \tag{9.16}$$

is a constant of the motion whereas charge conservation implies

$$\begin{aligned}Q &= \int (\bar{p}\gamma_4 p)\, d^3x = \int (p^\dagger p)\, d^3x \\ &= N(p) - N(\bar{p})\end{aligned} \tag{9.17}$$

But

$$B = \int N^\dagger \begin{pmatrix} I & 0 \\ 0 & I \end{pmatrix} N d^3x$$

and

$$Q = \int N^\dagger \begin{pmatrix} I & 0 \\ 0 & 0 \end{pmatrix} N d^3x \tag{9.18}$$

Hence any linear combination of B and Q is also a constant of motion. In particular consider

$$Q - \frac{B}{2} = \frac{1}{2} \int N^\dagger \begin{pmatrix} I & 0 \\ 0 & -I \end{pmatrix} N d^3x \tag{9.19}$$

In this block form, $\begin{pmatrix} I & 0 \\ 0 & -I \end{pmatrix}$ has the same structure as σ_3. It is customary to define

$$\begin{aligned}T_3 &= Q - \frac{B}{2} = \frac{1}{2} \int N^\dagger \tau_3 N d^3x \\ \tau_3 \begin{pmatrix} p \\ 0 \end{pmatrix} &= \begin{pmatrix} p \\ 0 \end{pmatrix}, \quad \tau_3 \begin{pmatrix} 0 \\ n \end{pmatrix} = -\begin{pmatrix} 0 \\ n \end{pmatrix} \\ 2T_3 &= N(p) - N(\bar{p}) - N(n) + N(\bar{n})\end{aligned} \tag{9.20}$$

Note that for the antiproton state the isospin is *down* and for the antineutron state the isospin is *up*.

In analogy with the charge triplet case we define T_k for all three components

$$\mathbf{T} = \tfrac{1}{2}\int N^\dagger \boldsymbol{\tau} N d^3x$$

$$\tau_1 = \begin{pmatrix} 0 & I \\ I & 0 \end{pmatrix}, \qquad \tau_2 = \begin{pmatrix} 0 & -iI \\ +iI & 0 \end{pmatrix}, \qquad \tau_3 = \begin{pmatrix} I & 0 \\ 0 & -I \end{pmatrix} \tag{9.21}$$

where I is a 4×4 unit matrix that acts on the Dirac four-component spinor. An infinitesimal isospin rotation is given by

$$N \to \left(1 + \frac{i\boldsymbol{\varepsilon}\cdot\boldsymbol{\tau}}{2}\right)N \tag{9.22}$$

and the unitary operator with the property

$$UNU^{-1} = \left(1 + \frac{i\boldsymbol{\varepsilon}\cdot\boldsymbol{\tau}}{2}\right)N$$

turns out to be

$$U = 1 + i\mathbf{T}\cdot\boldsymbol{\varepsilon}$$

where $\mathbf{T}$ can be shown to be the same $\mathbf{T}$ as defined above. (Use the anticommutation relations

$$\{N^\dagger_{\alpha\tau_3}(\vec{x}, x_0), N_{\beta\tau'_3}(\vec{x}', x_0)\} = \delta_{\alpha\beta}\delta_{\tau_3\tau'_3}\delta^{(3)}(\vec{x} - \vec{x}').)$$

A finite isospin rotation can be obtained by compounding successively infinitesimal rotations of the form (9.22). We may note that the most general isospin rotation operator that acts on a two-component isospinor takes the form

$$\exp\left(i\frac{\boldsymbol{\tau}\cdot\hat{n}\theta}{2}\right) = \begin{pmatrix} \cos\frac{\theta}{2} + in_3\sin\frac{\theta}{2} & i(n_1 - in_2)\sin\frac{\theta}{2} \\ i(n_1 + in_2)\sin\frac{\theta}{2} & \cos\frac{\theta}{2} - in_3\sin\frac{\theta}{2} \end{pmatrix} \tag{9.23}$$

$$= \begin{pmatrix} \alpha & \beta \\ -\beta^* & \alpha^* \end{pmatrix}$$

where α and β are complex numbers satisfying $|\alpha|^2 + |\beta|^2 = 1$. The group associated with this transformation matrix is known as a unitary unimodular group in complex two dimensions. This is not the most general 2×2 unitary matrix. The most general unitary matrix can be written as $\exp(i\gamma)$ times the rotation matrix (9.23) where γ is real. Physically speaking, $\exp(i\gamma)$ generates a gauge transformation for nucleon conservation.

We shall not explicitly prove that the invariance of the Lagrangian under isospin rotations leads to the conservation of isospin current, since the proof is very similar to the one given in Chapter 8, Section 1 for electric charge conservation. If π and N interact with each other, the quantity that is divergenceless is the sum of the nucleon isospin current and the pion isospin current, i.e.

$$\partial_\mu \mathbf{J}_\mu = 0 \tag{9.24}$$

where

$$\mathbf{J}_\mu = i\bar{N}\frac{\boldsymbol{\tau}}{2}\gamma_\mu N - \boldsymbol{\pi} \times \partial_\mu \boldsymbol{\pi} \tag{9.25}$$

Isospin rotations considered in the usual formalism are space-time independent in character; $\boldsymbol{\epsilon}$ in (9.22) and (9.23) is not a function of space-time coordinates. This means that, although the principle of charge independence implies that the differentiation between a proton and a neutron is purely arbitrary in the absence of the electromagnetic coupling, once we decide on what we call a proton at one space-time point, what we should call a proton at some other space-time point is no longer arbitrary. We may naturally ask: What are the consequences of requiring that the Lagrangian be invariant under space-time dependent isospin rotations? This question was examined by Yang and Mills (1954). As in the electromagnetic case, in order to preserve the invariance, we must postulate the existence of a $T = 1$ vector field coupled to the isospin current. We shall discuss the possible manifestations of such a vector field in later parts of the monograph.

In order that charge independence be valid, the interaction Lagrangian must be invariant under isospin rotations. In theories where the pion is treated as a fundamental field, and nucleons as well as pions are treated relativistically, we have the *ps-ps* coupling (γ_5 coupling)

$$L_{\text{int}} = -iG\boldsymbol{\pi}\cdot\bar{N}\gamma_5\boldsymbol{\tau}N \tag{9.26}$$

or the *ps-pv* coupling

$$L_{\text{int}} = -i\frac{F}{\mu_\pi}\frac{\partial\boldsymbol{\pi}}{\partial x_\mu}\cdot\bar{N}\gamma_5\gamma_\mu\boldsymbol{\tau}N \tag{9.27}$$

In the renormalized γ_5 coupling theory there is, in addition, a term of the form

$$\lambda(\boldsymbol{\pi}\cdot\boldsymbol{\pi})^2 \tag{9.28}$$

which gives directly the amplitude for the scattering of a pion by another pion in the weak coupling limit.

In theories where the nucleon is treated as an infinitely heavy particle, we have the so-called Chew-Low static Hamiltonian

$$H_{\text{int}} = \frac{F}{\mu_\pi}\vec{\sigma}\cdot\vec{\nabla}\boldsymbol{\tau}\cdot\boldsymbol{\pi}\rho(|\vec{x}|) \tag{9.29}$$

where $\vec{\nabla}$ and $\vec{\sigma}$ are to be contracted in ordinary space, and $\boldsymbol{\pi}$ and $\boldsymbol{\tau}$ are to be contracted in isospin space. We shall come back to (9.29) in the next section. We can also write down non-Yukawa type interactions such as

$$\boldsymbol{\pi}^2\rho(|\vec{x}|) \quad \text{and} \quad \boldsymbol{\tau}\cdot\left(\boldsymbol{\pi} \times \frac{\partial \boldsymbol{\pi}}{\partial x_0}\right)\rho(|\vec{x}|)$$

which describe *phenomenologically* isospin independent and isospin dependent s wave π-N scattering respectively, in contrast to the Chew-Low Hamiltonian which leads to scattering in p states only.

It is instructive to write down the $\boldsymbol{\tau}\cdot\pi$ interaction in terms of the various charge states. We have (omitting γ matrices)

$$\begin{aligned}\boldsymbol{\pi}\cdot(\bar{N}\boldsymbol{\tau}N) &= (\bar{p}, \bar{n})\begin{pmatrix} \pi_3 & \pi_1 - i\pi_2 \\ \pi_1 + i\pi_2 & -\pi_3 \end{pmatrix}\begin{pmatrix} p \\ n \end{pmatrix} \\ &= \sqrt{2}\,(\bar{p}n\pi_{\text{ch.}}^{\dagger} + \bar{n}p\pi_{\text{ch.}}) + (\bar{p}p - \bar{n}n)\pi^0\end{aligned} \tag{9.30}$$

where

$$\begin{aligned}\pi_{\text{ch.}} &= \frac{1}{\sqrt{2}}\,(\pi_1 + i\pi_2) \\ \pi_{\text{ch.}}^{\dagger} &= \frac{1}{\sqrt{2}}\,(\pi_1 - i\pi_2) \\ \pi^0 &= \pi_3\end{aligned} \tag{9.31}$$

$\pi_{\text{ch.}}$ is a wave function for an emitted π^+ or an absorbed π^- in the c-number theory. In the q-number theory $\pi_{\text{ch.}}$ annihilates π^- and created π^+.

At first glance the coefficients $\sqrt{2}$ and -1 in (9.30) may look asymmetric, and one might doubt whether (9.30) is truely charge-independent. Actually the appearance of these coefficients is absolutely essential if charge independence is to be maintained. To illustrate this point, let us consider n-p, p-p, and n-n forces brought about by the exchange of a single pion. (See Fig. 9.3.) From the diagrams it is evident that the equality of the nuclear forces in the three states can be maintained only with factors like $\sqrt{2}$ and -1 in the appropriate places. It is easy to convince oneself that if we had $\bar{p}n\pi_{\text{ch.}}^{\dagger} + \bar{n}p\pi_{\text{ch.}} + (\bar{p}p + \bar{n}n)\pi^0$ the equality would not hold at all. Considerations along this line can be generalized to higher-order diagrams.

We now consider π-N scattering. It is here useful to construct the projection operators for $T = \frac{1}{2}$ and $T = \frac{3}{2}$. We combine one unit of isospin of the pion with a half unit of isospin of the nucelon to form isospin $\frac{1}{2}$ or isospin $\frac{3}{2}$. The whole thing is analogous to what we did for π^+-p scattering in the $p_{1/2}$ and $p_{3/2}$ states. Formerly we obtained (in Chapter 3, Section 2) the projection operators for p wave scattering

$$\begin{aligned}\Lambda(J = \tfrac{1}{2}) &= \tfrac{1}{3}(\vec{\sigma}\cdot\hat{q}_f)(\vec{\sigma}\cdot\hat{q}_i) \\ \Lambda(J = \tfrac{3}{2}) &= \tfrac{1}{3}[3\hat{q}_f\cdot\hat{q}_i - (\vec{\sigma}\cdot\hat{q}_f)(\vec{\sigma}\cdot\hat{q}_i)]\end{aligned}$$

In precisely the same way

$$\Lambda(T = \tfrac{1}{2}) = \tfrac{1}{3}\tau_\beta\tau_\alpha = \tfrac{1}{3}(\delta_{\beta\alpha} + \tfrac{1}{2}[\tau_\beta, \tau_\alpha])$$
$$\Lambda(T = \tfrac{3}{2}) = \tfrac{1}{3}(3\delta_{\beta\alpha} - \tau_\beta\tau_\alpha) = \tfrac{1}{3}(2\delta_{\beta\alpha} - \tfrac{1}{2}[\tau_\beta, \tau_\alpha]) \qquad (9.32)$$

where $\tau_\alpha(\tau_\beta)$ stand for the projection of the nucleon $\boldsymbol{\tau}$ matrix along the "direction" of the initial (final) pion isovector (in the "linear," rather than "circular," polarization language). (An alternative way to obtain (9.32) is to consider the eigenvalues of $R_{\beta\alpha} = \tau_\beta\tau_\alpha$ using the relation $(R^2)_{\beta\alpha} = 3R_{\beta\alpha}$). These projection operators are very useful since they enable us to

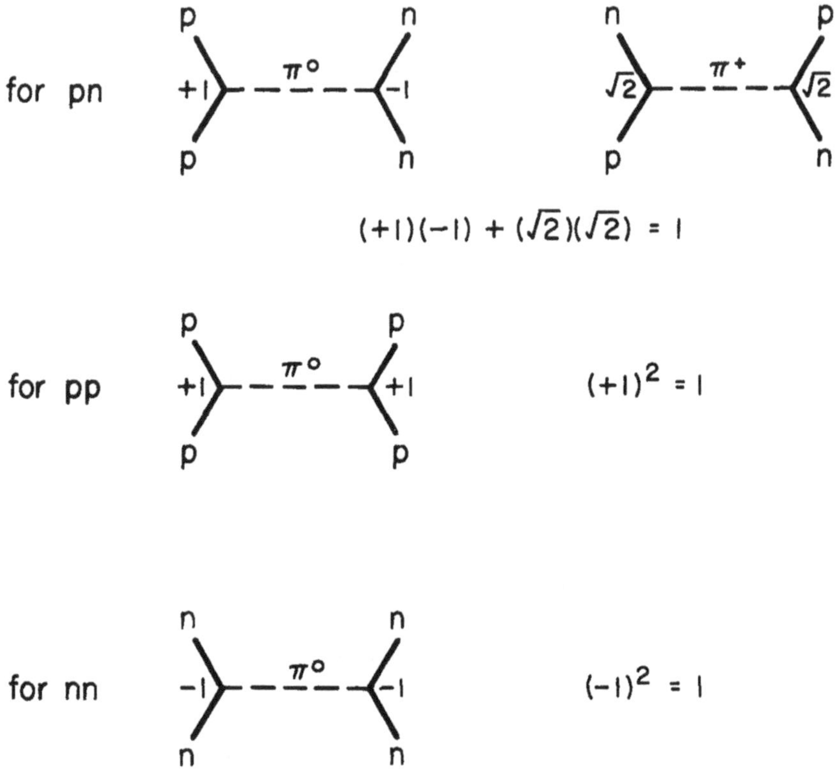

Figure 9.3. Charge independence of nuclear forces in the lowest order

separate the $T = \frac{3}{2}$ contributions from the $T = \frac{1}{2}$ contributions when the scattering matrix is given in invariant form in isospin space, i.e.

$$M_{\beta\alpha} = A\delta_{\beta\alpha} + \tfrac{1}{2}B[\tau_\beta, \tau_\alpha] \qquad (9.33)$$

A particular advantage of writing the scattering amplitude in this form is that the π-N scattering amplitude must satisfy the so-called crossing relations of Gell-Mann and Goldberger, the invariance under the substitutions

$$\vec{q}_i \rightleftarrows -\vec{q}_f, \qquad \omega_i \rightleftarrows -\omega_f, \qquad \alpha \rightleftarrows \beta \tag{9.34}$$

which we shall check directly for the lowest order calculations in the next section. The existence of this symmetry in all orders in perturbation theory can be most clearly seen in terms of Feynman diagrams. For any given diagram Fig. 9.4(a) there exists another diagram (b) which is obtained

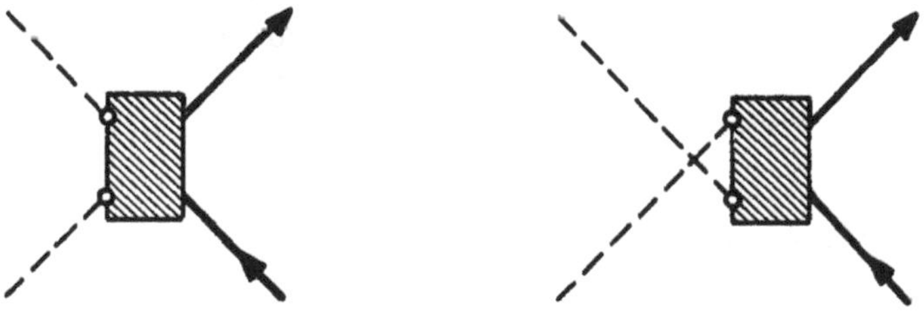

Figure 9.4. Crossing symmetry in π-N scattering

from the first by just interchanging the incoming and outgoing meson lines. The black box is left unchanged. Note also that we do not change the nucleon lines. Now the A term in (9.33) is symmetric between α and β whereas the B term is antisymmetric. Hence

$$\begin{aligned} A(\vec{q}_i, \vec{q}_f; \omega_i, \omega_f) &= A(-\vec{q}_f, -\vec{q}_i; -\omega_f, -\omega_i) \\ B(\vec{q}_i, \vec{q}_f; \omega_i\, \omega_f) &= -B(-\vec{q}_f, -\vec{q}_i; -\omega_f, -\omega_i) \end{aligned} \tag{9.35}$$

We should like to emphasize that neither $f_{3/2}$ nor $f_{1/2}$ exhibits simple behavior of evennesss or oddness under $\vec{q}_i \rightleftarrows -\vec{q}_f$, $\omega_i \rightleftarrows -\omega_f$, since

$$\begin{aligned} f_{1/2} &= A + 2B \\ f_{3/2} &= A - B \end{aligned} \tag{9.36}$$

In practice, it is useful to express A and B as linear combinations of $f(\pi^-p \to \pi^-p)$ and $f(\pi^+p \to \pi^+p)$ as follows

$$\begin{aligned} A &= \tfrac{1}{2}[f(\pi^-p \to \pi^-p) + f(\pi^+p \to \pi^+p)] \\ B &= \tfrac{1}{2}[f(\pi^-p \to \pi^-p) - f(\pi^+p \to \pi^+p)] \end{aligned} \tag{9.37}$$

In applying crossing symmetry, we must first understand what is meant by the scattering amplitude with a negative value of ω. It is apparent that crossing symmetry would be an empty statement unless it is supplemented by a knowledge on the analytic behavior of the scattering amplitude not just for positive real values of ω with $\omega > \mu_\pi$ but also for other values of ω in the complex ω plane. Fortunately from general postulates of field theory such as microscopic causality it is possible to prove that the real part of the scattering amplitude with momentum transfer fixed can be expressed as an integral over the imaginary part of the scattering amplitude with respect to real, but both positive and negative values of ω. Relations of this kind are called dispersion relations (see, e.g., Chapter 9 of Bogoliubov and Shirkov, 1959). In writing down the dispersion relations for π-N scattering we can use the information obtained from the crossing relations (9.35). Because of (9.36) and (9.37) simple dispersion formulas are satisfied not for $f_{1/2}, f_{3/2}$, nor for $f(\pi^\pm p \to \pi^\pm p)$ but only for the sum and difference of $f(\pi^+ p \to \pi^+ p)$ and $f(\pi^- p \to \pi^- p)$. As an example, we write down the π-N forward dispersion relations of Goldberger, Miyazawa, and Oehme (1955).

$$\tfrac{1}{2}\,\mathrm{Re}[f_-(k) + f_+(k)] - \tfrac{1}{2}\,\mathrm{Re}[f_-(0) + f_+(0)]$$
$$= 2f^2 \frac{k^2}{\omega^2 - (\mu_\pi^2/2M_N)^2}\,\frac{1}{2M_N} + \frac{k^2}{4\pi^2}\int_0^\infty dk'\,\frac{\sigma_-(k') + \sigma_+(k')}{k'^2 - k^2} \tag{9.38a}$$

$$\tfrac{1}{2}\,\mathrm{Re}[f_-(k) - f_+(k)] - \frac{\omega}{\mu_\pi}\,\tfrac{1}{2}\,\mathrm{Re}[f_-(0) - f_+(0)]$$
$$= -2f^2 \frac{k^2}{[\omega^2 - (\mu_\pi^2/2M_N)^2]}\,\frac{\omega}{\mu_\pi^2} + \frac{k^2\omega^2}{4\pi^2}\int_0^\infty \frac{dk'}{\omega'^2}\,\frac{\sigma_-(k') - \sigma_+(k')}{k'^2 - k^2} \tag{9.38b}$$

where k is the wave number in the laboratory system ($k^2 = \omega^2 - \mu_\pi^2$), and $f_\pm$ stands for the forward scattering amplitude for $\pi^\pm p \to \pi^\pm p$. Note that we have used the optical theorem $\sigma_{\text{tot.}} = (4\pi/k)\,\mathrm{Im}f$. The π-N dispersion relations were first tested experimentally by Anderson, Davidon, and Kruse (1955).

9.4. Classical Pion Physics

We are now in a position to discuss various attempts to understand the major features of low energy p wave π-N scattering, photo-pion production and the two-nucleon problem on the basis of the Yukawa concept. However, complete discussions on these topics would require another book; here we must content ourselves by merely sketching some of the results obtained, since the main purpose of the present monograph is to discuss applications of invariance principles which are independent of detailed dynamical considerations.

We start with the static model, which has been rather successful in *some* areas of low energy pion-nucleon physics. In writing down the static Hamiltonian, the following assumptions are made:

(1) The charge-triplet pseudoscalar pion can be emitted or absorbed singly by the charge-doublet nucleon in a charge-independent manner.

(2) The nucleon is an infinitely massive nonrelativistic particle; recoil effects, nucleon-pair effects, etc. are ignored.

The first assumption is the familiar Yukawa concept (Yukawa, 1935). The second is an obvious fiction since μ/M is $\sim\frac{1}{7}$, but it is nevertheless used because of its simplicity. Under these two assumptions the only simple interaction Hamiltonian density we can write down is

$$H_{\text{int.}} = \frac{F}{\mu}\,\vec{\sigma}\cdot\vec{\nabla}\boldsymbol{\pi}\cdot\boldsymbol{\tau}\rho(|\vec{x}|) \tag{9.29}$$

$\rho(|\vec{x}|)$ is the density of the "source," the dimension of which is assumed to be of the order of the nucleon Compton wavelength. The linearity in the pion field is the direct consequence of the assumption that pions can be emitted or absorbed *singly*. Equation (9.29) can be regarded as the nonrelativistic limit of the *ps-ps* or *ps-pv* interaction in the fully relativistic Lagrangian formalism, but it is worth emphasizing that the nonrelativistic limit of any interaction compatible with the assumptions listed above is expected to be of the form (9.29). For instance, Fermi and Yang (1949) considered a theory in which the pion is not an "elementary" particle, but is actually a tightly bound state of a nucleon and an antinucleon in the $^1S_0$, $T = 1$ state. Even in such a theory, the "effective" Hamiltonian useful in low-energy pion physics is expected to be of the form (9.29).

Since the gradient of the pion field operator $\boldsymbol{\pi}$ is nonvanishing at the origin only for p states, the static Hamiltonian (9.29) leads to only p state interactions (unless the de Broglie wavelength of the pion is comparable to the radius of the source, in which case the static model must break down anyway). This is precisely what one expects from simple parity and angular momentum considerations for an infinitely heavy nucleon. (The oddness due to the pseudoscalarity of the pion must be compensated by the odd parity associated with the p-state.) This means that we cannot predict anything about s wave scattering on the basis of the static model except that s wave scattering should be unimportant if the theory is to approximate reality. Empirically the s wave phase shifts *roughly* follow the relations

$$\delta_1 = 0.178\eta = 10.2^\circ\eta$$
$$\delta_3 = -0.087\eta = -5.0^\circ\eta$$

up to about 150 Mev. ($\eta = 1.4$) where $\eta = p_\pi/\mu_\pi$. It may be mentioned that naïve perturbation calculations based on the relativistic *ps-ps* theory predict s wave phase shifts that are ten times as large in magnitude with practically no isospin dependence.

For calculational purposes, it is more convenient to look at (9.29) in momentum space. By Fourier-transforming (9.29) we obtain

$$H_{\text{int.}} = i\frac{F}{\mu}\, v(q)\tau_\alpha \frac{\vec{\sigma}\cdot\vec{q}}{\sqrt{2\omega_q}}\, a_{\alpha\vec{q}} + \text{H.C.} \tag{9.39}$$

where

$$v(q) = \int \exp\,(-i\vec{q}\cdot\vec{x})\rho(|\vec{x}|)\, d^3x = \begin{cases} \approx 1 \text{ for } q < q_{\max} \\ \approx 0 \text{ for } q > q_{\max} \end{cases}$$

and $a_{\alpha\vec{q}}$ is the annihilation operator for the pion. If $\rho(|\vec{x}|)$ were δ function-like, $v(q)$ would be a constant independent of q, and the cut-off momentum $q_{\max}$ would be infinite; calculations based on the static Hamiltonian with a point source can be shown to lead to divergent observables (except in lowest-order calculations). In the static theory a single parameter $q_{\max}$ conveniently characterizes various complicated high-energy effects due to nucleon recoil, strange particles, nucleon pairs, etc.

Let us be simple-minded and calculate π-N scattering in the lowest order Born approximation. Using standard techniques of the second-order perturbation theory we obtain

$$\begin{aligned} H_{fi} &= \frac{F^2v^2(p)}{\mu^2}\left(\frac{1}{2\omega_i\omega_f}\right)^{1/2} \frac{\tau_f\vec{\sigma}\cdot\vec{p}_f\tau_i\vec{\sigma}\cdot\vec{p}_i}{(M+\omega_i)-M} + \frac{\tau_i\vec{\sigma}\cdot\vec{p}_i\tau_f\vec{\sigma}\cdot\vec{p}_f}{(M+\omega_i)-(M+\omega_i+\omega_f)} \\ &= \frac{F^2v^2(p)}{\mu^2 2\omega^2}[\tau_f\vec{\sigma}\cdot\vec{p}_f,\ \tau_i\vec{\sigma}\cdot\vec{p}_i] \end{aligned} \tag{9.40}$$

It is evident that the crossing relation discussed in the previous section is satisfied in this approximation. With the help of the projection operators (9.32), we have

$$\begin{aligned} [\tau_f\vec{\sigma}\cdot\vec{p}_f,\ \tau_i\vec{\sigma}\cdot\vec{p}_i] = -p^2[&(-\Lambda_{T=\frac{1}{2}} + 2\Lambda_{T=\frac{3}{2}})(-\Lambda_{J=\frac{1}{2}} + 2\Lambda_{J=\frac{3}{2}}) \\ &-9\Lambda_{T=\frac{1}{2}}\Lambda_{J=\frac{1}{2}}] \end{aligned} \tag{9.41}$$

This means that the expectation values of $[\tau_f\vec{\sigma}\cdot\vec{p}_f,\ \tau_i\vec{\sigma}\cdot\vec{p}_i]$ are $-4p^2$ in the $T = \frac{3}{2}$, $J = \frac{3}{2}$ state, $+2p^2$ in the $T = \frac{1}{2}$, $J = \frac{3}{2}$ state, etc. By taking account of the various normalization and kinematical factors, we can finally express the phase shifts in given T, J states as follows:

$$\begin{aligned} \delta_\alpha &= \lambda_\alpha \frac{p^3v^2(p)}{\omega} \\ \lambda_\alpha &= \frac{f^2}{3}\begin{cases} -8 \text{ for } T = \frac{1}{2}, J = \frac{1}{2} \\ -2 \text{ for both } T = \frac{1}{2}, J = \frac{3}{2} \text{ and } T = \frac{3}{2}, J = \frac{1}{2} \\ \ \ 4 \text{ for } T = \frac{3}{2}, J = \frac{3}{2} \end{cases} \\ f^2 &= \frac{F^2}{4\pi} \end{aligned} \tag{9.42}$$

where the approximation $\exp(i\delta)\sin\delta = \delta$ has been made. Recall that $H_{fi} < 0$ implies an attractive interaction which, if the Born approximation is valid, means that the phase shift δ is positive.

Now, do these numbers represent reality in any sense? As we have seen, empirically the dominant p wave interaction is in the $T = \frac{3}{2}$, $J = \frac{3}{2}$ state (3-3 state). According to the Born approximation, the scattering is most pronounced in the $T = \frac{1}{2}$, $J = \frac{1}{2}$ state, contrary to observation. Moreover, if there is anything like a resonance (i.e., δ going through 90°), the lowest-order Born approximation result (which implies that the scattering amplitude is real) has no connection with reality.

However, there is one rather important feature of the Born-type results worth examining. According to (9.40) and (9.41) the 3-3 state is the only state that is attractive. It is generally the case that a state that is attractive (repulsive) according to Born-type calculations becomes more attractive (less repulsive) in more exact treatments. So we may hope that the 3-3 state which is the only state with a positive phase shift gets enhanced so that δ_{33} eventually passes through 90°.

Detailed considerations by various workers—in particular by Chew and Low (1956a)—have shown that such a resonant behavior is indeed expected for the 3-3 state on the basis of the static theory with cut-off. They have derived an effective range formula of the form

$$\frac{p^3\cot\delta_\alpha}{\mu^2\omega^*} = \frac{1}{\lambda_\alpha}\left(1 - \frac{\omega^*}{\omega_\alpha}\right) \tag{9.43}$$

$$\omega^* = \sqrt{p^2 + \mu^2} + \frac{p^2}{2M}$$

$$\omega_\alpha = \omega_r\begin{cases} -\frac{5}{4} \text{ for } T = \frac{1}{2}, J = \frac{1}{2} \\ -\frac{5}{4} \text{ for } T = \frac{1}{2}, J = \frac{3}{2} \text{ and } T = \frac{3}{2}, J = \frac{1}{2} \\ 1 \text{ for } T = \frac{3}{2}, J = \frac{3}{2} \end{cases}$$

$$\omega_r = \frac{\pi\mu^2}{4f^2\omega_{\max}}$$

A resonance is expected if the right-hand side of (9.43) vanishes. This is impossible in all but the 3-3 state; for the 3-3 state the resonance does occur at $\omega_r < \omega_{\max}$ provided that the coupling constant is sufficiently large. If $p^3\cot\delta_{33}/\omega^*$ is plotted as a function of ω^*, one indeed finds a roughly straight line up to $E_{\text{kin}}^{(\text{lab})} \approx 250$ Mev. The intercept at $\omega^* = 0$ (using the 3-3 amplitudes corresponding to the energy intervals 10–20, 20–30, 30–40 Mev.) gives

$$f^2 = 0.081 \pm 0.007$$

as shown in Fig. 9.5, and the empirical value $\omega_r = 2.1\mu$ implies $\omega_{\max} \approx 6\mu$.

No successful tests for the effective range formula have been made for

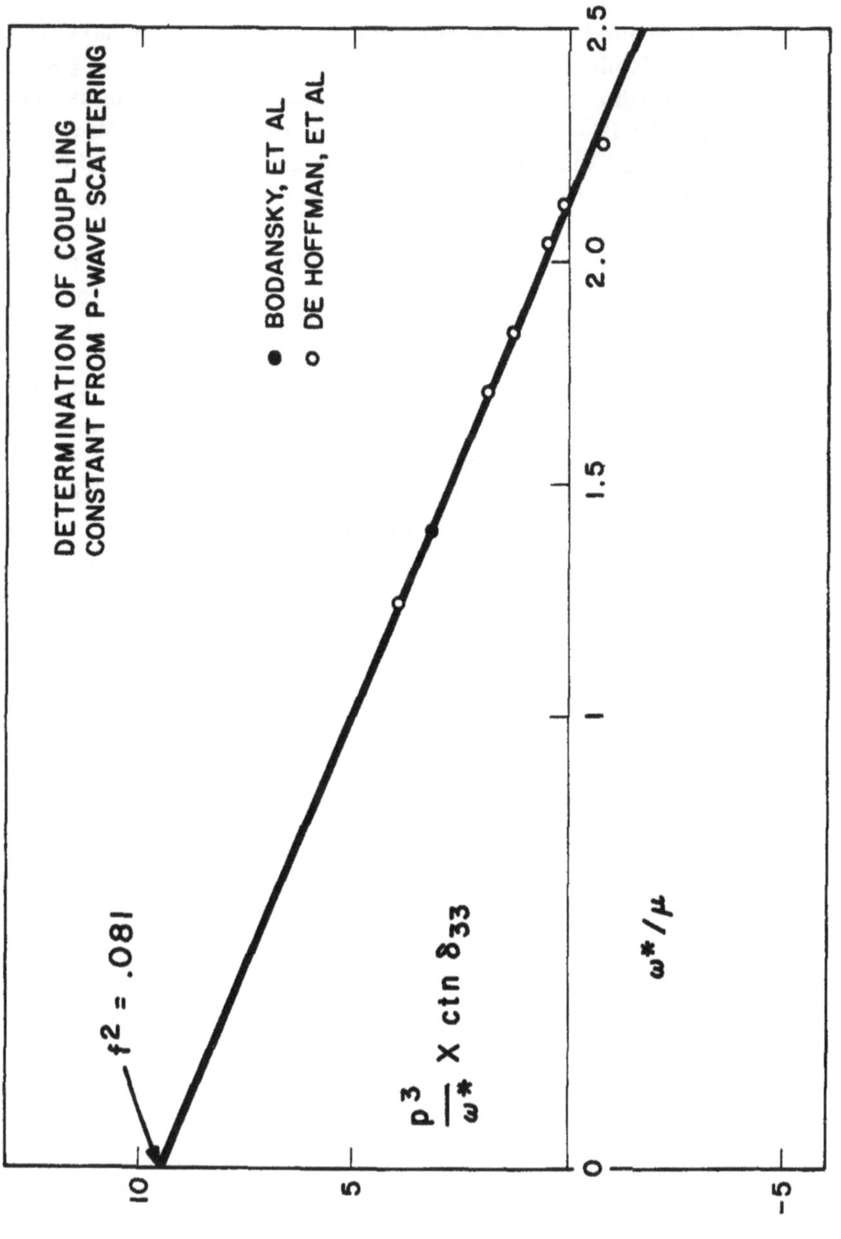

Figure 9.5. Chew-Low method for determining the coupling constant

the so-called "small" p wave phase shifts, which are at most of the order 10^0 in magnitude. It is expected that for these small phase shifts, various corrections that have to be made, for instance, corrections due to pion-pion interactions are comparable to the quantities predicted by the effective range formula (Bowcock, Cottingham, and Lurié, 1960).

For further reading on the static theory applied to p wave π-N scattering, the reader may consult Wick (1955), Edwards and Matthews (1957) and Weisskopf (1959), as well as Chew and Low (1956a).

Note added in proof. Detailed discussions of the static theory can also be found in two new textbooks on field theory: Schweber (1960) and Henley and Thirring (1962).

One may ask why the simple, nonlocal static model works so well. Some light is thrown on this question if we compare the predictions of the static theory with those of the dispersion relations which are deduced from a fully relativistic local field theory. Once we assume that p states are dominant, dispersion theory essentially reproduces the Chew-Low integral equation from which the effective range formula was obtained. Detailed discussions along this line are beyond the scope of the present book (cf. Chew *et al.*, 1957a). The coupling constant f^2 can also be determined from the "complete" dispersion relations of the form (9.38) (i.e., not the "truncated" dispersion relations, in which the explicit assumption that the 3-3 resonance state dominates the dispersion integrals is made), and we obtain $f^2 = 0.073 \pm 0.007$.

Another important reason for the success of the static model is that, as far as the dominant p wave interaction in the 3-3 state is concerned, we can neglect pion-pion interactions. Physically this means that the gross feature of low energy p wave scattering can be understood on the basis of the Yukawa concept in which a pseudoscalar pion is emitted or absorbed by the nucleon in p states, rather than on the basis of the so-called $\pi\pi$ mechanism in which the incident pion interacts with pions in the "clouds" surrounding the nucleon core, or the pion and the nucleon exchange a correlated pair of pions. If the pion-pion interactions played dominant roles in low-energy p wave scattering, the predictions of the static theory would be completely erroneous. It is expected that the pion-pion interactions will play important roles in low energy s wave scattering and also in multiple pion processes (e.g., pion production by pions) as well as in p wave scattering in states other than the 3-3 state. It is not likely, therefore, that the techniques that have been rather successful in understanding the 3-3 resonance will be useful in attacking problems in pion-nucleon physics at energies above the 3-3 resonance.

A discussion on the success story of the static model would be incomplete without mentioning something about photopion production. According to the arguments presented in Chapter 8, the static Hamiltonian (9.29) has to be modified in the presence of the radiation field $\vec{A}$ via the prescription $\vec{\nabla}\pi_{\text{ch.}} \to (\vec{\nabla} - ieA)\pi_{\text{ch.}}$. This modification results in an additional term in the interaction Hamiltonian density of the form

$$H_{\text{int.}} = \sqrt{2}e\frac{F}{\mu_\pi}\vec{\sigma}\cdot\vec{A}(\tau_+\pi_{\text{ch.}} + \tau_-\pi_{\text{ch.}}^\dagger)\rho(|\vec{x}|)$$
$$= e\frac{F}{\mu_\pi}\vec{\sigma}\cdot\vec{A}(\tau_1\pi_1 - \tau_2\pi_2)\rho(|\vec{x}|) \tag{9.44}$$

where we have ignored a term needed to preserve gauge invariance within the "source." Note that this time there is no gradient operator acting on $\pi_{\text{ch.}}$ so that the s state wave function rather than the p state wave function is large at the origin. So we have a natural mechanism for the ejection of an s state charged pion in γ-p or γ-n collisions. Since we know that the s state π-N interaction is weak, the so-called final state interactions may be ignored to a first approximation. This means that (9.44) in momentum space is essentially the matrix element for the production of a π^+ or π^- near threshold. The matrix element obtained in this manner is indeed of the form $i\vec{\sigma}\cdot\vec{\varepsilon}$ characteristic of the $E1 \to S_{1/2}$ transition with coefficient proportional to ef.

The above considerations immediately offer another possibility of obtaining f^2, which was first pointed out by Kroll and Ruderman (1954). The physical significance of the procedure may be understood in analogy with Thomson scattering. In the Thomson case, if we perform a scattering of a very "soft" photon by an electrically charged particle, we can determine the constant e^2 that characterizes the coupling between the electromagnetic field and the charged particle of mass m via the relation $\sigma = \frac{8\pi}{3}(e^2/4\pi m)^2$. Similarly if we had a zero energy photon creating a zero *total* energy pion, we could measure the product ef. Because of the finite rest mass of the pion, the zero energy pion is not available in reality, but since μ/M is small, the procedure works. It turns out that the μ/M correction factor to the static-theoretic method of determining f^2 is calculable in a fully relativistic theory (in the absence of π-π interactions), and is roughly equal in magnitude but opposite in sign for the two reactions

$$\gamma + n \to p + \pi^- \tag{9.45a}$$

$$\gamma + p \to n + \pi^+ \tag{9.45b}$$

(Chew *et al.*, 1957b). The coupling constant f^2 has been determined from the average value of the reaction rates for the two processes (9.45a) and (9.45b). Using the most reliable data available to date one obtains

$$f^2 = 0.081 \pm 0.007$$

The reader may wonder as to how the reaction rate for (9.45a) is obtained when there are no free neutron targets available. The rate for π^- photoproduction is calculable using detailed balancing if we know the radiative capture rate for the inverse reaction $\pi^- + p \rightarrow n + \gamma$, which in turn can be calculated from the Panofsky ratio

$$R(\pi^- + p \rightarrow n + \pi^0)/R(\pi^- + p \rightarrow n + \gamma)$$

at rest and the charge exchange scattering length

$$\lim_{p\to 0}\,[\exp(i\delta_3)\sin\delta_3 - \exp(i\delta_1)\sin\delta_1]/p = (\delta_3 - \delta_1)/p = (a_3 - a_1)$$

(which gives the *absolute* rate for $\pi^- + p \rightarrow n + \pi^0$). Of course, a more direct method to obtain the negative-to-positive photoproduction ratio $\sigma(\gamma + n \rightarrow \pi^- + p)/\sigma(\gamma + p \rightarrow \pi^+ + n)$ is to deduce it from the measured negative-to-positive photoproduction ratio in deuterium, i.e., $\sigma(\gamma + d \rightarrow \pi^- + 2p)/\sigma(\gamma + d \rightarrow \pi^+ + 2n)$, using the impulse approximation method or, more recently, the Chew-Low extrapolation method (Chew and Low, 1959). A great deal of experimental and theoretical work has centered around this interesting topic in "classical" pion physics.

It is apparent from (9.44) that no analogous method for direct s state photoproduction exists for neutral pions within the framework of the static model. This is compatible with the intuitive picture that the electric dipole associated with the virtual process $p \rightarrow p + \pi^0$ is $\mu/(\sqrt{2}M)$ times that associated with $p \rightarrow n + \pi^+$. Experimentally the s state production of π^0 is indeed negligible. So the major contribution arises from the 3-3 resonance. A quantitative relation between p wave scattering in the 3-3 state and photoproduction in the 3-3 state has been derived by Chew and Low (1956b), who showed that the matrix element for π^0 photoproduction in γ-p collisions can be obtained from the matrix element for charge elastic π^0-p scattering by replacing $\vec{p}_{in}$ in the scattering case by $\vec{k} \times \vec{\varepsilon}$ in the photoproduction case, except for a calculable constant factor. Their static theory predicts

$$\begin{aligned}\sigma_{\text{tot.}}(\gamma p \rightarrow p\pi^0) &= \frac{k}{q}\frac{e^2}{f^2}\left(\frac{\mu_p + 1 - \mu_n}{4M}\right)^2 \sigma_{\text{tot.}}(\pi^0 p \rightarrow \pi^0 p)\\ &= \frac{2\pi}{q}\left(\frac{e^2}{f^2 M^2 \mu^2}\right)(\mu_p + 1 - \mu_n)^2 \frac{k}{q^3}\sin^2\delta_{33} \qquad (9.46)\end{aligned}$$

(where μ_p and μ_n are the *anomalous* static moments of the nucleon in units of $e/2M$, i.e., $\mu_p = 1.79$, $\mu_n = -1.91$), which is verified to a rather impressive degree of accuracy up to the resonance (Koester and Mills, 1957). One may then compute the scattering phase shift δ_{33} from photoproduction experiments (if one knows f^2), or knowing δ_{33} from scattering experiments, one may determine f^2 in an entirely independent manner.

A third way to obtain f^2 from photoproduction experiments is to look at the manifestations of the

$$ief\,\vec{\sigma}\cdot(\vec{q} - k)\vec{q}\cdot\vec{\varepsilon}[-(k - \omega)^2 + (\vec{k} - \vec{q})^2 + \mu^2]^{-1}$$

term corresponding to Fig. 9.6. This term is analogous to the photoelectric ejection of an electron in an atom, and contains the same kind of retardation factor $(1 - \beta \cos \theta)^{-1}$, $\beta = v_\pi/c$. Because of this term the production amplitude at some fixed energy regarded as a function of $\cos \theta$ has a pole at a nonphysical angle $\cos \theta = \beta^{-1} > 1$, the residue of which is determined by e^2f^2 (Taylor, Moravcsik, and Uretsky, 1959). The coupling constant determined in this manner has a larger error ($f^2 = 0.09 \pm 15\%$), but in substantial agreement with other values.

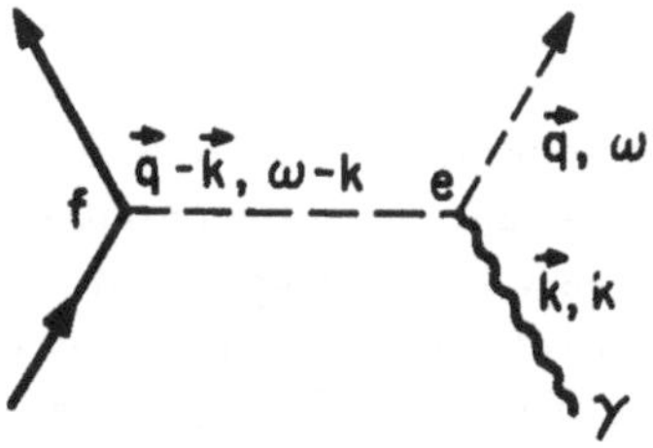

Figure 9.6. One pion-exchange mechanism in photoproduction

We now turn our attention to the oldest and most widely discussed topic in strong interaction physics; namely, the two-nucleon problem, for which meson theory was invented more than twenty-five years ago. As is well known, the original hypothesis of Yukawa is that the forces between two nucleons arise from exchanges of quanta, now known as pions. Because of the uncertainty principle, the range of nuclear forces due to the exchange of a single pion between two nucleons is expected to be of the order of $1/\mu = 1.41 \times 10^{-13}$ cm. and, in general, if n pions are exchanged, the corresponding range is of the order of $1/n\mu$ or smaller. This is rather encouraging. We may at least understand the *tail end* of the two-nucleon potential for distances large compared to $1/\mu$, say, $\gtrsim 2/\mu$ just by estimating the one-pion-exchange (OPE) contribution. In other words, the longest-range part of the two-nucleon potential is essentially the potential expected from the Born approximation.

Again we use the static Hamiltonian (9.29), and compute the OPE potential in a straightforward manner. We obtain (for $v(q) = 1$)

$$\begin{aligned} V_{\text{OPE}} &= -\frac{F^2}{\mu^2}\boldsymbol{\tau}_1\cdot\boldsymbol{\tau}_2\int \vec{\sigma}_1\cdot\vec{q}\vec{\sigma}_2\cdot\vec{q}\,\frac{\exp(i\vec{q}\cdot\vec{x})}{q^2+\mu^2}\,d^3q \\ &= \frac{F^2}{4\pi\mu^2}\boldsymbol{\tau}_1\cdot\boldsymbol{\tau}_2(\vec{\sigma}_1\cdot\vec{\nabla})(\vec{\sigma}_2\cdot\vec{\nabla})\,\frac{\exp(-\mu r)}{r} \qquad (9.47)\\ &= f^2\,\frac{\boldsymbol{\tau}_1\cdot\boldsymbol{\tau}_2}{3}\left\{\vec{\sigma}_1\cdot\vec{\sigma}_2 + S_{12}\left(\frac{3}{\mu^2r^2}+\frac{3}{\mu r}+1\right)\frac{\exp(-\mu r)}{r}\right\} \\ & \qquad (r = |\vec{x}|) \end{aligned}$$

(There is an additional δ function term which gives rise to a zero-range repulsion in even L states, but has been omitted here since the *OPE* potential approximates reality only for large ($\gtrsim 2/\mu_\pi$) distances.) The operator S_{12} is the well-known tensor operator

$$S_{12} = \frac{3(\vec{\sigma}_1 \cdot \vec{x})(\vec{\sigma}_2 \cdot \vec{x})}{r^2} - \vec{\sigma}_1 \cdot \vec{\sigma}_2 \tag{9.48}$$

Charge *independence* requires that the force due to the exchange of a $T = 1$ system (π in this case) is necessarily isospin *dependent*. $\boldsymbol{\tau}_1 \cdot \boldsymbol{\tau}_2$ is positive (negative) when the two isospins of the nucleons are parallel (antiparallel).

$$\begin{aligned} \boldsymbol{\tau}_1 \cdot \boldsymbol{\tau}_2 &= 1 \qquad \text{For } T = 1 \\ \boldsymbol{\tau}_1 \cdot \boldsymbol{\tau}_2 &= -3 \qquad \text{For } T = 0 \end{aligned} \tag{9.49}$$

Similarly for $\vec{\sigma}_1 \cdot \vec{\sigma}_2$. Now, as we have previously shown in Section 2, the exclusion principle requires

$T = 1$ (*pp*, *nn*, *pn*): singlet even, triplet odd
$T = 0$ (*pn*) : triplet even, singlet odd

where "even" and "odd" refer to L even and L odd, and "singlet" and "triplet" refer to $S = 0$ and $S = 1$.

Let us consider the S states. For both ^3S and ^1S we have $(\vec{\sigma}_1 \cdot \vec{\sigma}_2)(\boldsymbol{\tau}_1 \cdot \boldsymbol{\tau}_2) = -3$. This means that the central part of the OPE potential is the same and attractive in both S states. The tensor force (S_{12} term) has no effect on ^1S, but makes the ^3S state more attractive. Note also that the expectation value of S_{12} is largest when $\vec{x}$ is along the symmetry axis determined by the triplet spin direction so that $\boldsymbol{\tau}_1 \cdot \boldsymbol{\tau}_2 S_{12}$ for the deuteron is negative when $\vec{x}$ and the spin direction are parallel. So the nucleons in the deuterium nucleus have greater probabilities of spending their time in stretched out, "cigar-like" (rather than "orange-like") configurations. This accounts for the observed fact that the quadrupole moment of the deuteron is positive. Thus the OPE potential can qualitatively account for both the existence and the sign of the quadrupole moment of the deuteron. In fact the so-called symmetric pseudoscalar theory (a theory based on the charge-independent Yukawa coupling of the pseudoscalar, charge-triplet meson field to the nucleon) was preferred by many theoreticians long before the discovery of the pion.

Successful though these qualitative predictions are, it is impossible to calculate quantities such as the *s* wave scattering lengths, the effective range parameters, the deuteron binding energy and the deuteron quadrupole moment from the OPE potential alone. Some assumptions have to be made on the nature of the two-nucleon potential in the "inner" region. This is unfortunate, particularly because various workers have not agreed on even the two-pion-exchange contributions to the nuclear forces. All that can be said at this moment is that the various low energy parameters are consistent with the OPE potential with $f^2 = 0.08$, provided that we make some reasonable assumptions about the nature of the nuclear forces at short distances. (For earlier attempts along this line see, e.g. a series of review papers edited by Taketani (1956).)

There have been quantitative tests of the OPE contribution at *high* energies. Since many partial waves are likely to contribute at high energies, one may expect that analyses may become hopelessly complicated. But if the tail end of the N-N interactions is described by the OPE contribution, just from the range and centrifugal barrier arguments we see that the phase shifts with large L values must be correctly given by the OPE contribution, which is fortunately calculable in a fully covariant manner without recourse to the static potential language. Recently Moravcsik and collaborators have demonstrated that the inclusion of *all* higher angular momentum contributions (from G waves on) expected from the OPE matrix element with $f^2 \approx 0.06$ significantly improves the goodness-of-fit parameters in the previous phase shift analysis (which goes up to H waves), and eliminates many of the ambiguities that existed in the earlier purely phenomenological approaches (Cziffra *et al.*, 1959; MacGregor, Moravcsik, and Stapp, 1959). The correctness of the OPE tail has also been demonstrated by Breit *et al.* (1960b) using the "potential" approach.

As mentioned earlier, the two-pion-exchange contributions are already controversial. One may use the static Hamiltonian with f^2 and $q_{\max}$ determined from π-N scattering to construct a fourth-order potential in a straightforward manner (neglecting rescattering effects, nucleon pair effects, pion-pion correlation effects, etc.). This was done by Gartenhaus (1955), who succeeded in fitting the various low-energy parameters rather well. But his potential failed miserably in 100 Mev regions when attempts were made to fit quantities measured in double and triple scattering experiments.

It is now well established that a spin-orbit potential of some form is essential at least in the $T = 1$ state, as emphasized by Wolfenstein (1955), by Gammel and Thaler (1957) and by Signell and Marshak (1957). The need for a fairly short-ranged ($\sim 1/3\mu$) spin-orbit force is particularly evident from the pp phase shift analysis at 310 Mev., where the ordering of the triplet P phase shifts, $\delta(^3P_0) < 0$, $\delta(^3P_1) < 0$, $\delta(^3P_2) > 0$ differ remarkably from the ordering we expect from a dominant tensor force $\delta(^3P_1) < 0$, $0 < \delta(^3P_2) < \delta(^3P_0)$ (Stapp, Ypsilantis, and Metropolis, 1957). (If there are no bound states, and if the phase shifts are not large, the above argument based on the Born approximation is expected to be correct, at least qualitatively.) Since the static Hamiltonian gives only central plus tensor forces even in higher orders, the inadequacy of the static model applied to high-energy N-N scattering is rather apparent.

Another striking feature of the nature of nuclear forces at short distances is the existence of a short-ranged repulsion at distances $\sim 1/4\mu$, which can be most directly seen from the fact that the $^1S_0$ phase shift changes its sign (positive to negative) at ~ 250 Mev. This is the so-called repulsive core in the potential approach. At one time it was widely believed that this core was due to the δ function singularity of the OPE potential. Recent dispersion theoretic considerations have demonstrated that this "explanation" for the core is incorrect since the alleged effect is necessarily too short-ranged ($\sim 1/M$) (Noyes, 1960a). Moreover, various phenomenological analyses seem to indicate that the repulsive core is needed in all angular momentum and parity states while the δ function singularity of the OPE potential is repulsive or attractive, depending on whether the L value is even or odd.

To sum up, in spite of the tremendous amount of efforts spent on the nuclear force problem, we are still far from being able to understand the nature of the two-nucleon interaction from first principles. The only non-controversial part of the two-nucleon interaction is the OPE contribution which was written down long before the discovery of the pion. Even the two-pion-exchange contributions are moderately controversial so that the nature of the two-nucleon interactions between $0.5/\mu$ and $1.5/\mu$ is not established too well. (However, recently some feeling of optimism has been expressed concerning the usefulness and reliability of Mandelstam's double-dispersion representation in computing the two-pion-exchange contributions.) Regions with distances shorter than $0.5/\mu$ are completely speculative. In reviewing the history of nuclear forces, we cannot but concur with Goldberger's remark on this subject: "Scarcely ever has the world of physics owed so little to so many." (Goldberger, 1960.)

9.5. G Conjugation Invariance; N$\bar{\text{N}}$ Selection Rules, Nucleon Structure

Suppose we know experimentally that strong interactions are invariant under isotopic spin rotations and also under charge conjugation. In quantum mechanics whenever we have two symmetry operations that are "good," we immediately ask whether or not they commute. If they do not commute, we cannot characterize states as simultaneous eigenstates of these operators.

Let us ask whether **T** commutes with the charge conjugation operation C. Take T_3 for instance

$$T_3 = \tfrac{1}{2}(N(p) - N(\bar{p}) - N(n) + N(\bar{n})) \tag{9.50}$$

where it is to be understood that $N(p)$, etc. are number operators. Under charge conjugation

$$T_3 = \tfrac{1}{2}(-N(p) + N(\bar{p}) + N(n) - N(\bar{n})) \tag{9.51}$$

Hence

$$CT_3C^{-1} = -T_3 \tag{9.52}$$

So C and T_3 do not commute; rather they anti-commute.

Consider, however

$$G = C \exp(i\pi T_2) \tag{9.53}$$

i.e., 180° rotation in isospin space about the second axis followed by the charge conjugation operation. This operation was first considered by Michel (1953) (see also Amati and Vitale, 1955; Goebel, 1956; Lee and Yang, 1956a). Under $\exp(i\pi T_2)$ the nucleon isospinor transforms as follows

$$\begin{pmatrix} p \\ n \end{pmatrix} \to (\cos \pi/2 + i\tau_2 \sin \pi/2)\begin{pmatrix} p \\ n \end{pmatrix} = \begin{pmatrix} n \\ -p \end{pmatrix}$$

So under G

$$\begin{pmatrix} p \\ n \end{pmatrix} \xrightarrow{G} \begin{pmatrix} n^c \\ -p^c \end{pmatrix} \tag{9.54}$$

We may ask how **T** transforms under G

$$T_1 = \tfrac{1}{2}\int N^\dagger \begin{pmatrix} 0 & I \\ I & 0 \end{pmatrix} N d^3x = \tfrac{1}{2}\int (p^\dagger n + n^\dagger p)\, d^3x$$

$$\xrightarrow{G} \tfrac{1}{2}\int (-n^{c\dagger}p^c - p^{c\dagger}n^c)\, d^3x$$

$$T_2 = \tfrac{1}{2}\int N^\dagger \begin{pmatrix} 0 & -i \\ i & 0 \end{pmatrix} N d^3x = -\frac{i}{2}\int (p^\dagger n - n^\dagger p)\, d^3x$$

$$\xrightarrow{G} -\frac{i}{2}\int (-n^{c\dagger}p^c + p^{c\dagger}n^c)\, f^3x$$

$$T_3 = \tfrac{1}{2}\int (p^\dagger p - n^\dagger n)\, d^3x \xrightarrow{G} \tfrac{1}{2}\int (n^{c\dagger}n^c - p^{c\dagger}p^c)\, d^3x$$

Noting

$$:\int p^{c\dagger}p^c d^3x: \; = \; -:\int p^\dagger p d^3x:$$

we have

$$T_3 \xrightarrow{G} T_3$$

For T_1 and T_2 we note that the phase factors associated with C can be chosen in such a way that p and n transform in the same way under C, i.e., $\eta_C(p) = \eta_C(n)$. Then

$$T_1 \xrightarrow{G} T_1, \qquad T_2 \xrightarrow{G} T_2$$

where we have used

$$:\int p^{c\dagger}n^c d^3x: \; = \; -:\int n^\dagger p d^3x: \text{ etc.}$$

Thus we have accomplished our goal

$$[\mathbf{T}, G] = \left[\int \bar{N}\gamma_4 \frac{\boldsymbol{\tau}}{2} N d^3x, G\right] = 0 \tag{9.55}$$

This means that we have found a symmetry operation that commutes with isospin rotations.

For the isospin-current four vector we have

$$i\bar{N}\gamma_\mu \boldsymbol{\tau} N \xrightarrow{G} i\bar{N}\gamma_\mu \boldsymbol{\tau} N.$$

In general

$$\bar{N}\Omega\boldsymbol{\tau}N \xrightarrow{G} -\omega_C \bar{N}\Omega\boldsymbol{\tau}N \tag{9.56}$$

in contrast to

$$\bar{N}\Omega N \xrightarrow{C} \omega_C \bar{N}\Omega N$$

where

$$\begin{aligned} \omega_C &= 1 \text{ for } S, A, P \\ \omega_C &= -1 \text{ for } V, T \end{aligned}$$

as the reader may readily verify.

We may also obtain the commutation relation between the baryon number and G in a straightforward manner.

$$\begin{aligned} B = \int N^\dagger N d^3x &= \int (p^\dagger p + n^\dagger n)\, d^3x \\ &\xrightarrow{G} \int (n^{c\dagger} n^c + p^{c\dagger} p^c)\, d^3x \\ &= - \int (n^\dagger n + p^\dagger p)\, d^3x \end{aligned}$$

$$GBG^{-1} = -B \tag{9.57}$$

So the baryon number does *not* commute with G. This is not surprising since G involves the charge conjugation operation which interchanges the nucleon state and the corresponding anti-nucleon state.

It is worth noting that

$$\begin{pmatrix} p \\ n \end{pmatrix} \xrightarrow{G} \begin{pmatrix} n^c \\ -p^c \end{pmatrix} \xrightarrow{G} \begin{pmatrix} -p \\ -n \end{pmatrix}$$

or

$$N \xrightarrow{G^2} -N \tag{9.58}$$

A many-nucleon state with the total baryon number B can be obtained by applying successively the creation operator $a^\dagger$ (which has the same symmetry property as the field operator N) to the vacuum state $|0\rangle$ B times. Since

$$G^2 a^\dagger(B) G^{-2} G^2 a^\dagger(B-1) \ldots G^2 a^\dagger(1) G^{-2} G^2 = (-1)^B a^\dagger(B) \ldots a^\dagger(1)$$

we have

$$G^2 = (-1)^B \tag{9.59}$$

where B is the total baryon number. Actually this relation is not completely general if there are strange particles. Instead

$$G^2 = (-1)^{B+S}$$

where S stands for strangeness which will be defined in the next chapter.

Because of

$$[G, \mathbf{T}] = 0, \qquad \{G, B\} = 0$$

the G conjugation operation applied to a nucleon state gives an anti-nucleon state with the same T and T_3

$$G|B, T, T_3\rangle = |-B, T, T_3\rangle \tag{9.60}$$

except for a possible phase factor arising from $\eta_C(N)$. So unless the total baryon number is zero, systems consisting of nucleons and anti-nucleons cannot be eigenstates of G. This is exactly analogous to the charge conjugation case where only systems with zero total electric charge can be eigenstates of C. Eigenvalue of G is often referred to as G conjugation parity or G parity.

Let us now look at the pion field. For the total isospin we have

$$\mathbf{T} = -\int \left(\frac{\partial \boldsymbol{\pi}}{\partial x_0} \times \boldsymbol{\pi}\right) d^3x$$

with the components of $\boldsymbol{\pi}$ Hermitian. We want $\mathbf{T} \to \mathbf{T}$ under G. It may appear that this can be accomplished with either $\boldsymbol{\pi} \xrightarrow{G} \boldsymbol{\pi}$ or $\boldsymbol{\pi} \xrightarrow{G} -\boldsymbol{\pi}$. (Since the components of $\boldsymbol{\pi}$ must be Hermitian, $\boldsymbol{\pi} \to \eta\boldsymbol{\pi}$ with η complex will not do.) We now show that only the second possibility is admissible for the pion field occurring in nature.

Recall first that π_3 corresponds to π^0. But π^0 is even under charge conjugation since $\pi^0 \to 2\gamma$ is allowed. This means that $\pi_3 \to \pi_3$ under C without any phase ambiguity. Under the 180° rotation about the second axis in isospin space we have

$$\begin{pmatrix} \pi_1 \\ \pi_2 \\ \pi_3 \end{pmatrix} \xrightarrow{\exp(i\pi T_2)} \begin{pmatrix} -\pi_1 \\ \pi_2 \\ -\pi_3 \end{pmatrix}$$

So for π_3,

$$\pi_3 \xrightarrow{G} -\pi_3$$

If we are to have

$$\pi_1 \xrightarrow{G} -\pi_1$$

$$\pi_2 \xrightarrow{G} -\pi_2$$

we must require

$$\pi_1 \xrightarrow{C} \pi_1$$

$$\pi_2 \xrightarrow{C} -\pi_2$$

The fact that π_1 and π_2 should behave oppositely under *charge* conjugation is not surprising since

$$\pi_{\text{ch.}} = \frac{1}{\sqrt{2}}(\pi_1 + i\pi_2) \xrightarrow{C} \eta_C \pi_{\text{ch.}}{}^{\dagger} = \eta_C(\pi_1 - i\pi_2)\sqrt{2}$$

Our new point is that the phase of π_1 under C can be chosen to conform to the phase of π_3 under C so that all three components of $\boldsymbol{\pi}$ behave in the same manner under G.

To sum up

$$\boldsymbol{\pi} \xrightarrow{G} -\boldsymbol{\pi} \tag{9.61}$$

The operation G flips all three components of the pion field. The pion field may be regarded as a *polar* vector in isospin space if we interpret G as the inversion operation of all three axes in isospin space. As in ordinary space this inversion operation cannot be brought about by compounding proper rotations. The commutation relation

$$[G, \mathbf{T}] = 0$$

expresses the fact that isospin rotation and isospin inversion commutes.

Since G is the product of charge conjugation and a particular kind of isospin rotation, any interaction that is invariant under both C and T rotations is necessarily invariant under G. So the introduction of G may look redundant. But the whole point in introducing G is that this G operation can be diagonalized simultaneously with T^2 and T_3 whereas C cannot be diagonalized with T^2 and T_3. In fact, working with G makes it easier to discover selection rules which are not at all obvious if we just work with C, T^2 and T_3.

Before we discuss $N\bar{N}$ selection rules that follow from G conjugation invariance, we first write down the isospin functions for antinucleons. This is done in Table 9.2. The $-$ sign for the $T_3 = -\frac{1}{2}$, $B = -1$ state is

TABLE 9.2

Isospin Functions for N and $\bar{N}$

	$B = 1$	$B = -1$
$T_3 = \frac{1}{2}$	$\|p\rangle$	$\|n^c\rangle$
$T_3 = -\frac{1}{2}$	$\|n\rangle$	$-\|p^c\rangle$

essential if the nucleon state and the corresponding antinucleon state transform in a consistent manner under isospin rotations. For instance, under an infinitesimal rotation about the second axis we have

$$|N\rangle \to \left(1 + i\tau_2 \frac{\lambda}{2}\right)|N\rangle$$

which means

$$
|p\rangle \to |p\rangle + \frac{\lambda}{2}|n\rangle
$$
$$
|n\rangle \to |n\rangle - \frac{\lambda}{2}|p\rangle \tag{9.62}
$$

In order that the antinucleon isospinor be transformed in the same way, i.e.

$$
|\bar{N}\rangle \to \left(1 + i\tau_2 \frac{\lambda}{2}\right)|\bar{N}\rangle
$$

we must have

$$
|\bar{N}\rangle = \begin{pmatrix} |n^c\rangle \\ -|p^c\rangle \end{pmatrix} \tag{9.63}
$$

so that

$$
|p^c\rangle \to |p^c\rangle + \frac{\lambda}{2}|n^c\rangle
$$
$$
|n^c\rangle \to |n^c\rangle + \frac{\lambda}{2}(-|p^c\rangle) \tag{9.64}
$$

in agreement with (9.62). Note that $|\bar{N}\rangle$ given by (9.63) is indeed the G conjugate of $|N\rangle$. This is reasonable because under G we do not change T and T_3.

Let us consider a system consisting of one nucleon and one antinucleon. The isospin wave functions look as follows:

$$
\begin{aligned}
T = 1 \begin{cases} T_3 = 1 & |p\rangle|n^c\rangle \\ = 0 & \tfrac{1}{\sqrt{2}}[|p\rangle(-|p^c\rangle) + |n\rangle|n^c\rangle] \\ = -1 & |n\rangle|p^c\rangle \end{cases} \\
T = 0,\ T_3 = 0 \qquad \tfrac{1}{\sqrt{2}}[|p\rangle(-|p^c\rangle) - |n\rangle|n^c\rangle]
\end{aligned} \tag{9.65}
$$

For an actual system we have to consider the position and spin variables. Suppose we have a $T = 1$, $T_3 = 1$ nucleon-antinucleon system with the nucleon variables $\vec{x}_1$, $\vec{\sigma}_1$, and the antinucleon variables $\vec{x}_2$, $\vec{\sigma}_2$. Then

$$
|p, \vec{x}_1, \vec{\sigma}_1\rangle|n^c, \vec{x}_2, \vec{\sigma}_2\rangle \xrightarrow{G} |n^c, \vec{x}_1, \vec{\sigma}_1\rangle(-|p, \vec{x}_2, \vec{\sigma}_2\rangle)
$$

since under G we just change the internal quantum numbers leaving the spatial-spin properties unchanged. But this can be rewritten as

$$
|n^c, \vec{x}_1, \vec{\sigma}_1\rangle(-|p, \vec{x}_2, \sigma_2\rangle) = (-1)^{L+S+1}|p, \vec{x}_1, \sigma_1\rangle|n^c, \vec{x}_2, \vec{\sigma}_2\rangle
$$

(Note that there is a minus sign arising from the interchange of the n^c and the p operator which cancels with the minus sign in front of $-|p\rangle$.) Thus the G-conjugation parity is $(-1)^{L+S+1}$ for $T = 1$, $T_3 = 1$. It is easy to check that this is also true for $T = 1$, with $T_3 = 0, -1$. For $T = 0$, we get $G = +(-1)^{L+S}$. In general we have

$$G = (-1)^{L+S+T} \tag{9.66}$$

for a nucleon-antinucleon system with orbital angular momentum L, spin S, and isospin T.

Another way to obtain (9.66) goes as follows. We first note that for $T_3 = 0$ systems, $\exp(i\pi T_2)$ just gives $(-1)^T$ because the isospin function for a $T_3 = 0$ system transforms like the spherical harmonic Y_{T0} which changes into $(-1)^T\, Y_{T0}$ under a 180° rotation about the second axis. Meanwhile we know from Chapter 5, Section 2 that for $p\bar{p}$ and $n\bar{n}$ systems which must necessarily have $T_3 = 0$, the charge conjugation parity is given by $C = (-1)^{L+S}$. It then follows that for a linear combination of $p\bar{p}$ and $n\bar{n}$ which is an eigenstate of $\mathbf{T}^2$, we must have $G = (-1)^{L+S+T}$. Since the G quantum numbers are independent of T_3 by construction (otherwise G and $\mathbf{T}$ would not commute), this result must hold not only for $T_3 = 0$ systems but also for $T_3 = \pm 1$ systems. Quite generally speaking, this $G = C(-1)^T$ rule for $T_3 = 0$ systems enables us to determine the G quantum number of a multiplet whenever the C quantum number of the neutral $T_3 = 0$ member is known.

As for the pion, a state of one pion is necessarily odd under G. This follows readily since the creation and annihilation operator of pions must have the same symmetry properties as the field operator $\boldsymbol{\pi}$, which is odd under G. More generally for any state with n pions we have

$$G = (-1)^n \tag{9.67}$$

regardless of its spatial configuration. We can combine this with our previous result on the nucleon-antinucleon system with orbital angular momentum L, total spin S, total isospin T to get a selection rule for

$$N + \bar{N} \to n\pi$$

The rule is amazingly simple. States with even $L + S + T$ can decay only into an even number of pions and states with odd $L + S + T$ can decay only into an odd number of pions. For example, consider a $\bar{p}$-n system in the ${}^1S_0$ state. It is necessarily in $T = 1$, since $T_3 = -1$; we have

$$G = (-1)^{0+0+1} = -1$$

so this system cannot decay into 2π, 4π, 6π, etc. The annihilation selection rules are summarized in Table 9.3 (taken from Lee and Yang, 1956a).

Equation (9.67) also implies that any diagram with odd number of external pion lines with no other external lines must vanish. This is analogous to Furry's theorem discussed in Chapter 5. For instance, in pion-pion collisions, the processes $\pi + \pi \to 3\pi$, 5π, etc. are forbidden.

TABLE 9.3

Selection Rules for $\bar{p} + p \rightarrow N\pi$

State	Spin Parity	C	T	G	$2\pi^0$	$\pi^+ - \pi^-$	$3\pi^0$	$\pi^+ + \pi^- + \pi^0$	$4\pi^0$	$\pi^+ + \pi^- + 2\pi^0$	$2\pi^+ + 2\pi^-$	$5\pi^0$	$\pi^+ + \pi^- + 3\pi^0$	$2\pi^+ + 2\pi^- + \pi^0$
$^1S_0$	0−	+	0	+	×	×	−	−				−	−	−
			1	−	×	×				−	−			
$^3S_1$	1−	−	0	−	×	−	×		×	−	−	×		
			−	+	×		×	−	×			×		
$^1P_1$	1+	−	0	−	×	×	×		×	−	−	×		
			1	+	×	×	×	−	×			×	−	−
$^3P_0$	0+	+	0	+			×	×				−	−	−
			1	−	−	−	×	×	−	−	−			
$^3P_1$	1+	+	0	+	×	×	−	−				−	−	−
			1	−	×	×			−	−	−			
$^3P_3$	2+	+	0	+			−	−				−	−	−
			1	−	−	−			−	−	−			

Selection Rules for $\bar{p} + n \rightarrow N\pi$

State	Spin Parity	T	G	$\pi^- + \pi^0$	$2\pi^- + \pi^+$	$\pi^- + 2\pi^0$	$2\pi^- + \pi^+ + \pi^0$	$\pi^- + 3\pi^0$	$3\pi^- + 2\pi^+$	$2\pi^- + \pi^+ + 2\pi^0$	$\pi^- + 4\pi^0$
$^1S_0$	0−	1	−	×			−	−			
$^3S_1$	1−	1	+		−	−			−	−	−
$^1P_2$	1+	1	+	×	−	−			−	−	−
$^3P_0$	0+	1	−	−	×	×	−	−			
$^3P_1$	1+	1	−	×			−	−			
$^3P_2$	2+	1	−	−			−	−			

The × means strictly forbidden, and the − means forbidden so far as the isotopic spin is a good quantum number.

As another application of G conjugation invariance let us consider the electromagnetic structure of the nucleon. Since the physical nucleon has a finite-size charge distribution and an anomalous magnetic moment, the vertex that characterizes the interaction of the photon and the nucleon is not just $iA_\mu \bar{u}_p(p')\gamma_\mu u(p)$ (as would be the case if the proton were a "pure" Dirac particle) but $M = iA_\mu J_\mu$ where

$$J_\mu = \bar{u}_N(p')\{\gamma_\mu F_1^{(N)}((p' - p)^2) - i\sigma_{\mu\nu}(p' - p)_\nu F_2^{(N)}((p' - p)^2)\}u_N(p) \tag{9.68}$$

Note that terms like $\bar{u}_N(p')\gamma_5\gamma_\mu u_N(p)$ and $\bar{u}_N(p')(p' + p)_\mu u_N(p)$ cannot appear if parity conservation and charge conservation $[(p' - p)_\mu J_\mu = 0]$ are to be satisfied. (Note also that we can write $\bar{u}_N(p')(p' - p)_\mu u_N$ as a linear combination of the γ_μ and $\sigma_{\mu\nu}(p - p)_\nu$ terms.) $F_1(0)$ and $F_2(0)$ give the total electric charge and the anomalous magnetic moment respectively. Clearly we expect J_μ to be different for the proton and for the neutron. Hence F_1 and F_2 (which are called the Dirac form factor and the Pauli form factor respectively) depend on the third component of the nucleon isospin. So we write

$$F_{1,2} = F_{1,2}^{(S)} + \tau_3 F_{1,2}^{(V)} \tag{9.69}$$

where S and V refer to isoscalar and isovector.

$$\begin{aligned} F_{1,2}^{(S)} &= \tfrac{1}{2}(F_{1,2}^{(p)} + G_{1,2}^{(n)}) \\ F_{1,2}^{(V)} &= \tfrac{1}{2}(F_{1,2}^{(p)} - F_{1,2}^{(n)}) \end{aligned} \tag{9.70}$$

The form factors have simple physical interpretations. Take the proton form factor $F_1^{(p)}$, for instance. In analogy with electron diffraction scattering from a crystal we have

$$\begin{aligned} F_1^{(p)}(q^2) &= \int \exp\,(i\vec{q}\cdot\vec{x})\rho(r)\,d^3x \\ &= \frac{4\pi}{q}\int_0^\infty \rho(r)\sin\,(qr)\,rdr \\ &= e - \frac{e\langle r^2\rangle q^2}{6} + \frac{e\langle r^4\rangle q^4}{120}\cdots \end{aligned} \tag{9.71}$$

where $\rho(r)$ is the charge density in $\vec{x}$ space. The leading term e is, of course, the total charge. If a very low energy electron ($\lambda\!\!\!^{-} \gg 10^{-13}$ cm.) is scattered by the proton, only the total charge is relevant; so in spite of the finite size of the proton, the situation is indistinguishable from the "point" proton $F_1^{(p)}(q^2) = e$ for all q^2. In electron-proton and electron-deuteron scattering experiments at energies $\gtrsim$400 Mev., considerable deviations from the point charge model have been observed especially at large angles (i.e. for large momentum transfer). In this manner, physicists at Stanford started exploring the finite size of the proton. More quantitatively the scattering cross section for high energy electrons by protons is given by the Rosenbluth (1950) formula

$$\frac{d\sigma}{d\Omega} = \frac{1}{4E_{\text{lab}}} \frac{\cos^2 \theta/2}{\sin^4 \theta/2} \frac{1}{1 + (2E_{\text{lab}}/M_N) \sin^2 \theta/2}$$

$$\times \frac{1}{(4\pi)^2}\left\{F_1^{(N)2} + \frac{q^2}{4M^2}[2(F_1^{(N)} + 2M_N F_2^{(N)})^2 \tan^2 \theta/2 + (2M_N F_2^{(N)})^2]\right\} \quad (9.72)$$

where

$$q^2 = \frac{4E_{\text{lab}}^2 \sin^2 \theta/2}{(1 + 2E_{\text{lab}}/M_N) \sin^2 \theta/2}$$

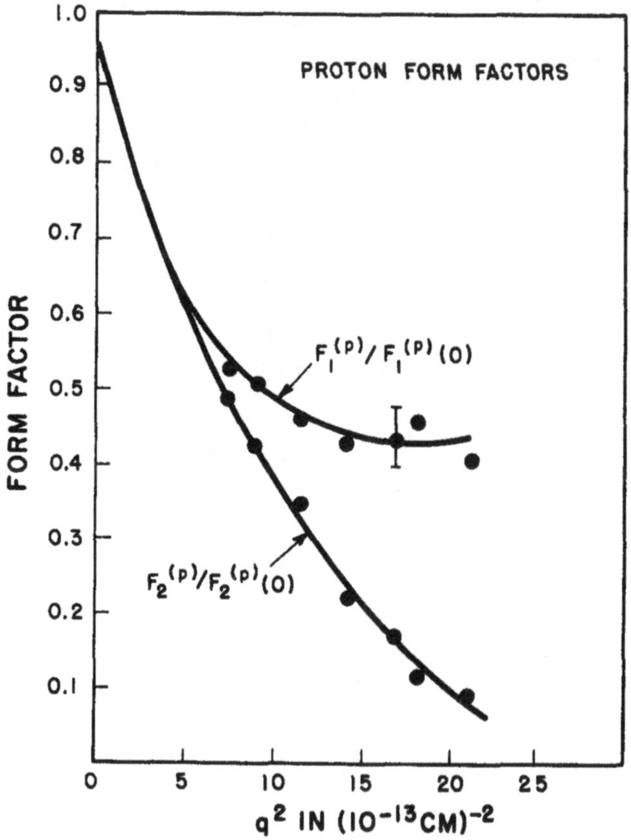

Figure 9.7. Charge and magnetic form factors of the proton

Note that by fixing q^2 but varying θ one can distinguish "charge scattering" from "magnetic moment scattering." There is no theoretical reason to expect $F_1^{(p)}(q^2)/F_1^{(p)}(0) = F_2^{(p)}(q^2)/F_2^{(p)}(0)$; in fact experimentally they do differ for large values of q^2 as shown in Fig. 9.7. (Hofstadter, Bumiller, and

Croissiaux, 1960, see also a review paper by Hofstadter (1957) which summarizes the earlier works).

It has been conjectured (though no rigorous proof has been given except in perturbation theory (Nambu, 1957b)) that the following integral representations exist

$$\begin{aligned}
F_1^{(S)}(q^2) &= \frac{e}{2} - \frac{q^2}{\pi}\int \frac{d\xi^2\sigma_1^{(S)}(\xi^2)}{\xi^2(q^2+\xi^2)} \\
F_1^{(V)}(q^2) &= \frac{e}{2} - \frac{q^2}{\pi}\int \frac{d\xi^2\sigma_1^{(V)}(\xi^2)}{\xi^2(q^2+\xi^2)} \\
F_2^{(S)}(q^2) &= \frac{1}{\pi}\int d\xi^2\,\frac{\sigma_2^{(S)}(\xi^2)}{\xi^2+q^2} \\
F_2^{(V)}(q^2) &= \frac{1}{\pi}\int d\xi^2\,\frac{\sigma_2^{(V)}(\xi^2)}{\xi^2+q^2}
\end{aligned} \tag{9.73}$$

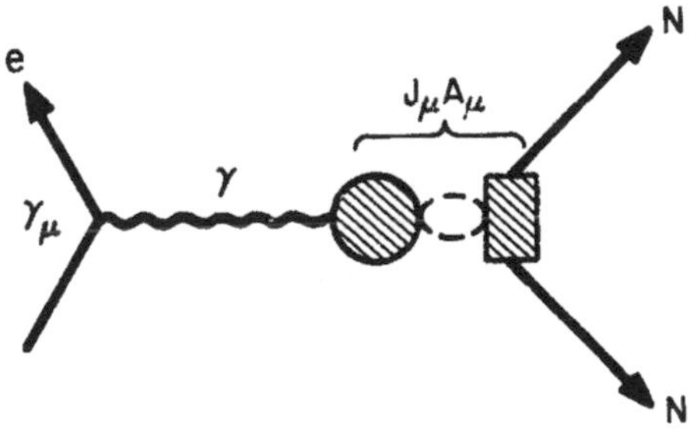

Figure 9.8. *e-N* scattering

The σ's which are real are the absorptive parts of F in the sense of dispersion theory. For a negative value of q^2 which is *not* accessible in *e-p* scattering the relation

$$F_1^{(S)}(q^2) = \frac{e}{2} - \frac{q^2}{\pi}\int d\xi^2\,\frac{\sigma_1^{(S)}(\xi^2)}{\xi^2(q^2+\xi^2-i\varepsilon)} \quad \text{etc.}$$

holds so that

$$\sigma(\xi^2) = ImF(-\xi^2)$$

F develops an imaginary part whenever there exists an energy-conserving process in which the photon of "mass" $\sqrt{-\xi^2}$ can create a system of real pions (or $K\bar{K}$, nucleon pairs, etc.) with total energy $\sqrt{-\xi^2}$ which is in turn coupled to the $N\bar{N}$ system. ImF is essentially the product of matrix elements $\langle\gamma|2\pi, \text{ or } 3\pi\ldots\rangle\langle 2\pi, \text{ or } 3\pi\ldots|N\bar{N}\rangle$.

Instead of looking at Feynman diagrams from "down" to "up" as we normally do in discussing *e-p* scattering, we look at Feynman diagrams corresponding to the interaction of the electromagnetic field with the physical nucleon from "left" to "right," i.e., we regard the photon creating some intermediate states of zero baryon number which in turn create nucleon pairs. The variable ξ^2 in $ImF(-\xi^2) = \sigma(\xi^2)$ represents the square of the mass of the various intermediate states through which the $\gamma \to N\bar{N}$ transition takes place. As an illustrative example, the reader may verify that, if a state with mass m completely dominated the dispersion integral for F_2 (as might be the case if there is a sharp π-π resonance) in the sense that the corresponding σ_2 is a δ-function at $\xi^2 = m^2$, then the mean square radius $\langle r^2 \rangle$ would be given by $6/m^2$. The density distribution in configuration space ($\rho(r)$ in (9.71)) would be of the Yukawa form $\exp(-mr)/r$. In general, the dominance of a low-mass state in the dispersion integral leads to a "spread out" distribution, as is evident from the uncertainty principle.

The intermediate states that appear must have the same symmetry properties as the photon. We can visualize the photon having sometimes isospin 1 and sometimes isospin zero. The quantum numbers of the relevant intermediate states can be shown to be $T = 1, J = 1^-$ for $F^{(V)}$ and $T = 0, J = 1^-$ for $F^{(S)}$. Consider now

$$\bar{u}_N \gamma_\mu F_1^{(S)} u_N$$

This is odd under G. Hence only those intermediate states which are odd under G contribute. Since the photon (odd charge conjugation parity) cannot create a single neutral pion (even charge conjugation parity), the lower limit of integration for $F_1^{(S)}$ is $(3\mu_\pi)^2$. On the other hand

$$\bar{u}_N \gamma_\mu \tau_3 F_1^{(V)} u_N$$

is even under G. Hence $\sigma_1^{(V)}$ starts with $(2\mu)^2$. Since $\bar{u}\gamma_\mu u$ and $\bar{u}\sigma_{\mu\nu}(p' - p)_\nu u$ have the same symmetry property under C, the integrations for $F_2^{(S)}$ and $F_2^{(V)}$ also extend from $(3\mu_\pi)^2$ and $(2\mu_\pi)^2$ respectively. To sum up, states with odd (even) number of pions contribute to the isoscalar (isovector) form factors. $K\bar{K}$ states and $N\bar{N}$ states can contribute both to the isoscalar and isovector form factors.

Experimentally

$$F_2^{(S)}(0) = \frac{(\mu_p + \mu_n)}{2} \frac{e}{2M_N} = -0.06 \frac{e}{2M_N}$$

is small whereas

$$F_2^{(V)}(0) = \frac{(\mu_p - \mu_n)}{2} \frac{e}{2M_N} = 1.8 \frac{e}{2M_N}$$

is large. So we may be tempted to conclude that the two pion states are mainly responsible for the electromagnetic structure of the nucleon. But this view faces serious difficulties when we consider the mean square radii of the proton and the neutron given by

$$\langle r^2\rangle_1^{(p)} \equiv -\frac{6}{e}\frac{dF_1^{(p)}(q^2)}{dq^2} \approx \frac{0.32}{\mu_\pi^2} \approx (0.8 \times 10^{-13}\ \text{cm})^2$$

$$\langle r^2\rangle_1^{(n)} \equiv -\frac{6}{e}\frac{dF_1^{(n)}(q^2)}{dq^2} \approx 0$$

which means that the isoscalar charge radius is as large as the isovector radius.

More detailed considerations by many workers (Chew *et al.*, 1958; Federbush, Goldberger, and Treiman, 1958) show that we do not even understand the isovector form factors on the basis of the usual Yukawa picture in which the nucleon emits and absorbs *uncorrelated* pions. The calculated isovector moment $F_2^{(V)}(0)$ turns out to be much smaller than the observed moment. Clearly some kind of enhancement mechanism is needed, as emphasized by Drell (1958). Dispersion theoretic calculations carried out by Frazer and Fulco (1959) indicate that a fairly sharp resonance in the $T = 1, J = 1^-$ state of the two pion system at a total energy of $\sim 4\mu_\pi$ may account for the observed isovector form factors.

It has been conjectured by Nambu (1957a) and Chew (1960) that the isoscalar form factors may be explained by a similar enhancement mechanism. A strongly interacting neutral vector meson (ω^0) or a sharp *three* pion resonance in the $T = 0, J = 1^-$ state can not only account for the unexpected large isoscalar charge radius but can also throw light on the nature of the two nucleon interactions at short distances—particularly the origins of the repulsive core and the spin-orbit interactions (Breit, 1960; Sakurai, 1960a, b).

Such pion resonances (or vector mesons) may be looked for in more direct processes, e.g. in multiple pion production. Perhaps the clearest search along these lines will involve reactions such as

$$e^- + e^+ \rightarrow 2\pi, 3\pi\ (\text{or } \omega^0)$$

which may become feasible with clashing electron-positron beams in the near future.

Note added in proof. The conjectured vector mesons have been discovered in recent bubble chamber experiments, as will be discussed in Section 7.

PROBLEM 11. Using G conjugation invariance, show that the OPE potential must change sign as we go from the NN system to the $N\bar{N}$ system. What can we say about the two-pion exchange contributions?

9.6. Isospin and Weak Interactions; Conserved Vector Theory

In spite of the fact that isospin is not conserved in weak decays, the concept of isospin plays a decisive role in quantitative discussions of weak interactions.

In discussing the universal Fermi interaction hypothesis, we have seen that the vector coupling constant C_V in beta decay agrees very well with the coupling constant g_{VA} in muon decay. This is somewhat surprising for the following reason. A physical neutron can sometimes be found in a "dissociated" state consisting of π^- and p; $n \rightleftarrows \pi^- + p$. During this time, i.e. while the physical neutron is in this dissociated state, the beta decay of the physical neutron is not expected to take place. So we expect that even if the bare beta decay coupling constant (i.e. the beta decay constant we would measure if we could "switch off" the strong interaction) is the same as the muon decay constant, the renormalized decay constant which we actually measure, and which characterizes the decay of the physical neutron, is not expected to be the same. So this agreement between C_V in beta decay and g_{VA} in muon decay is miraculous unless there is some kind of mechanism that guarantees this equality.

We note that we actually have a very similar situation in electrodynamics. The electron is believed to be a simple Dirac particle with no charge distribution and no anomalous moment either whereas the proton as we know today is a very complicated object with a pion cloud around the core. Virtual interactions of the form $p \rightleftarrows n + \pi^+$ are constantly taking place. Yet the total charge of p, which we measure in electron-proton scattering at very low energies ($\lambdabar \gg$ charge radius of p), or by Thomson scattering by a very soft γ ray is the same as the proton charge we would measure if it were not for the virtual pion interactions. This means that other interactions (e.g., pion interactions) are arranged in such a way that the equality between the physical electric charge and the bare electric charge is not disturbed; the electric charge of the physical proton is the same as the electric charge of e^+, provided that the bare proton charge is the same as the e^+ charge.

How is this equality achieved in electrodynamics? First of all electric charge conservation holds in the process $p \rightleftarrows n + \pi^+$. Secondly even while the proton is in this "dissociated" state the electromagnetic field can interact with the π^+ with the same strength. The fact that A_μ is coupled to the conserved current which consists of the sum of both the charge current of proton and the charge current of $\pi^\pm$,

$$j_\mu = i\bar{p}\gamma_\mu p + i\left(\frac{\partial\pi^\dagger_{\text{ch.}}}{\partial x_\mu}\pi_{\text{ch.}} - \pi^\dagger_{\text{ch.}}\frac{\partial\pi_{\text{ch.}}}{\partial x_\mu}\right)$$

guarantees the equality of the "physical electric charge" one measures in the low energy limit to the electric charge we would measure in the absence of the strong interactions. (This argument can be readily generalized in the presence of strongly interacting strange particles.)

The experimental near equality of C_V in beta decay to g_{VA} in muon decay strongly suggests that an analogous mechanism is at work in the beta decay case. Needless to say, $\bar{e}\gamma_\mu(1 + \gamma_5)\nu$ takes the place of A_μ.

Now the bilinear covariant constructed out of p and n that appears in the vector part of the beta decay interaction can be written as

$$i\bar{p}\gamma_\mu n = i\bar{N}\tau_+\gamma_\mu N \tag{9.74}$$

where

$$\tau_+ = \tfrac{1}{2}(\tau_1 + i\tau_2) = \begin{pmatrix} 0 & 1 \\ 0 & 0 \end{pmatrix}$$

Is this current conserved in the presence of strong interactions? No; in the usual Yukawa theory in which both the nucleon and the pion are "fundamental," that which is conserved is the sum of the isospin current for the nucleon and the isospin current for the pion:

$$J_\mu^{(+)} = i\bar{N}\tau_+\gamma_\mu N - \left(\frac{\partial \boldsymbol{\pi}}{\partial x_\mu} \times \boldsymbol{\pi}\right)_+ \tag{9.75}$$

where

$$\left(\frac{\partial \boldsymbol{\pi}}{\partial x_\mu} \times \boldsymbol{\pi}\right)_+ = \left(\frac{\partial \boldsymbol{\pi}}{\partial x_\mu} \times \boldsymbol{\pi}\right)_1 + i\left(\frac{\partial \boldsymbol{\pi}}{\partial x_\mu} \times \boldsymbol{\pi}\right)_2$$
$$= \sqrt{2}\left(\frac{\partial \pi^\dagger_{\text{ch.}}}{\partial x_\mu}\pi^0 - \pi^\dagger_{\text{ch.}}\frac{\partial \pi^0}{\partial x_\mu}\right)$$

In order that the vector part of the beta decay constant be unchanged by the strong couplings, the leptonic current $\bar{e}\gamma_\mu(1 + \gamma_5)\nu$ should be coupled to the total $J_\mu^{(+)}$ given by (9.75) and not just to $i\bar{N}\tau_+\gamma_\mu N$. More rigorous proofs of this argument can be given within the framework of field theory using a generalized form of the identity relation in quantum electrodynamics obtained by Ward (1951). (See, e.g., Okubo, 1959; Bernstein, Gell-Mann, and Michel, 1960). The hypothesis that the leptonic current be coupled to $J_\mu^{(+)}$ is called the conserved vector current hypothesis, and was first suggested by Gershtein and Zeldovich (1955). At that time people believed that the beta interaction was of the form S, T, and P, so this suggestion was only of academic interest. A few years later it was taken up again by Feynman and Gell-Mann (1958).

How can we test the conserved vector current hypothesis? Perhaps the most direct way to test it is to measure the transition rate

$$\pi^+ \rightarrow \pi^0 + e^+ + \nu.$$

Note that this process could also occur in the old theory via intermediate nucleon pairs. But in the new conserved vector theory, the rate is fixed. Unfortunately this process occurs very rarely; the $(\pi^0 e^+\nu)/(\mu^+\nu)$ branching ratio is expected to be of the order of 10^{-8}.

Note added in proof. The branching ratio for $\pi^+ \rightarrow \pi^0 + e^+ + \nu$ has recently been determined to be $(1.1^{+1.0}_{-0.5}) \times 10^{-8}$ by Dunaitsev *et al.* (1962) at Dubna, $(1.7 \pm 0.5) \times 10^{-8}$ by Depommier *et al.* (1962) at CERN, and

$(2.0 \pm 0.6) \times 10^{-8}$ by Bacastow *et al.* (1962) at Berkeley. The theoretically expected ratio (according to the conserved vector hypothesis) is $(1.00 - 1.06) \times 10^{-8}$.

There is another place where we might be able to test the conserved vector current hypothesis. Recall that the total electromagnetic current consists of two parts:
the isoscalar part

$$J_\mu^{(S)} = \frac{i}{2} \bar{N}\gamma_\mu N$$

and the isovector part

$$J_\mu^{(3)} = \frac{i}{2} \bar{N}\tau_3\gamma_\mu N - \left(\frac{\partial \pi}{\partial x_\mu} \times \pi\right)_3$$

Consider a magnetic dipole transition with $T = 1 \rightarrow T = 0$. Only the isovector part of the electromagnetic interaction is operative. Let us suppose that the other members of the $T = 1$ multiplet beta decay to the same $T = 0$ state (see Fig. 9.9). If the conserved vector current hypothesis

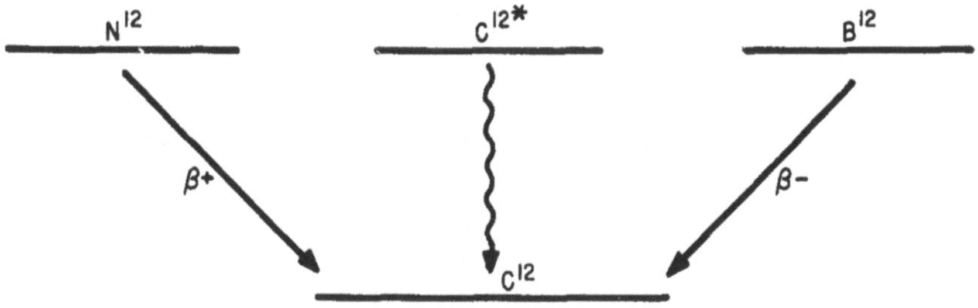

Figure 9.9. Analogous β and γ transitions

holds, the analogous β^- and β^+ transition must contain matrix elements which can be obtained from the matrix element for the γ transition under the substitution

$$eA_\mu \rightarrow C_V \bar{e}\gamma_\mu(1 + \gamma_5)\nu$$

$$J_\mu^{(3)} \rightarrow J_\mu^{(1)} + iJ_\mu^{(2)} \tag{9.76}$$

Now the γ transition goes via a phenomenological interaction of the form

$$-\frac{\mu e}{2M} (\vec{\nabla} \times \vec{A}) \cdot \vec{I}$$

evaluated at the nucleus where $\vec{I}$ is the nuclear spin. If the conserved vector hypothesis holds so that the substitution (9.76) really generates the beta decay interaction, we must have

$$-\frac{\mu}{2M} C_V[\vec{\nabla} \times (\bar{e}\vec{\gamma}(1 + \gamma_5)\nu)] \cdot \vec{I} \tag{9.77}$$

This is a non-allowed vector-type interaction with dependence on lepton momenta. The essential point is that the rates of these beta transitions are calculable since the same μ appears. The transition moment μ is due to very complicated effects of pions, but as long as the interactions that produce μ are charge independent, the same coefficient is directly taken over to the β decay case. This effect is called "weak magnetism."

It turns out that the dominant part of $B^{12} \xrightarrow{\beta^-} C^{12}$ and $N^{12} \xrightarrow{\beta^+} C^{12}$ decay is an allowed Gamow-Teller transition. In spectra we expect interference between the dominant allowed A and the small "weak magnetic" V. Such V, A interference would cause the Kurie plot to deviate from a straight line, and the sign of the V, A interference is opposite between the β^- decay of B^{12} and the β^+ decay of N^{12} (see Chapter 5, Section 7). So the ratio of the B^{12} and N^{12} spectra (with Coulomb corrections and corrections due to the difference in the end point energies taken into account) should deviate from a constant. The magnitude of this deviation is calculable from the known transition rate of $C^{12*} \rightarrow C^{12}$. It has been estimated that this effect is of the order of 20%. The experimental status of this effect is not clear; to date there has not been any evidence for weak magnetism. This interesting experiment was proposed by Gell-Mann (1958) to test the conserved vector current hypothesis.

Note added in proof. Recently Mayer-Kuckuk and Michel (1961) at Caltech made a comparison between the B^{12} spectrum and the N^{12} spectrum. If the ratio of the two spectra is written as $1 + aE$, then the value of a predicted by the conserved vector hypothesis is $1.08 \pm 0.22\%$ per Mev. This is to be compared with the experimental value $1.13 \pm 0.25\%$ per Mev. It is to be emphasized that without the weak magnetism effect, the constant a is expected to be about five times smaller.

One parenthetical remark: if the pion is to be regarded as a bound state of a nucleon and an antinucleon, and if there are no "fundamental" particles with isospin other than the nucleon (as in the case of the model proposed by Sakata (1956) in which only p, n and Λ are "fundamental"), then the consequences of the conserved vector current hypothesis follow automatically *without any additional postulate*. This is because $J_\mu^{(1)} + iJ_\mu^{(2)}$ is nothing but $i\bar{p}\gamma_\mu n$ itself.

It is instructive to write down the matrix element for neutron decay by explicitly taking into account the fact that the nucleon is "dressed" rather than bare. We have

$$M = M_\mu \bar{u}_e \gamma_\mu (1 + \gamma_5) u_\nu + M_{5\mu} \bar{u}_e i\gamma_5 \gamma_\mu (1 + \gamma_5) u_\nu$$

where

$$M_\mu = \bar{u}_p (f_V \gamma_\mu + g_V \sigma_{\mu\nu} q_\nu) u_n$$
$$M_{5\mu} = \bar{u}_p (f_A i\gamma_5 \gamma_\mu + g_A q_\mu \gamma_5) u_n \tag{9.78}$$

q_λ is the four momentum transfer to the lepton field. If the nucleon were "bare," we would have our old familiar results

$$f_V = C_V, \qquad g_V = 0$$
$$f_A = C_A, \qquad g_A = 0$$

The f's and g's are functions of q^2 but in beta decay where the momentum transfer is small they can be regarded as constants. Strictly speaking, they are "form factors" and can be handled in the same way as the electromagnetic form factors (Goldberger and Treiman, 1958a).

In the conserved vector current theory "weak magnetism" is responsible for the g_V term. We have

$$g_V(0) = \frac{(\mu_p - \mu_n)}{2M_N} C_V$$
$$f_V(0) = C_V \tag{9.79}$$

which are of the same form as $F_2^V(0)$ and $F_1^V(0)$ discussed in the previous section.

As for the axial vector part we cannot construct an exactly conserved current. So we do not expect $f_A(0) = C_A^{(\text{bare})}$, and this is probably the reason that the observed C_A/C_V ratio departs from -1. The g_A term cannot be expressed in terms of the known parameters. But the major contribution to the g_A term can be shown to arise from the diagram shown in Fig. 9.10. The black box is the same black box we had for the π^- decay.

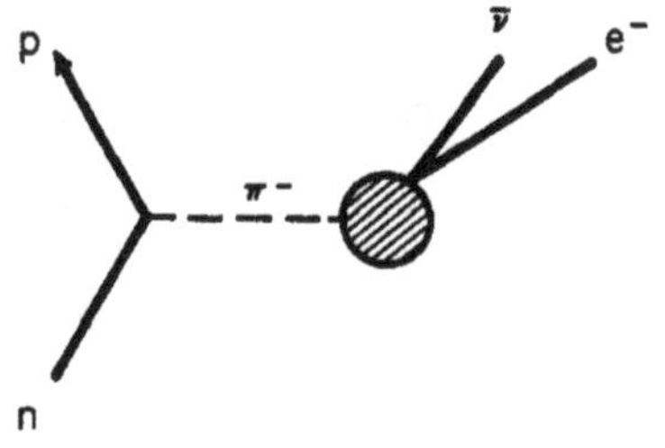

Figure 9.10. Induced pseudoscalar

This additional contribution arises because the neutron is part of the time a π^- and a proton, and the π^- can beta decay $\pi^- \to e^- + \bar{\nu}$ (since September 1958). Using the equations of motion for the leptons we can reduce the g_A term to an effective pseudoscalar term

$$\bar{u}_p(g_A q_\mu \gamma_5)u_N \bar{u}_e i\gamma_5\gamma_\mu(1 + \gamma_5)u_\nu = m_e g_A \bar{u}_p \gamma_5 u_n \bar{u}_e \gamma_5(1 + \gamma_5)u_\nu$$

So we have an effective pseudoscalar interaction with $C_P^{(\text{eff.})} = m_e g_A$. This should not be confused with a "fundamental" pseudoscalar interaction in the Lagrangian which should *not* be present at all if the πe-$\pi\mu$ ratio is to be given correctly. This contribution is called "induced pseudoscalar." It can be shown that the presence of this induced pseudoscalar term is much too small to be observed in beta decay. But in muon capture we have $C_P^{(\text{eff.})} \approx m_\mu g_A$. Since m_μ is large we might expect a large contribution. We can estimate g_A from the known π^+ lifetime and the pion-nucleon coupling constant. It turns out that $C_P^{(\text{eff.})} \approx 8|C_A|$ (Wolfenstein, 1958). Note, however, that $\bar{u}_p\gamma_5 u_n$ is intrinsically small. In any case the diagram shown cannot *alone* give a sufficiently large contribution to account for the major part of the μ^- capture. However, it gives rise to about a 20% effect.

We may ask why $M_{5\mu}$ does not contain an additional term of the form $ih_A\gamma_\mu\gamma_5 q_\nu$. The reason is that $M_{5\mu}$ and the bare $C_A^{(\text{bare})}\bar{u}_p\gamma_5\gamma_\mu u_n$ must have the same symmetry properties if $M_{5\mu}$ is to be obtained from the bare axial vector interaction by "dressing" the nucleon. Tensor behaves oppositely from axial vector under G conjugation; hence the h_A term should be absent. The relevance of G conjugation invariance to the matrix elements that appear in weak interactions is discussed by Weinberg (1958b).

9.7 Quantum Numbers of ρ, ω, and η (added in proof)

In this section we discuss the quantum numbers of the recently discovered mesons: ρ, ω, and η.

(a) *ρ meson*

The first conclusive evidence for what is now known as the ρ meson, or the 750 Mev., $T = 1$, $\pi\pi$ resonance, was presented by a Wisconsin group (Erwin *et al.*, 1961) in the reaction

$$\pi^- + p \to \pi^+ + \pi^- + n, \quad \pi^- + \pi^0 + p \tag{9.80}$$

at a π^- momentum of 1.9 Bev./c lab. When the invariant mass of the final-state dipion system

$$\sqrt{(E_1 + E_2)^2 - (\vec{p}_1 + \vec{p}_2)^2}$$

was plotted, pronounced peaks were found in both the $\pi^+\pi^-$ mass distribution and the $\pi^-\pi^0$ mass distribution. In other words, a substantial fraction of the events in (9.80) could be interpreted as

$$\pi^- + p \to \rho^0 + n, \quad \rho^- + p$$

where the short-lived isobar ρ subsequently decayed into two pions. The center of the ρ peak seemed to be at a total mass of about 750–770 Mev.

The existence of the 750 Mev. ρ peak in the dipion system has subsequently been confirmed by numerous groups in a variety of reactions:

$$\bar{p} + p \rightarrow \pi^+ + \pi^- + \pi^0, \quad 2\pi^+ + 2\pi^- + \pi^0,$$
$$\pi^+ + p \rightarrow \pi^+ + \pi^0 + p, \quad 2\pi^+ + \pi^- + p, \text{ etc.}$$

as well as in (9.80) at different energies. The observed width seems to vary somewhat from experiment to experiment, but the most likely value appears to be $\Gamma \approx$ 100–130 Mev., where Γ stands for the full width at half maximum. Such a large value of Γ implies that the decay of the ρ meson into two pions must proceed via strong interactions. From the absence of $\pi^+\pi^+$ and $\pi^-\pi^-$ peaks at 750 Mev. in reactions such as $\pi^+ + p \rightarrow \pi^+ + \pi^+ + n$, the isospin of the ρ meson has been determined to be unity

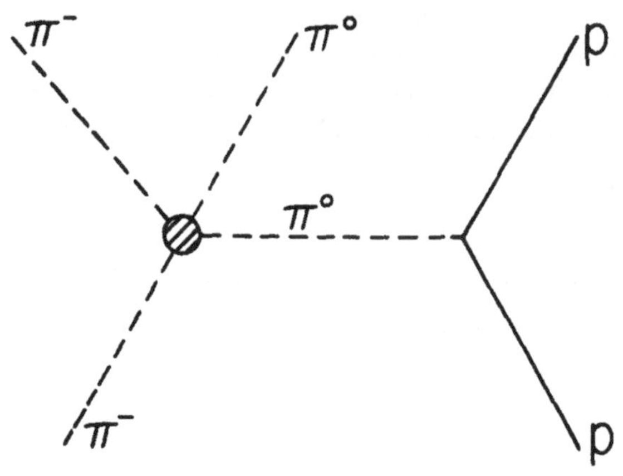

Figure 9.11. Peripheral model used to study pion-pion scattering

(Stonehill *et al.*, 1961). The G conjugation parity of the ρ must be even since $\rho \rightarrow 2\pi$ is a "strong decay" which conserves G as well as isospin.

Since the $T = 1$ isospin wave function is necessarily anti-symmetric (cf. Appendix A), Bose-statistics requires that the orbital angular momentum of the pion pair in ρ decay be odd. To determine the spin of the ρ meson, the angular distribution of the pion pair has been studied in the reaction

$$\pi^{\pm} + p \rightarrow \pi^{\pm} + \pi^0 + p.$$

If we select those events in which the momentum transfer to the nucleon is small, the energy distribution of the recoil proton turns out to agree rather well with the predictions of the peripheral model of Chew and Low (1959) and Goebel (1958), according to which the one-pion-exchange diagram of Fig. 9.11 is assumed to be dominant. Such a model also predicts a two-to-one ratio for $(\pi^- p \to \rho^0 n)/(\pi^- p \to \rho^- p)$, in agreement with observation. More quantitatively, the peripheral model gives

$$\frac{\partial^3 \sigma(\pi^- p \to \pi^- \pi^0 p)}{\partial \Delta^2 \partial \omega^2 \partial(\cos\theta_{\pi\pi})} = \frac{f_{\pi N}^2}{2\pi} \frac{\omega(\omega^2 - 4\mu^2)^{1/2}}{2q^2} \frac{\Delta^2}{(\Delta^2 + \mu^2)^2 \mu^2} \frac{d\sigma(\pi\pi)}{d(\cos\theta_{\pi\pi})} \tag{9.81}$$

where Δ denotes the four-momentum transfer to the nucleon, ω the total c.m. energy of the $\pi\pi$ system, $\theta_{\pi\pi}$ the angle of $\pi\pi$ scattering in the rest system of the dipion, and q the laboratory momentum of the incident π^- in the $\pi^- p$ collision. Using the formula (9.81), we can study the angular distribution of pion-pion scattering. This was done by Carmony and Van de Walle (1962) at Berkeley and also by the Saclay-Orsay-Bologna-Bari group (Alitti *et al.*, 1962). (See also Pickup, Robinson, and Salant, 1961.) The differential cross section of pion-pion scattering turns out to be almost of a pure $\cos^2\theta$ form in the neighborhood of 750 Mev. ($\omega^2/\mu^2 \approx 29$). This, of course, agrees with the $J = 1^-$ assignment for the ρ meson since the 3^- assignment would lead to a $(25\cos^6\theta - 30\cos^4\theta + 9\cos^2\theta)$ distribution which is very strongly peaked forward and backward. Moreover, the forward-backward asymmetry in $\pi^\pm + \pi^0 \to \pi^\pm + \pi^0$ changes its sign as the energy goes through 750 Mev., which indicates that we indeed have a p wave scattering (together with a small s wave background) whose phase shift goes through 90°. This point is illustrated in Fig. 9.12.

Another argument in favor of the $J = 1^-$ rather than the $J = 3^-$ assignment comes from the total pion-pion cross section obtained by assuming the validity of the peripheral model. The total cross section for elastic pion-pion scattering exhibits a peak at $\approx$ 750 Mev. Its maximum value is close to (about $\frac{2}{3}$ of) $12\pi\lambda\!\!\!^{-2}$ (expected for the $J = 1^-$ assignment), and is quite far from $28\pi\lambda\!\!\!^{-2}$ (expected for the $J = 3^-$ assignment).

To conclude, the ρ meson is a $T = 1$, $J^{PG} = 1^{--}$ meson which decays via strong interactions into a p wave pion pair, and appears as a p wave resonance in pion-pion scattering, $\pi + \pi \to \rho \to \pi + \pi$. Its mass is in the neighborhood of 750 Mev., and its full width at half maximum is 100–130 Mev.

(b) *ω meson*

The ω meson that decays predominantly into $\pi^+ + \pi^- + \pi^0$ was first discovered in the reaction

$$\bar{p} + p \to \pi^+ + \pi^- + \pi^+ + \pi^- + \pi^0 \tag{9.82}$$

by Maglić *et al.* (1961) at Berkeley. When the invariant mass of three of the five pions

$$\sqrt{(E_1 + E_2 + E_3)^2 - (\vec{p}_1 + \vec{p}_2 + \vec{p}_3)^2}$$

was plotted for various charge combinations, a pronounced peak was observed in the $\pi^+\pi^-\pi^0$ mass distribution at a total mass of about 780

Mev. whereas no such peak was observed in the $\pi^+\pi^+\pi^0$, $\pi^-\pi^-\pi^-$, $\pi^+\pi^+\pi^-$, $\pi^-\pi^-\pi^0$ mass distributions. At the $\bar{p}$ momentum of 1.6 Bev./c where (9.82) was first studied, about 30% of the reaction (9.82) can be interpreted as

$$\bar{p} + p \rightarrow \pi^+ + \pi^- + \omega, \quad \text{followed by } \omega \rightarrow \pi^+ + \pi^- + \pi^0,$$

where ω stands for a short-lived state responsible for the 780 Mev. peak in the $\pi^+\pi^-\pi^0$ mass distribution. From the non-existence of charged ω's, the isospin of the ω meson has been inferred to be zero.

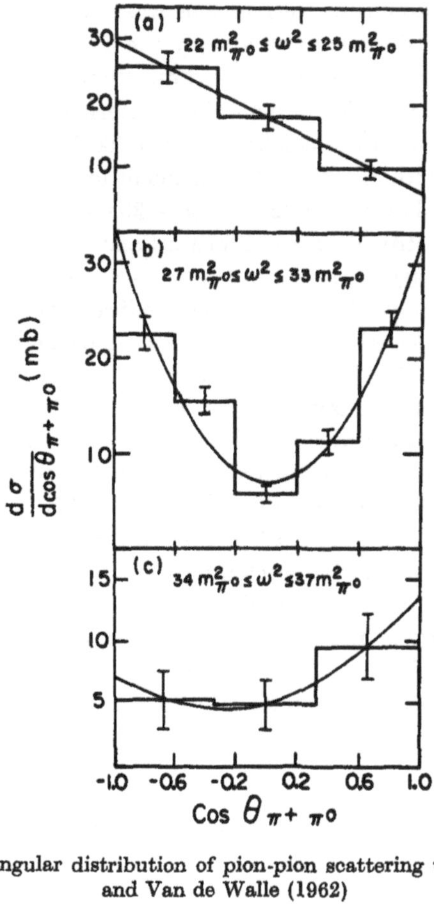

Figure 9.12. The angular distribution of pion-pion scattering taken from Carmony and Van de Walle (1962)

Subsequently the 780 Mev. ω meson has been observed in many other reactions such as

$$\bar{p} + p \rightarrow 3\pi^+ + 3\pi^- + \pi^0, \quad K + \bar{K} + \pi^+ + \pi^- + \pi^0$$

$$K^- + p \rightarrow \Lambda + \pi^+ + \pi^- + \pi^0$$

$$\pi^+ + d \rightarrow \pi^+ + \pi^- + \pi^0 + 2p$$

$$\pi^+ + p \rightarrow \pi^+ + \pi^- + \pi^0 + \pi^+ + p$$

In all these reactions the width of the observed ω peak is roughly equal to the width of the experimental resolution function. For this reason, only an upper limit on the Γ width has been established. The best value available seems to be $\Gamma \lesssim 15$ Mev.

Before we discuss the spin-parity of the ω meson, we first make some general remarks on the properties of a phase space plot, or a Dalitz plot, which was invented earlier in connection with τ decay (Dalitz, 1953). Let us consider the decay of the ω meson into $\pi^+ + \pi^- + \pi^0$ in the rest system of the parent ω meson. Let E_i, T_i, and $\vec{p}_i$ be the total energy, the kinetic energy and the momentum of the i th pion in the ω rest system. We have

$$\sum_i^3 E_i = m_\omega, \quad \sum_i^3 \vec{p}_i = 0, \quad \sum_i^3 T_i = Q = m_\omega - 3m_\pi$$

The proposal of Dalitz is to represent an event as a point in a two-

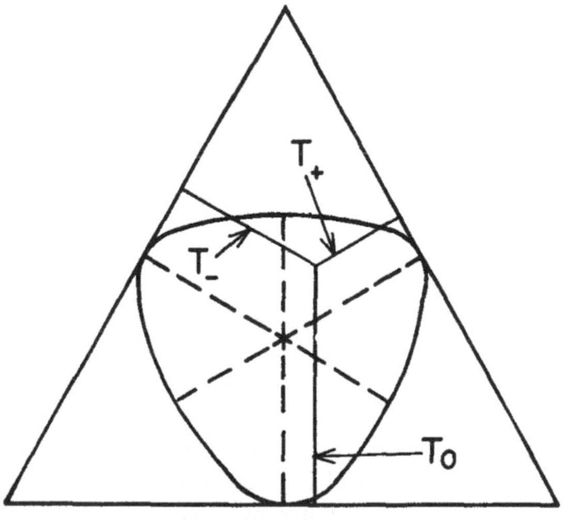

Figure 9.13. Dalitz plot

dimensional plot as shown in Fig. 9.13. The kinematically allowed configurations must satisfy the triangular relation

$$(p_1 + p_2 - p_3)(p_2 + p_3 - p_1)(p_3 + p_1 - p_2)(p_1 + p_2 + p_3) > 0$$

where the equality holds only when the three decay pions are collinear. If the decay pions were nonrelativistic (as in the τ decay case), the kinematical boundary (on which collinear events are represented) would be the circle inscribed by an equilateral triangle of height Q; relativistic kinematics, however, distorts the circular boundary, as shown in Fig. 9.13. (For details, see Fabri, 1954.)

The Dalitz plot has the remarkable property that equal areas in the plot correspond to equal volumes of covariant phase space. In other words, if the decay distribution is given by the Lorentz invariant phase space, then the density of points in the Dalitz plot must be uniform. The proof goes as follows. The standard three-body phase space, when combined with the $\delta^{(4)}$ function coming from energy-momentum conservation and with $(1/\sqrt{2E_i})^2$ coming from each of the meson wave functions, takes the Lorentz invariant form

$$\frac{1}{(2\pi)^9}\frac{d^3p_1d^3p_2d^3p_3}{2m_\omega 2E_1 2E_2 2E_3}(2\pi)^4\delta^{(3)}(\textstyle\sum \vec{p}_i)\delta(m_\omega - \sum E_i) \tag{9.83}$$

We can immediately integrate over $\vec{p}_3$ using momentum conservation. Moreover, since the absolute orientation of the decay configuration is of no significance, we can integrate over all angles except the angle θ between $\vec{p}_1$ and $\vec{p}_2$. We are then left with only two independent variables which we may choose to be E_1 and E_2. So the phase space function $\rho(E_1, E_2)$ is given by

$$\rho(E_1, E_2)dE_1dE_2 = \frac{4\pi p_1^2dp_1 2\pi p_2^2 dp_2}{(2\pi)^5 16 m_\omega E_1E_2E_3}\frac{d(\cos\theta)}{d\,(\sum E_i)} = \frac{dE_1dE_2}{64\pi^3 m_\omega} \tag{9.84}$$

where we have used $p_idp_i = E_idE_i$ and

$$d(\textstyle\sum E_i) = dE_3 = d(\sqrt{m_3^2 - (\vec{p}_1 + \vec{p}_2)^2} = \dfrac{p_1p_2}{E_3}d\,(\cos\theta)$$

for fixed E_1, E_2. Since E_1 and E_2 are linearly related to $(T_1 - T_2)/\sqrt{3}$ and T_3, which are the horizontal and the vertical variables of the Dalitz plot, the constancy of $\rho(E_1, E_2)$ implies that equal areas in the Dalitz plot correspond to equal volumes in the covariant phase space. In other words, the density of points in the Dalitz plot is directly proportional to $|\mathcal{M}|^2$, where $\mathcal{M}$ is the Lorentz invariant matrix element in the sense of Feynman (i.e., defined without kinematical factors such as $1/\sqrt{2E_i}$ etc.). (At first sight the use of T_1 etc. in the Dalitz plot may seem noncovariant. But note that E_1 is linearly related to the invariant mass squared of the $\pi_2\pi_3$ system.)

In discussing the decay mode $\omega \rightarrow \pi^+ + \pi^- + \pi^0$, we first assume that the decay $\omega \rightarrow \pi^+ + \pi^- + \pi^0$ is allowed by strong interactions. The final 3π system must then have $T = 0$. But the $T = 0$ 3π isospin wave function is totally antisymmetric. (The only quantity which is linear in each of the vectors $\vec{a}$, $\vec{b}$, and $\vec{c}$ and invariant under rotations is $\vec{a}\cdot(\vec{b}\times\vec{c})$ which is necessarily antisymmetric under the interchange of any pair. Alternatively, we may note that a $T = 0$ 3π system can be regarded as a $T = 1$ pion plus a dipion which must necessarily be in the $T = 1$ (antisymmetric state.)) Bose statistics then requires that the spatial wave function must also be totally antisymmetric. Hence the matrix element squared must be symmetric under the interchange of any pair of the three pions. In the Dalitz plot this would imply that the density of points must be invariant under the interchange of any of the three symmetry axes denoted by dotted lines in Fig. 9.13. Experimentally this "sextant symmetry" seems to be very well satisfied by the available data (based on more than 1000 ω events.) We now discuss the spin-parity of the ω meson, following Maglić *et al.* (1961) and Dalitz (1962). First, let us note that the 0^+ (scalar meson) assignment is ruled out by angular momentum and parity conservation (cf. our earlier discussion of the $\tau - \theta$ puzzle). As for the 0^- (pseudoscalar meson) case, let $\vec{q}$ be the relative momentum of any pair of pions, which we call π_1 and π_2, and $\vec{p}$ be the momentum of the third pion π_3 in the rest system of the decaying ω. Because of the odd intrinsic parity of the pion which introduces $(-1)^3$, the spatial wave function in momentum space must transform like a scalar. The only scalar we can construct which is odd under $\pi_1 \rightleftarrows \pi_2$ is of the form $\vec{p}\cdot\vec{q}$ times a scalar function of $(\vec{p}\cdot\vec{q})^2$. But such a wave function vanishes whenever $\vec{p}$ is perpendicular to $\vec{q}$, or equivalently whenever the π_1 and the π_2 have the same energies. On the Dalitz plot, this implies that the density must vanish on the symmetry axes shown in broken lines. In particular, the density must go to zero very strongly at the center of the Dalitz plot where the three symmetry axes intersect.

For higher spins, let us focus our attention on what we may call the symmetry configuration in which all pion energies are the same. The distinctive feature of this configuration is that the momentum of any one of the three pions makes an angle of 120° with respect to the momentum of each of the other two pions. Suppose we rotate this configuration about the axis normal to the decay plane by 120°, and then make two interchanges, $\pi_1 \rightleftarrows \pi_2$ followed by $\pi_2 \rightleftarrows \pi_3$; this sequence of operations restores the original configuration. If the angular momentum is quantized with respect to the axis of rotation, this sequence of operations amounts to multiplying the three-pion function by

$$\exp{(2\pi i m/3)}(-1)^2$$

which must be equal to unity by the uniqueness of integral J wave functions. We see that for ω spin < 3, this is possible only if $m = 0$. We now consider a 180° rotation about the axis normal to the decay plane followed by space inversion. Its effect on the *spatial* wave function is

$$\exp{(im\pi)}w_\omega(-1)^3 = -w_\omega \text{ for } m = 0$$

where w_ω stands for the parity of the ω meson; i.e., the parity of the 3π system. (Recall that the parity of the 3π system is $(-1)^3(-1)^{\ell+L}$ where ℓ and L are defined as in the τ decay case.) Again this sequence of operations restores the original configuration, and the uniqueness of the spatial wave function demands an odd parity for the ω. In other words, the symmetry configuration is forbidden for 1^+ and 2^+. We can show that this symmetry configuration is impossible also for 2^- by considering a 180° rotation about the axis along the momentum of one of the pions; this is left as an exercise.

Experimentally the symmetry configuration in which the pions have equal energies (corresponding to the center of the Dalitz plot) is fully allowed. For ω spin < 3, then, the only possible assignment appears to be 1^-. Does the distribution agree with the 1^- (vector meson) assignment? Generally speaking, if the parity of the 3π system is equal to $(-1)^J$, then configurations in which the three pions are collinear must be forbidden, as we have seen in the τ decay case. It is not difficult to see that collinear configurations are mapped onto points on the boundary of the Dalitz plot. The pioneering work of Maglić *et al.* (1961) already indicated that the density of points in the ω Dalitz plot indeed tends to go to zero at the boundary, which made them assert that the ω is a 1^- (vector) meson. Subsequent experiments (based on more than 1000 events) have substantiated this conclusion beyond any doubt. As an example, we show in Fig. 9.14 the folded Dalitz plot and the "distance-from-the-center plot" (Stevenson *et al.*, 1962) of the ω meson obtained by CERN, Paris groups (Armenteros, 1962a) in the reaction

$$\bar{p} + p \rightarrow K + \bar{K} + \pi^+ + \pi^- + \pi^0$$

(Although the total number of ω events is not large, we have chosen to show the Dalitz plot of the CERN-Paris collaboration experiment because the number of background (non-ω) events seems essentially zero in this reaction). The curves drawn for comparison are the predictions based on the simplest matrix elements

$$\begin{aligned} &1^- : \vec{p}_i \times \vec{p}_j \\ &1^+ : E_+(\vec{p}_- - \vec{p}_0) + E_-(\vec{p}_0 - \vec{p}_+) + E_0(\vec{p}_+ - \vec{p}_-) \\ &1^- : (E_+ - E_0)(E_- - E_0)(E_0 - E_+) \end{aligned}$$

constructed by Gell-Mann. (These simplest matrix elements are expected to represent reality if there are no strong final state interactions. These, of course, satisfy the symmetry requirements discussed earlier.)

Thus far we have assumed that $\omega \rightarrow \pi^+ + \pi^- + \pi^0$ is a G conserving, T conserving strong reaction. In other words, The $T = 0$, $G = -1$ parent ω meson has been assumed to decay into a $T = 0$, $G = -1$ 3π final state. Since the width of the ω meson is very slight, it is conceivable that this decay proceeds via virtual electromagnetic interactions which need not conserve G and T (Duerr and Heisenberg, 1962). For this reason we may consider the possibility that a $T = 0$, $G = +1$ ω meson decays into a $T = 1$, $G = -1$, 3π system by violating isospin and G conservation.

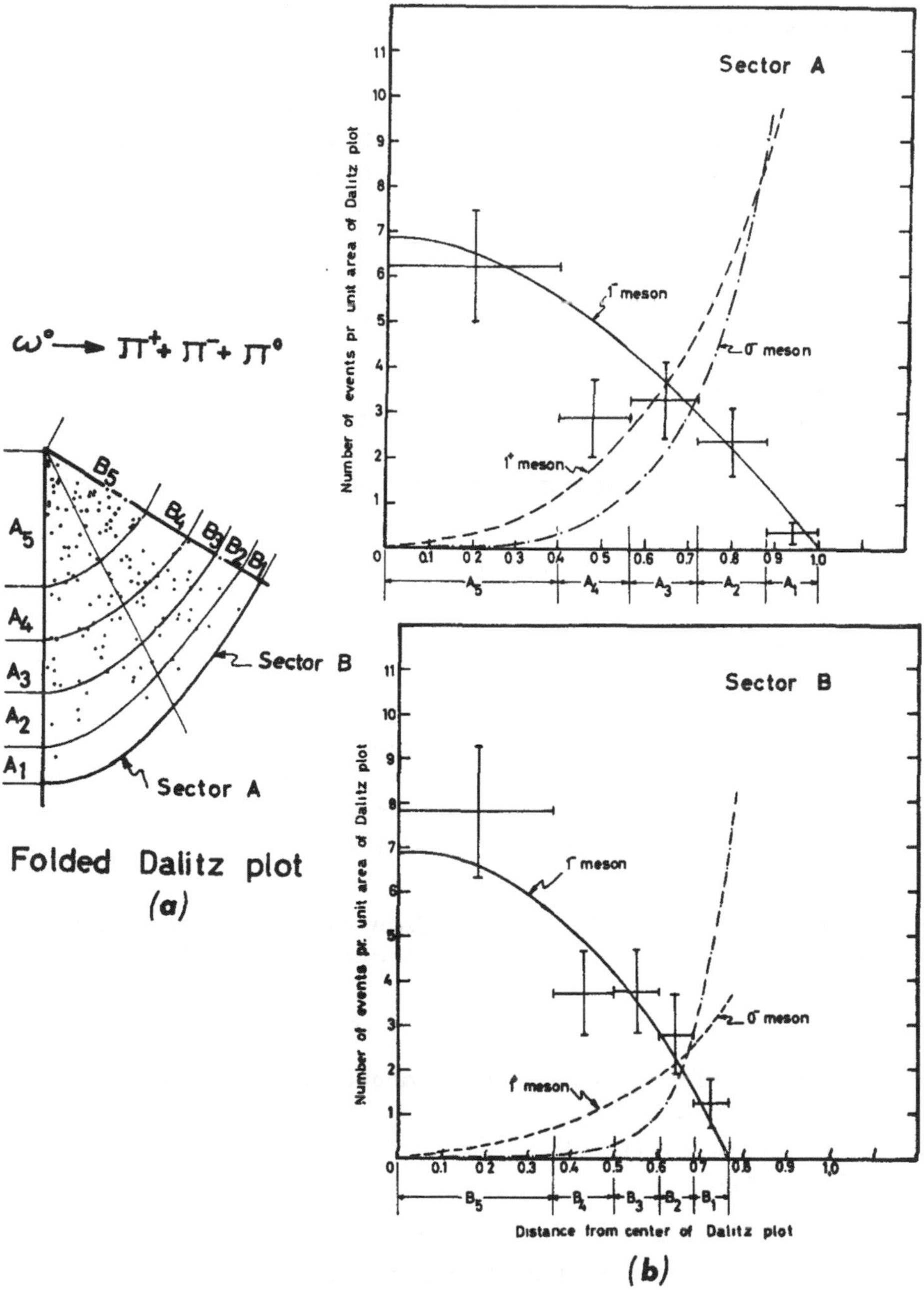

Figure 9.14

(a) The folded Dalitz plot of the ω meson obtained in $\bar{p} + p \rightarrow K + \bar{K} + \pi^+ + \pi^- + \pi^0$

(b) The distance-from-the-center plot of the ω meson.

Various assignments with $G = +1$, however, run into difficulty with the following experimental facts.

(i) With $G = +1$, the G allowed "strong transition" $\omega \rightarrow 4\pi$ is expected to dominate the G forbidden electromagnetic transition $\omega \rightarrow 3\pi$ with the exception of the $J^{PG} = 0^{-+}$ case. (The transition $0^{-+} \rightarrow 4\pi$ is not favored because two d states and one p state are needed.) Experimentally $(\omega \rightarrow 4\pi)/(\omega \rightarrow 3\pi) < 17\%$ (Xuong and Lynch, 1962).

(ii) Even if G parity is violated in ω decay, the C parity must still be conserved. Because of the $G = C(-1)^T$ rule for $T_3 = 0$ systems, the 3π state must be in $T = 1$ for G violating decays. (If G changes, T must necessarily change by an odd integer. Since $\Delta T \geq 2$ is forbidden for virtual electromagnetic transitions that go as e^2 in the amplitude, we must have $\Delta T = 1$.) This would allow $\omega \rightarrow 3\pi^0$ at a rate comparable to $\pi^+ + \pi^- + \pi^0$. In fact, the $(\pi^0\pi^0\pi^0)/(\pi^+\pi^-\pi^0)$ ratio would be 3/2 for totally symmetric configurations. Experimentally the combined data of various groups give something like

$$\frac{\omega \rightarrow \text{neutrals}}{\omega \rightarrow \pi^+ + \pi^- + \pi^0} \approx 10 - 20\%$$

(iii) The observed Dalitz plot does not favor any of the simplest matrix elements for positive G ω mesons; in particular it definitely excludes the 0^{-+} assignment.

We thus conclude that the only reasonable assignment for the $T = 0$ ω meson is $J^{PG} = 1^{--}$, as proposed in the original paper of Maglić *et al.* (1961). With this assignment, the final state decay products in $\omega \rightarrow$ neutrals are expected to be predominantly $\pi^0 + \gamma$.

(c) η *meson*

The η meson of mass 550 Mev. was discovered by Pevsner *et al.* (1961) in a Johns-Hopkins-Northwestern collaboration experiment in which the reaction

$$\pi^+ + d \rightarrow \pi^+ + \pi^- + \pi^0 + 2p$$

was studied. The invariant mass distribution of the $\pi^+ + \pi^- + \pi^0$ system in this reaction is shown in Fig. 9.15.

Subsequently the properties of the η meson have been studied in other reactions, e.g.,

$$K^- + p \rightarrow \Lambda^0 + \pi^+ + \pi^- + \pi^0, \quad \Lambda^0 + \text{neutrals},$$
$$\pi^+ + p \rightarrow \pi^+ + \pi^- + \pi^0 + \pi^+ + p$$

The isospin of the η meson has been determined to be zero by demonstrating the non-existence of $\eta^\pm$; more quantitatively, Carmony, Rosenfeld, and Van de Walle (1962) have shown that the triangular relation

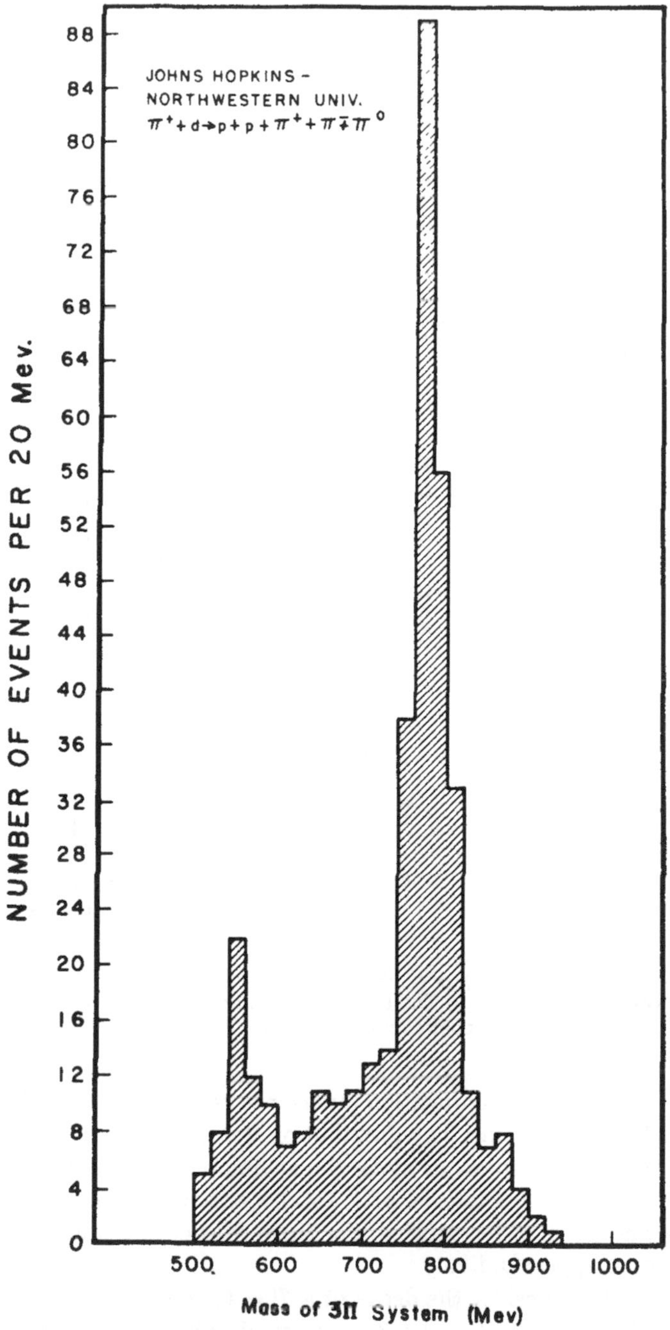

Figure 9.15. Evidence for the 780 Mev. ω meson and the 550 Mev. η meson in the reaction $\pi^+ + d \rightarrow \pi^+ + \pi^- + \pi^0 + 2p$ (Johns-Hopkins group)

expected for $T = 1$ η's is not satisfied by the reactions $\pi^+ + n \to \eta^0 + p$, $\pi^\pm + p \to \eta^\pm + p$. It is also known that

$$\frac{\eta \to \text{neutrals}}{\eta \to \pi^+ + \pi^- + \pi^0} \approx 3$$

(Bastien *et al.*, 1962).

Fig. 9.16 shows the combined Dalitz plot of many groups compiled by

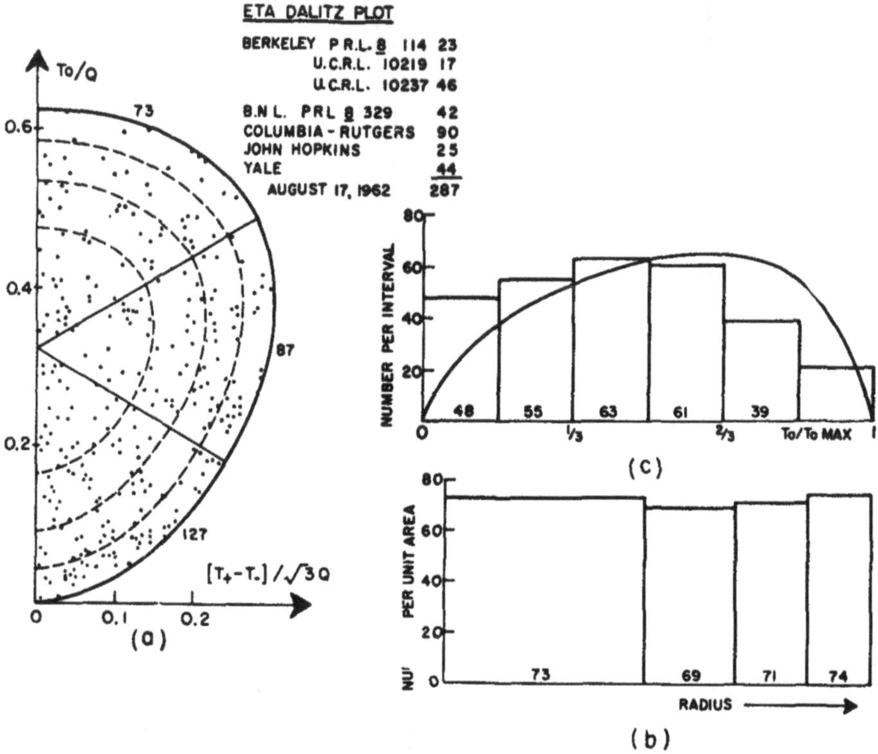

Figure 9.16. Dalitz plot for $\eta \to \pi^+ + \pi^- + \pi^0$ compiled by Alff *et al.* (1962); (a) The distribution of points, (b) The radial density, and (c) The projection of the points on the T_0 axis.

Alff *et al.* (1962). Charge conjugation invariance allows us to fold the plot with respect to the T_0 axis. The distribution does not seem to vanish at the boundary, at the center, or along the symmetry axes. Moreover, the sextant symmetry characteristic of a totally antisymmetric $T = 0$, 3π system does not seem to be satisfied. For these reasons odd G assignments for the η meson appear unlikely.

So we are led to consider the decay of a $T = 0$, $G = C = +1$ η meson into a $T = 1$, $G = -1$, $C = +1$ 3π system via virtual electromagnetic interactions that violate isospin and G conservation. Now let ℓ be the relative orbital angular momentum of the $\pi^+\pi^-$ system, and L the orbital angular momentum of the π^0 in the c.m. system. Since the C parity of the π^0 is even, the $\pi^+\pi^-$ system must necessarily be in an even ℓ state. (Recall that $C = (-1)^\ell$ for $\pi^+\pi^-$.) For $J^{PG} = 0^{-+}$, we expect $\ell = L = 0$, so no part in the Dalitz plot is forbidden by the angular momentum barrier; this is consistent with observation. The 1^{++} assignment would lead to $(\ell, L) = (0, 1), (2, 1)$ etc., forbidding the configuration in which the π^0 has its minimum energy. (The π^0 at rest is necessarily in an $L = 0$ state.) Experimentally the bottom of the Dalitz plot where T_0 is zero is very well populated. The 1^{-+} assignment would require the distribution to vanish at the boundary, again contrary to observation. So the $J^{PG} = 0^{-+}$ assignment appears most likely.

Moreover, evidence for the $\eta \to 2\gamma$ mode has been reported in a New England collaboration experiment by Chrétien *et al.* (1962). Now a $G = C = -1$ meson is forbidden to decay into 2γ rays by charge conjugation invariance. Note also that a spin 1 meson is forbidden to decay into 2γ rays by angular momentum conservation. So the 2γ mode of the η meson contradicts the $J^{PG} = 1^{++}, 1^{-+}$ assignments as well as all possible $G = -1$ assignments.

To sum up, for spin < 2, the only possible assignment for the $T = 0$ η meson is $J^{PG} = 0^{-+}$. (Historically Bastien *et al.* (1962) at Berkeley were the first to suggest that this assignment is the most likely one on the basis of the η mesons produced in $K^- + p \to \eta + \Lambda$.) Its decay into a $T = 1$, $G = -1$ 3π system takes place via G violating virtual electromagnetic interactions. The simplest matrix element for this assignment would lead to a completely uniform distribution in the Dalitz plot, but a density distribution of the form $a + b(T_0/T_0^{(\max)})$ (which roughly fits the observed non-uniformity in the Dalitz plot with $b/a \approx -0.4$) is quite possible; in fact, the η Dalitz plot resembles the τ Dalitz plot, which suggests that final state interactions of the same kind are involved (Wali, 1962). Events of the type $\eta \to$ neutrals must be accounted for by both the 2γ mode and the $3\pi^0$ mode. That the $\pi^+ + \pi^- + \gamma$ mode (whose decay rate is of the order of e^2) is not as frequent as the $\pi^+ + \pi^- + \pi^0$ mode (whose decay rate is of the order of e^4) is somewhat surprising. In any case, within the framework of strong interactions, the η meson is a stable elementary particle in the same sense as the π^0 and Σ^0 are stable elementary particles. The width of the η meson is expected to be of the order of a kev. which is much smaller than the experimentally measured limit of 10 Mev.

CHAPTER 10

Isospin and Related Topics ($S \neq 0$)

10.1. History

The discovery of the pion in 1947 was a great triumph for theoretical as well as for experimental physics. But just about that time evidence was accumulating that there were other strongly interacting particles, the existence of which no theoreticians had expected. In 1947, a strange V type event was reported by Rochester and Butler (1947), which might be interpreted as the decay of a new heavy neutral particle with mass $\sim 1000m_e$. Subsequently, "strange particles" in cosmic rays were extensively studied by various cloud-chamber and emulsion groups at Bristol, Manchester, Padua, Milan, Pasadena, Indiana, etc. (for history, see, e.g., Franzinetti and Morpurgo, 1957; Powell, Fowler, and Perkins, 1960).

As statistics improved, there appeared to be a variety of strange particles with different masses. They had names like V_1^0, κ, χ, etc. Fermi was once asked whether he could remember all these names; he answered, "If I could, I would much rather be a biologist." Since 1953 "strange particles" came to be created artificially at Brookhaven, and since 1955 at Berkeley; this enabled us to study these particles more systematically. In general there are two classes of strange particles: (1) Fermions, with masses greater than the nucleon mass, which are called "hyperons" and which eventually decay into nucleons and pions, (2) bosons with masses between the π mass and the N mass, which are called "K particles."

In the earlier days of strange particle physics there were two contradictory features to be explained. First of all, these particles seemed to be created copiously in nucleon-nuclei collisions; the interaction constants that characterize the production processes are of the order of the pion-nucleon constant (perhaps slightly smaller than the latter, but certainly much larger than 1/137). This alone is not peculiar. But the second feature is that these particles are long lived, in fact the interaction constants that characterize the decays of these particles are of the order of the beta decay constant in dimensionless units. To explain copious production, the couplings between strange particles and ordinary particles must necessarily be strong. But if the couplings between strange particles and ordinary particles were strong, strange particles would immediately decay into ordinary particles in $\sim 10^{-23}$ sec. so that we could not possibly explain their observed long lifetimes, 10^{-10} sec.–10^{-8} sec. One of the major questions in the early 1950's was: How can we reconcile the copious production of strange particles with their long lifetimes?

After a few unsuccessful proposals such as strange particles having higher spins, it was suggested by Pais (1952) that elementary particles are characterized by a new kind of internal quantum number. Pais' quantum numbers take even values for ordinary particles (N and π) and odd values for strange particles. Let us assume that the interactions that conserve this new quantum number are strong or "fast" whereas the interactions that do not respect the conservation of this quantum number are weak or "slow." Then strange particles can be created copiously in collisions of ordinary particles provided that they are produced in pairs since

$$\text{“even”} + \text{“even”} \rightarrow \text{“odd”} + \text{“odd”}$$

is "fast." Thus Pais was led to the idea of "associated production." At the same time Pais' scheme also explains the longevity of strange particles since

$$\text{“odd”} \rightarrow \text{“even”} + \text{“even”}$$

must be "slow."

This idea of Pais had been advanced before associated production was confirmed experimentally. In fact, when Pais proposed this idea, experimentalists laughed at it: "Why should nature create two strange objects at a time? Isn't creating one at a time good enough?" When the Cosmotron at Brookhaven started working, however, the idea of associated production was brilliantly confirmed (Fowler *et al.*, 1957). Strange particles are always created in reactions like

$$\pi^- + p \rightarrow \Lambda^0 + K^0$$

$$p + p \rightarrow \Lambda^0 + p + K^+$$

To date there have been no exceptions to the rule that a strange particle is created out of ordinary particles only in association with some other strange particle.

Successful as was the idea of the associated production hypothesis, it soon became recognized that the idea of associated production *alone* was not enough. Consider

$$n + n \rightarrow 2\Lambda^0$$

which is certainly allowed by the associated production hypothesis; the kinetic energy necessary in the c.m. system for this reaction to proceed is only 2×177 Mev. $= 354$ Mev. (since $M_\Lambda - M_n = 177$ Mev.). It has been shown experimentally that strange particles are not produced at such low energies. It is necessary to supply c.m. energy $M_K + M_\Lambda - M_n = 670$ Mev. to create strange particles; this is the actual threshold for strange particles produced according to experiments (Ridgway, Berley and Collins, 1956).

10.2 Strangeness Scheme

We are now in a position to discuss the classification scheme of Gell-Mann and Nishijima which introduced a remarkable order to strange particle physics (Gell-Mann, 1953, 1956a; Nakano and Nishijima, 1953; Nishijima, 1955). We first go back to the notion of charge independence introduced in Chapter 9. Consider the process (real or virtual)

$$\pi^- + p \rightleftarrows \Lambda^0 + K^0 \tag{10.1}$$

Suppose this reaction did not conserve isospin. Since strange particles are coupled strongly to nucleons and pions, we would expect some violation of isospin conservation in pion-nucleon physics, e.g. in pion-nucleon scattering, nuclear forces, etc. So we are led to the idea that the strong interactions of strange particles must also be charge-independent.

Going back to (10.1), let us suppose that experimentalists tell us that there is no charged hyperon with mass close (say within 10 Mev.) to the Λ^0 mass. (Historically this empirical fact was not well established at the time Gell-Mann and Nishijima proposed their scheme.) Then we argue that the Λ^0 particle must be a charge singlet, and assign $T = 0$, $T_3 = 0$. The initial π^--p system can be in either $T = \frac{1}{2}$ or $T = \frac{3}{2}$; so if isospin is conserved, the K^0 particle must be a member of a charge multiplet with $T = \frac{1}{2}$ or $T = \frac{3}{2}$. Suppose $T = \frac{3}{2}$; then there would be $2 \times \frac{3}{2} + 1 = 4$ charge states for K particles. Experimentally doubly charged heavy mesons have never been observed. So we assign $T = \frac{1}{2}$, $T_3 = -\frac{1}{2}$ to K^0 (since for $\pi^- p$, $T_3 = -\frac{1}{2}$). This assignment necessarily implies that there must be a charged counterpart of K^0, denoted by K^+ corresponding to $T = \frac{1}{2}$, $T_3 = \frac{1}{2}$.

Note that we have done something rather drastic. In both low-energy nuclear physics and in pion physics, a state with integral ordinary spin is *always* associated with integral isospin, e.g., $d\ (J = 1,\ T = 0)$, $\alpha\ (J = 0,\ T = 0)$, $\pi(J = 0,\ T = 1)$. On the other hand a state with $\frac{1}{2}$ integral J is always associated with $\frac{1}{2}$ integral isospin; e.g., $p\ (J = \frac{1}{2},\ T = \frac{1}{2})$, $H^3\ (J = \frac{1}{2},\ T = \frac{1}{2})$, π-nucleon system ($J = \frac{1}{2}, \frac{3}{2}, \frac{5}{2}$, etc., $T = \frac{1}{2}, \frac{3}{2}$). But in (10.1) we have assigned $T = 0$ to a particle with ordinary spin $\frac{1}{2}$, and $T = \frac{1}{2}$ to a particle with ordinary spin zero. The major step taken by Gell-Mann and Nishijima was that they pointed out the possibility of decoupling isospin from ordinary spin in the sense that strange particles like Λ and K have unconventional relations between ordinary spin and isospin. This may sound trivial now, but in 1953 the idea was considered to be highly speculative and even somewhat "crazy."

We construct a table of strongly interacting particles using the idea of charge multiplet (Table 10.1). Now the relationship between T_3 and Q is such that they differ only by a constant which is common to all members of the multiplet. This constant is $Y/2$ where Y is given by Table 10.1.

$$Q = T_3 + Y/2 \tag{10.2}$$

TABLE 10.1

Gell-Mann-Nishijima Scheme

		T	Y	S
$B = 1$	$\Xi^{-,0}$	$\frac{1}{2}$	-1	-2
	$\Sigma^{\pm,0}$	1	0	-1
	Λ^0	0	0	-1
	n, p	$\frac{1}{2}$	$+1$	0
$B = 0$	K^+, K^0	$\frac{1}{2}$	1	1
	$K^-, \bar{K}^0$	$\frac{1}{2}$	-1	-1
	$\pi^\pm, \pi^0$	1	0	0

Y is twice the average charge of the various members of the multiplet. Y is nonvanishing when $T = \frac{1}{2}$ and is sometimes called "hypercharge" or "isofermion number." It is amusing to note that no "elementary" particles with $|Y| > 1$ have been discovered.

It is more conventional to consider S, called "strangeness," with $S = Y - B$ so that

$$Q = T_3 + (S + B)/2 \tag{10.3}$$

where B stands for baryon number. This explains the third column of the table. Note that S (or Y) of antiparticle is opposite to that of the corresponding "particle." We may remark that historically Σ^0 and Ξ^0 were discovered after Gell-Mann and Nishijima constructed classification schemes essentially identical to Table 10.1. The existence of Σ^0 and Ξ^0 was predicted by the Gell-Mann-Nishijima scheme just as some elements in Mendeleev's periodic table were filled in experimentally only afterwards.

There are non-trivial consequences of the Gell-Mann-Nishijima scheme. Since Q and B are conserved absolutely in any reaction, the conservation of T_3 in strong interactions is equivalent to the conservation of S in strong interactions. So we examine the consequences of S conservation.

First of all, we note that the associated production hypothesis still holds. If we start with ordinary particles, we have $S = 0$ initially; so a particle with $S = 1$ must be created in association with a particle with $S = -1$, e.g.

$$\pi^-(0) + p(0) \rightarrow \Lambda(-1) + K^0(+1)$$

$$n(0) + p(0) \rightarrow \Lambda^0(-1) + K^+(+1) + n(0) \text{ etc.}$$

where the numbers in the brackets stand for S. For reactions involving Ξ we must have two other strange particles with $S = 1$ since Ξ is "doubly strange," e.g.

$$\pi^- + p \rightarrow K^0(1) + K^+(1) + \Xi^-(-2)$$

Some of the reactions allowed by the associated production hypothesis but forbidden by the strangeness scheme are

$$n + n \rightarrow \Lambda(-1) + \Lambda(-1)$$
$$\pi^- + p \rightarrow \Sigma^+(-1) + K^-(-1)$$

These reactions had never been observed experimentally. In this sense the strangeness scheme is more satisfactory than the associated production hypothesis alone.

It is extremely important to note that the strangeness scheme requires *two distinct* boson doublets which are "anti-doublets" to each other, namely (K^+, K^0) and (K^-, $\bar{K}^0$). In particular, observe that the K^0 and the $\bar{K}^0$ are characterized by different quantum number (in contrast to the π^0 whose antiparticle is also π^0). More about neutral K particles will appear in Section 4. Note also that although Σ^+ and Σ^- have opposite electric charges and take opposite values of T_3, they are *not* antiparticles of each other. Rather they both belong to the same multiplet with baryon number unity. $\overline{\Sigma^+}$ and $\overline{\Sigma^-}$ which are the antiparticles of Σ^+ and Σ^- do have $B = -1$.

The strangeness scheme predicts that the interaction of K^- with nucleons is fundamentally different from the interaction of K^+ with nucleons in the following sense. For K^-

$$K^-(-1) + N \rightarrow \Sigma(-1) + \pi,\ \Lambda(-1) + \pi$$

can occur as well as scattering

$$K^- + N \rightarrow \bar{K} + N \qquad (\bar{K} \text{ stands for } \overline{K^0} \text{ or } K^-)$$

In contrast, for K^+, only scattering is possible; the K^+ with positive S cannot convert a nucleon into a hyperon with negative S. Several hundreds of meters of K^+ tracks have been analyzed in emulsions and bubble chambers, but not a single hyperon production event has been reported.

The ways in which K^- can be created are also different from those in which K^+ can be created. To create K^+ all we have to supply is the K mass + ΛN mass difference in the c.m. system, e.g.

$$p + n \rightarrow n + \Lambda(-1) + K^+(+1)$$

which has a threshold of 1.6 Bev. in the lab. system. On the other hand, K^- can be created only in association with K^+ or K^0, e.g.

$$p + n \rightarrow p + n + K^+(+1) + K^-(-1)$$

which has a lab. threshold of 2.6 Bev. At Cosmotron energies (maximum energy 3 Bev.) K^- particles have been found to be extremely rare in comparison with K^+ particles.

What happens to the strangeness rule in the presence of the electromagnetic interactions? Recall that the electromagnetic couplings do not respect isospin conservation. Indeed there are small mass differences within charge multiplets, which are believed to be brought about by combined effects of the electromagnetic couplings and the strong couplings. For instance,

$$m(\Sigma^-) - m(\Sigma^+) = 7.2 \pm 0.1 \text{ Mev. (out of 1190 Mev.)}$$
$$m(\bar{K}^0) - m(K^-) = 3.9 \pm 0.6 \text{ Mev. (out of 490 Mev.)}$$

The existence of such mass differences indeed indicates that isospin conservation is only approximate.

Although the full rotational invariance in isospin space is broken as we "switch on" the electromagnetic coupling, the cylindrical symmetry about the third axis, i.e. the conservation of T_3, is not affected provided that the electromagnetic coupling is "minimal." This is because the photon field is always coupled to the electric charge current which consists of the sum of charge-current densities which transform like isoscalar and charge-current densities which transform like the third component of isovector. Thus T_3 commutes with the electromagnetic coupling Lagrangian. We have the important result: Strangeness conservation still holds in the presence of the minimal electromagnetic interaction. (The assumption of minimal electromagnetic coupling is essential here. Gell-Mann (1956) has constructed an example of a nonminimal electromagnetic interaction that leads to S violation. The connection between the metastability of strange particles and the assumption of minimal electromagnetic couplings was first emphasized by Pais (1952).)

In associated photoproduction processes isospin is not a good quantum number. But strangeness is still conserved. This means, for instance,

$$\gamma + p \rightarrow \Sigma^0(-1) + K^+(1)$$

is allowed whereas

$$\gamma + n \rightarrow \Sigma^+(-1) + K^-(-1)$$
$$\gamma + p \rightarrow \Lambda^0(-1) + \pi^+$$

are forbidden.

As a consequence of strangeness conservation which holds even in the presence of the electromagnetic interaction, all strange particles except Σ^0 are stable against decay as long as we do not switch on the weak couplings. For instance, the decay mode

$$\Lambda^0(-1) \rightarrow n(0) + \gamma$$

is forbidden. In contrast, for Σ^0 we have

$$\Sigma^0(-1) \rightarrow \Lambda^0(-1) + \gamma$$

which is allowed by strangeness conservation, and is expected to occur as we switch on the electromagnetic coupling. Nobody knows how to calculate the Σ^0 lifetime exactly, but assuming that the relative $\Lambda\Sigma$ parity is even and the $\Lambda\Sigma$ transition moment is of the order of the nuclear magneton, we obtain

$$\tau_{\Sigma^0} \sim 5 \times 10^{-20} \text{ sec.}$$

This corresponds to a mass uncertainty

$$\frac{\hbar}{\tau} \approx 10 \text{ kev.}$$

so the Σ^0 is a respectable "elementary particle" unlike the 3-3 resonance state which "lives" only for 10^{-23} sec.

It is assumed that the interactions that violate strangeness conservation are as "weak" as beta decay interactions. In this way we can account for the long lifetimes of strange particles. In weak interactions $|\Delta S| = 1$ has been suggested, e.g.

$$\Lambda(-1) \rightarrow p(0) + \pi^-(0)$$
$$\Xi^-(-2) \rightarrow \Lambda(-1) + \pi^-(0)$$
$$K^+(1) \rightarrow \pi^+(0) + \pi^0(0)$$
$$K^+(1) \rightarrow \mu^+(0) + \nu(0)$$

We assume that leptons do not carry strangeness. It is of interest to set up a firm limit on the nonexistence of

$$\Xi^- \rightarrow n + \pi^-, \qquad \Xi^0 \rightarrow p + \pi^-, n + \pi^0$$

The rule $|\Delta S| = 1$ is formally equivalent to the rule $|\Delta T_3| = \frac{1}{2}$. Later we shall present another piece of evidence for $|\Delta S| \neq 2$ for weak interactions based on the K_1^0-K_2^0 mass difference.

10.3 Charge Independence in Strange Particle Reactions

The ten years of strange particle experiments with high energy accelerators have fully confirmed the idea that $\Delta S = 0$ is an exact law as far as strong and electromagnetic interactions are concerned. Recall that we have deduced $\Delta S = 0$ as a consequence of a particular kind of rotational invariance in isospin space, namely, rotational invariance around the third axis. We now examine the validity of the charge independence hypothesis itself, namely full three-dimensional rotational invariance in isospin space.

Consider the following K^--d reactions

$$K^- + d \rightarrow \Sigma^- + p$$
$$K^- + d \rightarrow \Sigma^0 + n$$

K^- has isospin $\frac{1}{2}$, and the d isospin is zero. The reactions go via the $T = \frac{1}{2}$ channel only. Σ and π have the same isospin assignment, namely, $T = 1$

and $Q = T_3$. So, in complete analogy with what we did with the pion-nucleon system, we obtain (see Appendix A)

$$|T = \tfrac{1}{2}, T_3 = -\tfrac{1}{2}\rangle = -\sqrt{\tfrac{2}{3}}|\Sigma^- p\rangle + \sqrt{\tfrac{1}{3}}|\Sigma^0 n\rangle$$

We expect

$$\sigma(\Sigma^- p)/\sigma(\Sigma^0 n) = 2$$

if charge independence holds.

Experimentally it has been observed that when K^- is absorbed in d, a pion as well as Σ and N is also emitted most of the time. (This is not surprising since if the nucleon that absorbs K^- were free, the elementary interaction would be of the form, $K^- + N \rightarrow \Sigma(\text{or } \Lambda) + \pi$.) However, even in the reactions

$$K^- + d \rightarrow \Sigma + N + \pi$$

$$K^- + d \rightarrow \Lambda + N + \pi$$

we can still test the charge independence hypothesis.

Problem 12. Show

$$\tfrac{1}{2}[\sigma(\Sigma^-\pi^+n) + \sigma(\Sigma^0\pi^-p) + \sigma(\Sigma^+\pi^-n)] = \sigma(\Sigma^0\pi^0n) + \sigma(\Sigma^-\pi^0p) \quad (10.4a)$$

$$\tfrac{1}{2}\sigma(\Lambda^0\pi^-p) = \sigma(\Lambda^0\pi^0n) \quad (10.4b)$$

$$\sigma(\Sigma^-\pi^0p) = \sigma(\Sigma^0\pi^-p) \quad (10.4c)$$

if charge independence holds in K^--d reactions.

Experimentally it is hard to separate Λ^0 events from Σ^0 events. Equation (10.4c) cannot be readily checked since $\Lambda^0\pi^-p$ events cannot be distinguished from $\Sigma^0\pi^-p$ events. As for (10.4a) and (10.4b), if we add them together the question of the $\Lambda^0\Sigma^0$ confusion does not arise. For K^- absorbed at rest by d, it was found (after correcting for unseen neutral modes using the branching ratios we know from other experiments) by the Berkeley hydrogen bubble chamber group that

$$\text{L.H.S. of (a)} + \text{L.H.S. of (b)} = (1132 \pm 26)/2$$

$$\text{R.H.S. of (a)} + \text{R.H.S. of (b)} = 533 \pm 26$$

so that the ratio is 1.06 ± 0.05 (Alvarez, 1959). This is just about as good an agreement as we can expect.

It has been pointed out, however, that this kind of check is really a test of charge independence for pion-hyperon interactions, and may be insensitive, under some circumstances, to the crucial question whether K^- and $\bar{K}^0$ really form a charge doublet in the sense of Gell-Mann and Nishijima. A more sensitive test which actually touches the heart of the question would bo a comparison between

$$\pi^+ + d \rightarrow K^0 + \Sigma^+ + p$$

and

$$\pi^- + d \rightarrow K^+ + \Sigma^- + n$$

These two reactions are to be compared energy by energy and angle by angle. We have to have them in a one-to-one ratio if charge independence holds. On the other hand, if K^0 and K^+ do not form a doublet, it would be miraculous if we always obtain such a one to one ratio.

In π^+-d collisions one may also study

$$\pi^+ + \text{“}n\text{”} \rightarrow \Lambda^0 + K^+$$

which can be compared to the more extensively studied reaction

$$\pi^- + p \rightarrow \Lambda^0 + K^0$$

where "n" stands for the neutron in the deuterium nucleus. There is some evidence that both the cross section and polarization in Λ^0K^+ production are roughly equal to those observed in Λ^0K^0 production (Cork *et al.*, 1960). This, of course, supports the Gell-Mann-Nishijima assignment for K particles.

Note added in proof. Recently the reaction

$$\pi^+ + d \rightarrow \Lambda \text{ (or } \Sigma^0) + K^+ + p$$

and its charge symmetric counterpart

$$\pi^- + d \rightarrow \Lambda \text{ (or } \Sigma^0) + K^0 + n$$

have been studied in a Berkeley-Northwestern-Johns Hopkins experiment at lab. π^+ momenta of 1.23 Bev./c (Button-Shafer *et al.*, 1962). The results obtained are

$$\begin{aligned} \sigma(K^+) &= 0.71 \pm 0.07 \text{ mb} \\ \sigma(K^\circ) &= 0.76 \pm 0.06 \text{ mb} \\ \alpha\bar{P}_\Lambda &= 0.55 \pm 0.10 \text{ for } K^+ \text{ reactions} \\ \alpha\bar{P}_\Lambda &= 0.47 \pm 0.07 \text{ for } K^\circ \text{ reactions} \end{aligned}$$

where P_Λ stands for the average polarization of the Λ hyperon. Moreover, the angular distributions of the two processes have also been shown to be the same to an accuracy of about 10%. This experiment offers a very striking proof that the K^+ and K^0 indeed form a charge-doublet in the sense of Gell-Mann and Nishijima.

In

$$\begin{aligned} \pi^+ + p &\rightarrow \Sigma^+ + K^+ \\ \pi^- + p &\rightarrow \Sigma^0 + K^0 \\ \pi^- + p &\rightarrow \Sigma^- + K^+ \end{aligned}$$

the relations between the two isospin amplitudes and the various amplitudes expressed in terms of charge states are exactly the same as the corresponding relations in $\pi^\pm N$ scattering. So

$$\begin{aligned} f(\Sigma^+) &= a(\tfrac{3}{2}) \\ f(\Sigma^0) &= \tfrac{\sqrt{2}}{3}[a(\tfrac{3}{2}) - a(\tfrac{1}{2})] \\ f(\Sigma^-) &= \tfrac{1}{3}[a(\tfrac{3}{2}) + 2a(\tfrac{1}{2})] \end{aligned}$$

from which follows

$$\sqrt{2}f(\Sigma^0) + f(\Sigma^-) - f(\Sigma^+) = 0 \tag{10.5}$$

The cross sections depend on the squares of the amplitudes, and this means that $\sqrt{2d\sigma(\Sigma^0)}$, $\sqrt{d\sigma(\Sigma^-)}$, and $\sqrt{d\sigma(\Sigma^+)}$ must form the three sides of a triangle (Feldman, 1956). Attempts have been made to check the triangular relations such as

$$\sqrt{2d\sigma(\Sigma^0)} + \sqrt{d\sigma(\Sigma^-)} \geq \sqrt{d\sigma(\Sigma^+)}$$

Recent combined data of the Berkeley, Saclay, and Yale bubble chamber groups show that the observed differential cross sections at 1.1 Bev. indeed satisfy the inequality relations. (See Battay *et al.*, 1961.)

We now turn our attention to the KN and $\bar{K}N$ interactions which have been extensively studied at various energies. We have

$$\sigma_{\text{total}}(K^+p) = \sigma_{\text{total}}(S = 1,\ T = 1)$$
$$\sigma_{\text{total}}(K^+n) = \tfrac{1}{2}[\sigma_{\text{total}}(S = 1,\ T = 1) + \sigma_{\text{total}}(S = 1,\ T = 0)]$$
$$\sigma_{\text{total}}(K^-p) = \tfrac{1}{2}[\sigma_{\text{total}}(S = -1,\ T - 1) + \sigma_{\text{total}}(S = -1,\ T = 0)]$$
$$\sigma_{\text{total}}(K^-n) = \sigma_{\text{total}}(S = -1,\ T = 1).$$

Included in σ_{total} may be inelastic processes such as

$$K + N \rightarrow K + N + \pi$$

and in cases of K^- interactions, hyperon production

$$K^- + N \rightarrow \Sigma(\text{or } \Lambda) + \pi$$

as well as usual scattering.

Recently measurements of these total cross sections have been pushed up to 13 Bev./c at CERN. K^+p and K^+n cross sections look rather flat at all energies

$$\sigma_{\text{total}} \sim 4\pi\left(\frac{1}{\mu_K}\right)^2 \approx 20 \text{ mb.}$$

There does not seem to be any spectacular "structure" (nothing comparable to the various peaks we saw in pion-nucleon scattering) in these cross sections. It turns out that

$$\sigma_{\text{total}}(K^+p) \approx \sigma_{\text{total}}(K^+n)$$

except possibly at low energies ($E_k < 150$ Mev.), so we conclude that both $T = 0$ and $T = 1$ contribute appreciably.

As for K^-p, the absorption cross sections (cross sections corresponding to reactions $K^- + p \rightarrow \Sigma$ (or Λ) $+ \pi$) seem to comprise a substantial fraction of $\sigma_{\text{total}}(K^-p)$. For K^- captured at rest we have experimentally (Alvarez, 1959)

$$\Sigma^- : \Sigma^+ : \Sigma^0 : \Lambda^0 = 44 : 20 : 28 : 8$$

where we expect

$$\Sigma^- : \Sigma^+ : \Sigma^0 : \Lambda^0 = 1 : 1 : 1 : 0$$

for pure $T = 0$, and

$$\Sigma^- : \Sigma^+ : \Sigma^0 : \Lambda^0 = 1 : 1 : 0 : x$$

for pure $T = 1$ (x can be anything since charge independence does not tell us how the Λ^0 cross section should be related to the Σ cross sections).

At low energies $\sigma_{\text{total}}(K^-p)$ is very large—about 200 mb. at $p_K \sim 80$ Mev./c. (6 Mev. K.E.); this is not surprising for exothermic reactions. (If the matrix element for K^- absorption is roughly constant, we expect the well known $1/v_k$ dependence since

$$\sigma_{\text{abs}} \propto \frac{1}{v_K} |M|^2 \rho(E_f)$$

and $\rho(E_f)$ does not change so rapidly as $1/v_K$ if the reaction is exothermic.)

The interactions of K^- with protons at low energies have been discussed within the framework of a "generalized" effective range theory (Jackson, Ravenhall, and Wyld, 1958; Dalitz and Tuan, 1960). It is assumed that

$$k \cot \delta_T = \frac{1}{a_T + ib_T} \tag{10.7}$$

is a slowly varying function of energy where δ_T is the complex phase shift for $\bar{K}N \to \bar{K}N$ in the appropriate isospin (T) channel. In the first approximation a_T and b_T are taken to be real constants (zero-range approximation). The elastic $\bar{K}N$ cross sections are given by

$$\sigma_{\text{el.}}\left(K^-p \begin{array}{l} \nearrow K^-p \\ \searrow \bar{K}^0 n \end{array}\right) = \pi \left| \frac{a_0 + ib_0}{1 - ik(a_0 + ib_0)} \pm \frac{a_1 + ib_1}{1 - ik(a_1 + ib_1)} \right|^2 \tag{10.8}$$

while the absorption cross section is given by

$$\sigma_{\text{abs.}}(T) = \frac{4\pi}{k} b_T \left| \frac{1}{1 - ik(a_T + ib_T)} \right|^2 \tag{10.9}$$

where we have ignored the very important effect due to the K^--$\bar{K}^0$ mass difference. Using

$$\sigma_{\text{abs.}}(T = 1)/\sigma_{\text{abs.}}(T = 0) = \frac{\sigma(\Sigma^+) + \sigma(\Sigma^-) - 2\sigma(\Sigma^0) + \sigma(\Lambda^0)}{3\sigma(\Sigma^0)} \tag{10.10}$$

etc., one may solve for a_1, a_0, b_1, and b_0. In general there are four sets of solutions, two of which can be eliminated by studying Coulomb interference.

It is well known from the two-nucleon problem ($b = 0$ in this case) that, although a is positive for a weak attractive interaction with no bound state, a large negative value of a is still compatible with a strong attractive interaction provided that the interaction is so strong that the system forms a bound state. (See, e.g., Chapter 10 of Bethe and Morrison,

1956.) Likewise, in the $\bar{K}N$ case, if a is large and negative, it is possible for the K^-p system to develop a quasi-bound state which, however, is unstable against decay into $\Lambda + \pi$. Experimentally this would appear as a resonance state of the pion-hyperon system below the $\bar{K}N$ threshold, as first discussed by Dalitz and Tuan (1959, 1960). A π-Λ resonance has been observed (see below), but it is not clear at this moment whether the observed resonance is a resonance of the Dalitz-Tuan type.

Note added in proof. It now appears that the 1380 Mev. $\pi\Lambda$ resonance (now called Y_1^*) is a $J = 3/2$ resonance (Ely *et al.*, 1961); hence it cannot be an $s_{1/2}$ $\bar{K}N$ bound state of the Dalitz-Tuan type. However, there is evidence for a $T = 0$ $\pi\Sigma$ resonance, denoted by Y_0^* (Alston *et al.*, 1961b) at 1405 Mev., which might be an $s_{1/2}$ $\bar{K}N$ bound-state resonance. In addition, there exists another $T = 0$, $S = -1$ resonance denoted by Y_0^{**} at 1520 Mev.; by studying the polarizations and the angular distributions of the reactions $\bar{K} + N \rightarrow \bar{K} + N$, $\pi + \Sigma$ in the neighborhood of the Y_0^{**} energy, Tripp, Watson, and Ferro-Luzzi (1962) conclude that the Y_0^{**} appears as a $d_{3/2}$ resonance in both the $\bar{K}N$ and the $\pi\Sigma$ channel, hence the $K\Sigma$ parity must be odd. (This method for determining the $K\Sigma$ parity was proposed by Capps (1962).)

At higher energies the total cross sections for K^--p and also K^--n do decrease monotonically except for a peak in K^--p at ~ 1 Bev./c; 95 mb. at 300 Mev./c, 30 mb.–25 mb. at 3 Bev./c–13 Bev./c. There are a lot of new exciting developments arising from high-energy K^-p data. By studying the Q values of $(\Lambda^0\pi^+)$ and of $(\Lambda^0\pi^-)$ in

$$K^- + p \rightarrow \Lambda^0 + \pi^+ + \pi^-$$

Alston *et al.* (1960) at Berkeley have established the existence of a $T = 1$ hyperon-pion resonance denoted by Y^*, at 1380 Mev. (total c.m. energy), 50 Mev. below the K^-p threshold, with width ~ 60 Mev. (or smaller). The existence of a $\bar{K}\pi$ resonance (K^*) at ≈ 880 Mev. (total cm. energy) in

$$K^- + p \rightarrow \begin{Bmatrix} \bar{K}^0 + \pi^- \\ K^- + \pi^0 \end{Bmatrix} + p$$

has also been reported by the same group (Alston *et al.*, 1961a).

A hypernucleus is a bound system made up of nucleons and a Λ hyperon. Experimentally ${}_\Lambda H^3$ (one p, one n, and one Λ) exists but not ${}_\Lambda He^3$; ${}_\Lambda H^4$ and ${}_\Lambda He^4$ both exist with comparable binding energies, (2.20 ± 0.14) Mev. for ${}_\Lambda H^4$, and (2.36 ± 0.12) Mev. for ${}_\Lambda He^4$. These empirical facts can be understood as direct consequences of the assignment $T = 0$ to Λ. ${}_\Lambda H^3$ is essentially a $T = 0$ deuteron plus a $T = 0\Lambda$ bound together, so it must be a charge singlet. ${}_\Lambda H^4$ (${}_\Lambda He^4$) can be regarded as a bound system of a $T = \frac{1}{2}$ H^3 (He^3) and a $T = 0$ Λ. ${}_\Lambda H^4$ and ${}_\Lambda He^4$ form a charge doublet with $S = -1$ just as H^3 and He^3 form a charge doublet with $S = 0$. The only assumptions we have made are: (1) The Λ is a charge singlet. (2) Forces between Λ and nucleon are charge independent. The existing hypernuclei are listed in Table 10.2.

TABLE 10.2

Table of Hypernuclei prepared by R. Levi-Setti. In computing, the binding energies $Q_\Lambda = 37.58$ Mev. for free Λ decay has been used.

A	Hypernucleus	Binding energy (Mev.)
2	${}_\Lambda H^2$	(not bound)
3	${}_\Lambda H^3$	0.12 ± 0.26
4	${}_\Lambda H^4$	2.20 ± 0.14
4	${}_\Lambda He^4$	2.36 ± 0.12
5	${}_\Lambda He^5$	3.08 ± 0.09
6	${}_\Lambda He^6$	unobserved
6	${}_\Lambda Li^6$	(not bound)
7	${}_\Lambda He^7$	3.0 ± 0.5
7	${}_\Lambda Li^7$	5.46 ± 0.33
8	${}_\Lambda Li^8$	6.11 ± 0.35
8	${}_\Lambda Be^8$	6.6 ± 0.6
9	${}_\Lambda Li^9$	7.2 ± 0.6
9	${}_\Lambda Be^9$	6.60 ± 0.31
11	${}_\Lambda B^{11}$	9.9 ± 0.6
12	${}_\Lambda B^{12}$	9.8 ± 0.4
13	${}_\Lambda C^{13}$	10.8 ± 0.5

We may now ask what kind of Yukawa couplings are allowed by charge independence. Omitting γ matrices we have

$$\bar{N}\boldsymbol{\tau}N\cdot\boldsymbol{\pi} = \sqrt{2}(\bar{p}n\pi^+ + \bar{n}p\pi^-) + (\bar{p}p - \bar{n}n)\pi^0$$

$$\bar{\boldsymbol{\Sigma}}\Lambda\cdot\boldsymbol{\pi} + \text{H.C.} = \overline{\Sigma^+}\Lambda\pi^+ + \overline{\Sigma^0}\Lambda\pi^0 + \overline{\Sigma^-}\Lambda\pi^- + \text{H.C.}$$

$$(\bar{\boldsymbol{\Sigma}} \times \boldsymbol{\Sigma})\cdot\boldsymbol{\pi} = (\overline{\Sigma^0}\Sigma^- - \overline{\Sigma^+}\Sigma^0)\pi^+ + (\overline{\Sigma^+}\Sigma^+ - \overline{\Sigma^-}\Sigma^-)\pi^0 + (\overline{\Sigma^-}\Sigma^0 - \overline{\Sigma^0}\Sigma^+)\pi^- \quad (10.11)$$

$$\bar{\Xi}\boldsymbol{\tau}\Xi\cdot\boldsymbol{\pi} = \sqrt{2}(\overline{\Xi^0}\Xi^-\pi^+ + \overline{\Xi^-}\Xi^0\pi^-) + (\overline{\Xi^0}\Xi^0 - \overline{\Xi^-}\Xi)\pi^0$$

where

$$\Sigma^\pm = \frac{\Sigma_1 \mp i\Sigma_2}{\sqrt{2}}, \qquad \Sigma^0 = \Sigma_3, \qquad \Xi = \begin{pmatrix}\Xi^0\\ \Xi^-\end{pmatrix}$$

Note that we cannot write down a Yukawa coupling of the form $\bar{\Lambda}\Lambda\boldsymbol{\pi}$ which is invariant under isospin rotation. This is evident since

$$\Lambda \to \Lambda + \pi$$

would not conserve isospin (π is a charge triplet). An immediate consequence of this is that the force between N and Λ must be short ranged, since it cannot be brought about via the exchange of a single pion. Figure 10.1 illustrates this point. The diagram (a) is forbidden whereas the diagram (b) is allowed. The range of force associated with Fig. 10.1(b) is of

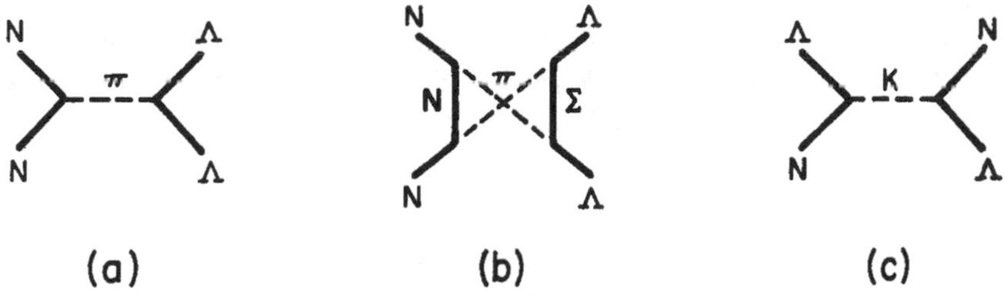

Figure 10.1. ΛN forces. (a) is forbidden by charge independence while (b) and (c) are fully allowed.

the order of $1/(2\mu_\pi)$. (The exchange of a single K particle does lead to an exchange force between Λ and N, but the range associated with this is also short $\sim 1/\mu_K$. See Fig. 10.1(c).)

As for the K particle couplings, we have

$$\bar{N}\Lambda K + \text{H.C.} = \bar{p}\Lambda K^+ + \bar{n}\Lambda K^0 + \text{H.C.}$$

$$\bar{N}\boldsymbol{\tau}\cdot\boldsymbol{\Sigma}K + \text{H.C.} = (\sqrt{2}\bar{n}\Sigma^- + \bar{p}\Sigma^0)K^+ + (\sqrt{2}\bar{p}\Sigma^+ - \bar{n}\Sigma^0)K^0 + \text{H.C.}$$

$$\bar{\Xi}\Lambda K^G + \text{H.C.} = \overline{\Xi^0}\Lambda\overline{K^0} - \overline{\Xi^-}\Lambda K^- + \text{H.C.} \qquad (10.12)$$

$$\bar{\Xi}\boldsymbol{\tau}\cdot\boldsymbol{\Sigma}K^G + \text{H.C.} = (\sqrt{2}\overline{\Xi^-}\Sigma^- + \overline{\Xi^0}\Sigma^0)\overline{K^0} - (\sqrt{2}\overline{\Xi^0}\Sigma^+ - \overline{\Xi^-}\Sigma^0)K^-$$

+ H.C. where

$$K = \begin{pmatrix} K^+ \\ K^0 \end{pmatrix}$$

in analogy with

$$N = \begin{pmatrix} p \\ n \end{pmatrix}$$

and

$$K^G = \left[i\tau_2 \begin{pmatrix} K^+ \\ K^0 \end{pmatrix} \right]^c = \begin{pmatrix} \bar{K}^0 \\ -K^- \end{pmatrix}$$

Note the use of K^G rather than $(\bar{K}^0, K^-)$. It is K^G that transforms in the same way as (K^+, K^0) under isotopic spin rotation (just as in the antinucleon case discussed in Chapter 9, Section 5). There are no interactions like $\bar{N}\Lambda K^G$, $\bar{\Xi}\Lambda K$. They may look invariant under isospin rotation but they conserve neither Q nor S.

We are now in a position to discuss the intrinsic parities of elementary particles assuming that parity conservation holds for strong but not for weak interactions. We take the pion-nucleon interaction to start with; although the relative *p-n* parity is arbitrary, once we fit it to be even by convention, charge independence requires that the intrinsic parities of the three pions have to be the same. Otherwise, it is easy to convince yourself that the equality between *p-p* and *p-n* force brought about by the exchange of a single pion, cannot be maintained. Charge independence enables us to choose the η_P's of various members of the same multiplet to be the same.

What about the relative ΛN parity? ΣN parity? There are no strong interactions like $\bar{N}\Lambda\pi$, $\bar{N}\Sigma\pi$. From weak processes such as

$$\Lambda \rightarrow p + \pi^-$$

we cannot determine the Λ parity, since parity is not conserved in these processes. (In fact, given two channels of opposite parity, we cannot tell which channel is parity conserving.)

We often *define* the Λ parity to be the same as the N parity. A nontrivial question is the K parity. If the relative $K\Lambda$ parity is even, we call K scalar, and in the Lagrangian formalism we must have $\bar{N}\Lambda K$. If the relative $K\Lambda$ parity is odd, we call K pseudoscalar, and we must have $i\bar{N}\gamma_5\Lambda K$ (or its equivalent pseudo-vector coupling). For reactions where this relative $K\Lambda$ parity can be determined, we refer the reader to Chapter 2, Section 4.

There are strong (and electromagnetic) interactions that connect Λ and Σ, e.g.

$$N + \Sigma \rightarrow N + \Lambda$$

$$\Sigma^0 \rightarrow \Lambda^0 + \gamma$$

So the relative $\Lambda\Sigma$ parity *is* a meaningful concept. We have

$$i\bar{\boldsymbol{\Sigma}}\gamma_5\Lambda\cdot\boldsymbol{\pi} \quad \text{for even } \Lambda\Sigma$$

$$\bar{\boldsymbol{\Sigma}}\Lambda\cdot\boldsymbol{\pi} \quad \text{for odd } \Lambda\Sigma$$

Once we determine the $K\Lambda$ parity and the $\Lambda\Sigma$ parity, the question of whether we insert 1 or $i\gamma_5$ for $\bar{N}\boldsymbol{\tau}\cdot\boldsymbol{\Sigma}K$ is already settled. For $-i(\bar{\boldsymbol{\Sigma}} \times \boldsymbol{\Sigma})\cdot\boldsymbol{\pi}$ and $\bar{\Xi}\boldsymbol{\tau}\Xi\cdot\boldsymbol{\pi}$ we must insert $i\gamma_5$.

Consider

$$\pi + N \rightarrow 2K + \Xi$$

$$\Xi^- + p \rightarrow 2\Lambda$$

These are strong interactions allowed by the conservation of strangeness. Studies of orbital angular momentum states involved in these reactions enable us to determine the relative $N\Xi$ parity in spite of the intrinsic ambiguity in the K parity or the Λ parity. (For possible experiments to determine the $N\Xi$ parity see, e.g., Treiman, 1959.) In the Lagrangian formalism the $N\Xi$ parity shows up in the question of whether we insert 1 or $i\gamma_5$ for $\bar{\Xi}\Lambda K^G$. Suppose the ΛK parity is odd, and suppose further that the $N\Xi$ parity is even. Then we must insert $i\gamma_5$ for both $\bar{N}\Lambda K$ and $\bar{\Xi}\Lambda K^G$. Note that the relative $N\Xi$ parity is meaningful solely because of the K couplings which indirectly connect N and Ξ via Λ (or Σ).

To sum up, apart from the π parity which we know to be odd, there are three independent physically meaningful parities to be determined from experiments

Relative ΛK parity (often called K parity)

Relative $\Lambda\Sigma$ parity (often called Σ parity)

Relative $N\Xi$ parity (often called Ξ parity)

We have assumed charge independence and taken advantage of the fact that the parities of the various members within the same charge multiplet can be chosen to be the same.

10.4. Neutral K Particles

The Gell-Mann-Nishijima scheme asserts that K^0 is distinct from its antiparticle $\bar{K}^0$ since K^0 and $\bar{K}^0$ are characterized by different strangeness quantum numbers. This situation is to be contrasted with π^0 whose antiparticle is π^0 itself. Now in decay interactions, strangeness is not conserved, and on the basis of $|\Delta S| = 1$, we expect both K^0 and $\bar{K}^0$ would go into the same final state products, e.g. two pions. It appears then that we can never test the strangeness scheme as far as neutral K particles are concerned because K^0 and $\bar{K}^0$ look the same, say, in a bubble chamber, so that we cannot really tell which is which. When Gell-Mann proposed the strangeness scheme, Fermi said to him: "I won't believe in your scheme until you have a way of telling K^0 from $\bar{K}^0$." This remark of Fermi stimulated Gell-Mann and Pais, whose subsequent investigations have led to one of the most far reaching ideas ever proposed in elementary particle physics.

Let us assume that CP invariance holds in weak decays. (The original argument of Gell-Mann and Pais (1955) used just C invariance.) Define

$$|K_1^0\rangle = \frac{1}{\sqrt{2}}(|K^0\rangle + CP|K^0\rangle)$$

$$|K_2^0\rangle = \frac{1}{\sqrt{2}}(|K^0\rangle - CP|K^0\rangle) \qquad (10.13)$$

Note

$$CP|K^0\rangle = \eta|\bar{K}^0\rangle \qquad |\eta| = 1$$

We can adjust the phase of $|K^0\rangle$ so that $\eta = 1$

$$|K_1^0\rangle = \frac{1}{\sqrt{2}}(|K^0\rangle + |\bar{K}^0\rangle), \qquad |K^0\rangle = \frac{1}{\sqrt{2}}(|K_1^0\rangle + |K_2^0\rangle)$$

$$|K_2^0\rangle = \frac{1}{\sqrt{2}}(|K^0\rangle - |\bar{K}^0\rangle), \qquad |\bar{K}^0\rangle = \frac{1}{\sqrt{2}}(|K_1^0\rangle - |K_2^0\rangle)$$

(10.14)

$|K^0\rangle$ and $|\bar{K}^0\rangle$ are eigenstates of S with eigenvalues $+1$ and -1 respectively but not eigenstates of CP. $|K_1^0\rangle$ and $|K_2^0\rangle$ are eigenstates of CP with eigenvalues $+1$ and -1 respectively, but not eigenstates of S.

In strong interactions S is conserved. So when neutral K particles are created, they are either K^0 or $\bar{K}^0$. The production of K^0 in a strong reaction can be regarded as the production of a fifty-fifty mixture of K_1^0 and K_2^0 with a definite phase relation. This situation is rather analogous to regarding a linearly polarized beam as a superposition of a right circularly polarized beam and a left circularly polarized beam with equal amounts and with a definite phase relation.

What happens in the decay interaction? Let us consider the decay modes under the assumption that CP is good. A state of 2π with a definite J in the center of mass system is necessarily even under CP. To see this, consider $\pi^+ + \pi^-$; the wave function must be even under total exchange, which can be accomplished by interchanging the space coordinates and then interchanging the charge indices. The operation CP is precisely such an operation. For $2\pi^0$ CP is the same as P which gives $(-1)^J$, where J must be even from Bose statistics. Thus both $\pi^+ + \pi^-$ and $2\pi^0$ are necessarily even under CP. If CP invariance holds

$$K_1^0 \to \pi^+ + \pi^-, 2\pi^0$$

but

$$K_2^0 \not\to \pi^+ + \pi^-, 2\pi^0$$

Thus certain decay modes are allowed for K_1^0 but not for K_2^0. Since the lifetime is determined by the matrix elements between K_1^0 (K_2^0) and its decay products, the K_1^0 and K_2^0 have different lifetimes. As far as the decay interactions are concerned, K_1^0 and K_2^0, rather than K^0 and $\bar{K}^0$, behave like "particles," if by a "particle" we mean some object with definite lifetime.

Now a system of three π^0's with $J = 0$ is necessarily odd under CP (C gives $(+1)^3 = 1$, and P gives -1 just as in the τ^+ decay); $\pi^+ + \pi^- + \pi^0$ is odd or even under CP depending on particular angular momentum states. We expect from phase space considerations that the decay rate via the 3π mode is smaller than that via the 2π mode. Moreover, CP must again be odd for the simplest $\pi^+\pi^-\pi^0$ configuration $\ell=0$, $L=0$, where ℓ refers to the relative angular momentum of the $\pi^+\pi^-$ pair and L stands for the orbital angular momentum of the π^0 relative to the c.m. of the $\pi^+\pi^-$ system. Transitions of K_1^0 into more complicated $\pi^+\pi^-\pi^0$ configurations with $CP = +1$ are expected to be highly suppressed because of the centrifugal barrier. The final state products of the leptonic modes, $\pi^+e^-\nu$ etc., are not eigenstates of CP, but these three-body modes must again be slower. Thus the dominant decay mode of K_1^0 should be 2π. K_2^0, which is forbidden to decay into 2π, should live much longer than K_1^0.

Historically, the neutral K particle decaying into 2π with a lifetime of $\sim 10^{-10}$ sec. was first observed. This is to be identified with K_1^0. Gell-Mann and Pais "predicted" that there must be a long-lived neutral K, namely K_2^0, which has a lifetime $\gg 10^{-10}$ sec., and which cannot decay into 2π. Their prediction was brilliantly confirmed by the Columbia cloud chamber group (Landé *et al.*, 1956; Bardon, Landé, and Lederman, 1958), which established the existence of neutral K particles decaying into $\pi^{\pm}e^{\mp}\nu$, $\pi^{\pm}\mu^{\mp}\nu$, and $\pi^+\pi^-\pi^0$ with lifetime $\sim 10^{-7}$ sec. No two-body disintegrations have been found in the long-lived neutral K beam.

Suppose that we create K^0's in a bubble chamber in π-p collisions

$$\pi^- + p \rightarrow \Lambda^0 + K^0$$

In a 12 inch bubble chamber the typical observation time is about 5×10^{-10} sec., a time scale long compared to the K_1^0 lifetime but much too short compared to the K_2^0 lifetime. It is possible under such circum-

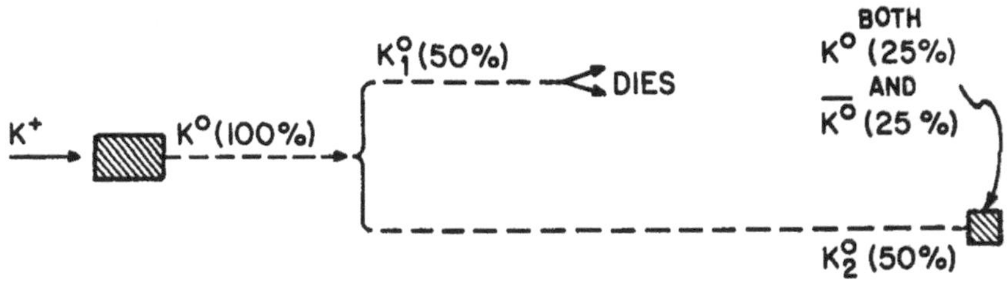

Figure 10.2. Conversion of K^0 into $\bar{K}^0$

stances to check the theoretical prediction that exactly half of K^0 produced in associated production should decay via the fast K_1^0 channel into two pions. Experimentally this prediction has been checked to an accuracy of 10%. (Eisler *et al.*, 1957a; Crawford *et al.*, 1959a; Brown *et al.*, 1959.)

Again suppose that we create pure K^0's but this time we observe the neutral K beam far away. If we wait long enough, the K_1^0 component dies out. At distances far away from the places of production we have a pure beam of K_2^0 with half of the original K^0 intensity. But the K_2^0 beam is to be regarded as a fifty-fifty mixture of K^0 and $\bar{K}^0$; when these neutral particles interact "strongly," it is strangeness that characterizes their states. Since $\bar{K}^0$ has $S = -1$, it can create hyperons. To sum up we have started with a K^0 ($S = 1$) beam which cannot create hyperons, and just by waiting long enough we have obtained a $\bar{K}^0$ ($S = -1$) beam which can create a hyperon. (See Fig. 10.2.)

The production of hyperons by long-lived neutral K particles has been observed (Fry, Schneps, and Swami, 1956). K^- can also be made by charge exchange scattering of a long-lived neutral K particle. This is of interest since K^- is much harder to make directly than K^+ at Cosmotron energies because of the higher threshold for K^-. Thanks to the peculiar behaviour of neutral K particles we have a way of obtaining K^- at relatively low prices. Such a K^- is called a "poor man's K^-."

Another interesting effect pointed out by Pais and Piccioni (1955) is the following. We create K^0, say by associated production in pion-nucleon collisions. The K_1^0 component dies out, and after a while we have a pure beam of K_2^0. Let this K_2^0 beam go through nuclear matter, e.g., Pb plate. The neutral K particle that interacts in the Pb plate is either K^0 or $\bar{K}^0$. But once the strong interactions project the K^0 or $\bar{K}^0$ state, then K^0 or $\bar{K}^0$ can be in turn analyzed in terms of K_1^0 and K_2^0 (Fig. 10.3). So with K_2^0 on

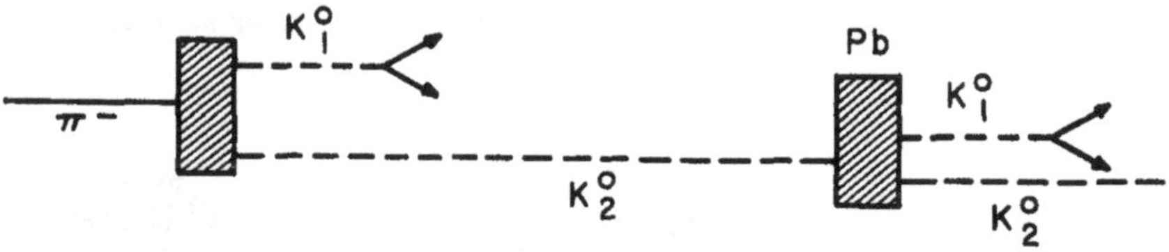

Figure 10.3. Regeneration of K_1^0

Pb, we get K_1^0 as well as K_2^0. This is called the regeneration effect (or the Pais-Piccioni effect), since the K_1^0 component which, we thought, had died out a long time ago reappears.

Note that the situation is quite analogous to the Stern-Gerlach experiment discussed in Chapter 2. We start with a pure $s_z = \frac{1}{2}$ beam; by applying an inhomogeneous magnetic field in the x direction, the beam splits into $s_x = -\frac{1}{2}$ and $s_x = +\frac{1}{2}$. Take $s_x = +\frac{1}{2}$; if we apply an inhomogeneous magnetic field in the z direction, we get both $s_z = \frac{1}{2}$ and $s_z = -\frac{1}{2}$. By measuring s_x we have lost our previous information on s_z. Similarly in the neutral K case we may start with K_2^0, with $CP = -1$, but by measuring its strangeness S via a strong interaction, we lose information on CP. S and CP do not commute any more than s_x and s_z.

Let us see how a neutral K beam changes with time when initially we have K^0 only, i.e.

$$|\psi(0)\rangle = |K^0\rangle = \frac{1}{\sqrt{2}}(|K_1^0\rangle + |K_2^0\rangle)$$

At later times we have

$$|\psi(t)\rangle = \left[\frac{1}{\sqrt{2}}(|K_1^0\rangle \exp(-i\mu_1 t)\exp(-\lambda_1 t/2) + |K_2^0\rangle \exp(-i\mu_2 t)\exp(-\lambda_2 t/2)\right] \quad (10.15)$$

λ_1 and λ_2 and μ_1 and μ_2 stand for the decay rates and the masses of K_1^0 and K_2^0 respectively; μ_1 and μ_2 can be slightly different because K_1^0 and

K_2^0 have different weak interactions so that the self energy effects associated with weak interactions are different in the two cases. From dimensional considerations the mass difference is probably of the order of

$$|\mu_1 - \mu_2| \approx \lambda_1 = \frac{1}{\tau_1} = 7 \times 10^{-12}\,\text{Mev.}$$

since both the self energy effect due to the weak interaction and the decay probability (not amplitude) involve the weak interaction vertex twice. (A more detailed discussion of the K_1-K_2 mass difference appears in the next section.) We obtain for the probability of observing K^0

$$\begin{aligned}|\langle K^0|\psi(t)\rangle|^2 &= |\tfrac{1}{2}\exp(-i\mu_1 t)\exp(-\lambda_1 t/2) \\ &\quad + \tfrac{1}{2}\exp(-i\mu_2 t)\exp\left(\frac{-\lambda_2 t}{2}\right)|^2 \\ &= \tfrac{1}{4}\Big\{\exp(-\lambda_1 t) + \exp(-\lambda_2 t) \\ &\quad + 2\exp\left(\frac{-\lambda_1 t - \lambda_2 t}{2}\right)\cos[(\mu_1 - \mu_2)t]\Big\}\end{aligned} \tag{10.16}$$

Similarly for $\bar{K}_0$

$$\begin{aligned}|\langle \overline{K^0}|\psi(t)\rangle|^2 &= |\tfrac{1}{2}\exp(-i\mu_1 t)\exp(-\lambda_1 t/2) \\ &\quad - \tfrac{1}{2}\exp(-i\mu_2 t)\exp\left(\frac{-\lambda_2 t}{2}\right)|^2 \\ &= \tfrac{1}{4}\Big\{\exp(-\lambda_1 t) + \exp(-\lambda_2 t) \\ &\quad - 2\exp\left(\frac{-\lambda_1 t - \lambda_2 t}{2}\right)\cos[(\mu_1 - \mu_2)t]\Big\}\end{aligned} \tag{10.17}$$

Recall again that the intensity of the $\bar{K}^0$ component can actually be "measured" by studying the rate of hyperon production. If $|\mu_1 - \mu_2|$ is moderately large compared to λ_1, we expect a rapid oscillation between K^0 and $\bar{K}^0$ so that the strangeness identity would be lost even before the K_1^0 component dies out. This is not the case experimentally. For $|\mu_1 - \mu_2| \approx \lambda_1$, we expect the results shown in Fig. 10.4. With $|\mu_1 - \mu_2| \approx \lambda_1$, it is clear that there definitely exists the possibility of measuring the fantastically small mass difference between K_1^0 and K_2^0 as pointed out by Treiman and Sachs (1956) and by Fry and Sachs (1958). The very fact that we do not see $\bar{K}^0$ very near the point of the production of K^0 shows that $|\mu_1 - \mu_2|$ cannot be too large. Recent bubble chamber data seem to point out that $|\mu_1 - \mu_2|$ is indeed of the order of $\lambda_1 = \frac{1}{\tau_1}$. For instance, Birge *et al.* (1960) studied the rate of hyperon production as a function of time by neutral K particles originally produced in charge exchange scattering $K^+ + n \rightarrow K^0 + p$. Their results show $0.5/\tau_1 < |\mu_1 - \mu_2| < 3/\tau_1$.

Note added in proof. More recent data of the same group (Camerini *et al.*, 1962) give $|\mu_1 - \mu_2| = (1.5 \pm 0.2)/\tau_1$ (based on 140 $\bar{K}^0$ interactions).

Another method for measuring the $K_1^0 - K_2^0$ mass difference has been proposed by Good (1958). Consider a parallel beam of K_2^0 incident on a plate. Since the K^0 component and the $\bar{K}^0$ component diminish with different rates, the phase relation between K^0 and $\bar{K}^0$ is no longer that of the pure K_2^0 beam. This means that the incident parallel beam of K_2^0 can

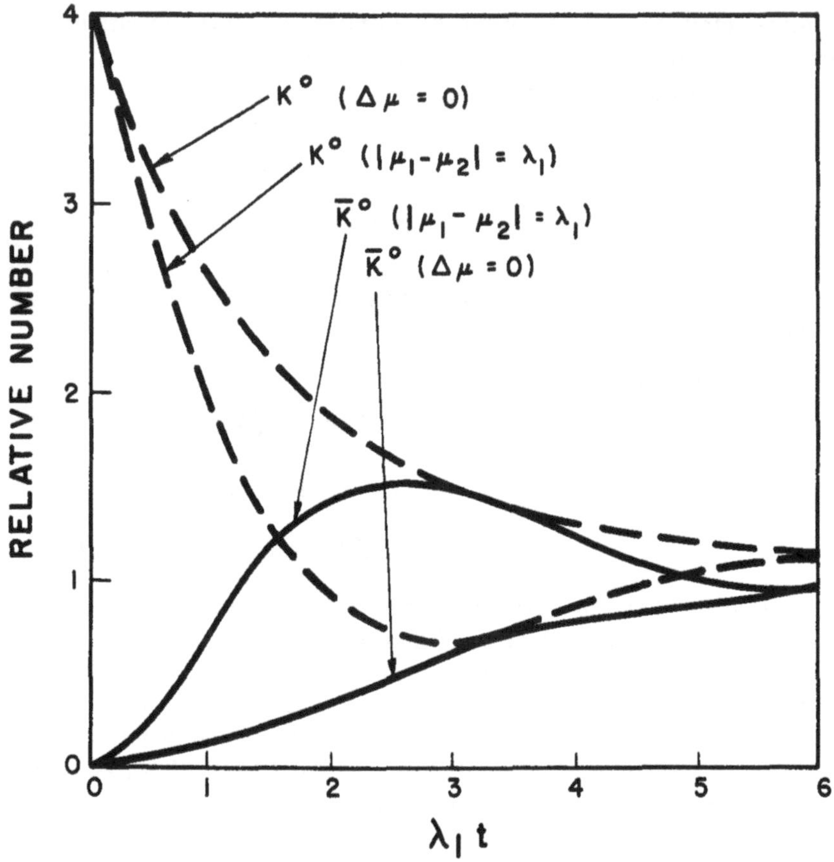

Figure 10.4. Intensities of K^0 and $\bar{K}^0$ as functions of time

regenerate a *parallel* beam of K_1^0 (not to be confused with a spherical wave of K_1^0) as the beam goes through matter. This phenomenon is called "transmission regeneration," and is to be distinguished from other phenomena whereby a particle transforms into another particle, e.g. charge exchange scattering of π^-. Meanwhile, K^0 and $\bar{K}^0$ can be diffracted by nuclei. Diffracted K^0 and $\bar{K}^0$ can in turn be analyzed in terms of K_1^0 and K_2^0. So K_1^0 can be regenerated also by diffraction scattering of K_2^0. Moreover, the ratio of the intensity of the K_1^0's regenerated by transmission and that of the K_1^0's regenerated in the forward direction by diffraction scattering depends only on the plate thickness, the K_1^0 decay rate and the K_1^0-K_2^0 mass difference. (This statement can be proved by solving a set of coupled differential equations the K_1^0 and K_2^0 amplitude must satisfy when both decay interactions and strong interactions (absorption and scattering processes) take place simultaneously. See Case (1956) and Good (1957).)

Using Good's idea, Muller *et al.* (1960) carried out an experiment to measure the mass difference. Neutral K particles produced in π-p collisions which traveled 22.5 ft. were incident on an iron plate in a propane bubble chamber. In addition to a diffraction peak in regions $\cos\theta > 0.985$ ($\theta < 10°$), they observed a spectacular peak confined entirely to $\cos\theta > 0.999$ ($\theta < 2.5°$) which cannot be explained by diffraction and must be attributed to transmission regeneration. The mass difference $|\mu_1 - \mu_2| = (0.85\ {}^{+0.4}_{-0.25})/\tau_1$ has been established from this experiment.

It is amusing to note that the foregoing discussions of neutral K particles would collapse if the $\bar{K}^0$ had a gravitational mass of sign opposite to the K^0 gravitational mass. Under the so-called "antigravity" hypothesis in which antimatter has a gravitational mass opposite in sign to its inertial mass, the K^0 state must be multiplied by $\exp(-i\mu_K\Phi t)$, while the $\bar{K}^0$ state by $\exp(i\mu_K\Phi t)$, where Φ is the gravitational potential on the surface of the earth. Note that

$$\mu_K\Phi \approx 4 \times 10^{-7}\ \text{Mev.} \gg \frac{1}{\tau_1} \approx 7 \times 10^{-12}\ \text{Mev.}$$

If we now repeat the sort of argument we used in discussing possible oscillations between K^0 and $\bar{K}^0$, we see that K_1^0 and K_2^0 (rather than K^0 and $\bar{K}^0$) would turn themselves periodically into each other with a period of oscillation much shorter than the K_1^0 lifetime by many orders of magnitude. Then all neutral K particles would decay via the short-lived channel, in disagreement with observation. More quantitatively, it is possible to show that the ratio of the gravitational K^0-$\bar{K}^0$ mass difference to the K mass must be smaller than a few parts in 10^{-10} if a K_1^0-K_2^0 mass difference of the order of $1/\tau_1$ is to be detected. This result is a striking confirmation of the equivalence principle which demands that the gravitational mass and the inertial mass must be equal. The connection between gravity and neutral K particles was first pointed out by Good (1961).

Thus far we have assumed CP invariance. It can be shown that even if CP invariance did not hold, we would expect two neutral particle states $|K_1^0\rangle$ and $|K_2^0\rangle$ characterized by distinct lifetimes. $|K_1^0\rangle$ and $|K_2^0\rangle$ would still be some orthogonal linear combinations of $|K^0\rangle$ and $|\bar{K}^0\rangle$, but they would no longer be eigenstates of CP. For details we refer to the paper by Lee, Oehme, and Yang (1957). Weinberg (1958a) has shown that the following empirically well-established facts taken together support CP invariance in K decays:

(1) There are no K_2^0's decaying into 2π.
(2) $\lambda_2 \ll \lambda_1$ (by a factor of $\sim 10^3$).
(3) K_1^0 decays into both $\pi^+ + \pi^-$ and $2\pi^0$, i.e. there are *two* rapid decay channels.

However, there is an important exception to Weinberg's conclusion; if the $|\Delta \mathbf{T}| = \frac{1}{2}$ rule to be discussed in the next section is exact so that K_1^0 decays *always* into 2π in $T = 0$, then no information about the validity of CP invariance can be drawn (since the 2π decay mode can be completely characterized by one amplitude).

To sum up the section on neutral K particles we should like to emphasize that not only the existence of K_2^0 but also the peculiar and rather subtle properties of neutral K particles predicted by Gell-Mann and Pais have been brilliantly confirmed by experiments. This is one of the greatest achievements of theoretical physics; all we have used are symmetry arguments plus a little of the superposition principle in quantum mechanics.

Note added in proof.

Recently, the particle mixture theory of Gell-Mann and Pais has been applied to $\bar{p}p$ annihilation processes. First, note that if the $\bar{p}$ captures at rest take place predominantly from atomic s states, as predicted by Day, Snow, and Sucher (1960), then the reaction

$$\bar{p} + p \rightarrow K^0 + \bar{K}^0$$

must necessarily proceed from the $C = -1$, $P = -1$, ${}^3S_1$ state. (The $C = 1$, $P = -1$, ${}^1S_0$ $\bar{p}p$ system is forbidden to turn into an s wave $K^0\bar{K}^0$ pair because of parity conservation.) This requires the final $K\bar{K}$ system to be in the $C = -1$, $J = 1^-$ state, which is an allowed configuration since $C = P = (-1)^\ell$ for a $K^0\bar{K}^0$ system (just as for the $\pi^+\pi^-$ system.) But a pure $C = -1$ $K^0\bar{K}^0$ system must necessarily be of the form

$$K_a^0\bar{K}_b^0 - \bar{K}_a^0K_b^0 = \frac{(K_{1b}^0 + K_{1a}^0)}{\sqrt{2}}\frac{(K_{1b}^0 - K_{2b}^0)}{\sqrt{2}} - \frac{(K_{1a}^0 - K_{2a}^0)}{\sqrt{2}}\frac{(K_{1b}^0 + K_{2b}^0)}{\sqrt{2}}$$

$$= -K_{1a}^0K_{2b}^0 + K_{2b}^0K_{1b}^0$$

where a and b stand for the space indices of the K particles. Thus we obtain the far reaching conclusion that if the capture takes place from an s state, then

$$\bar{p} + p \rightarrow K_1^0 + K_1^0, K_2^0 + K_2^0$$

are forbidden, whereas

$$\bar{p} + p \rightarrow K_1^0 + K_2^0$$

is allowed, as pointed out by Schwartz (1961) and d'Espagnat (1961). This selection rule implies that in a bubble chamber in which the decay of a K_2^0 is very unlikely, we should not see "double V" K pairs. (K_1^0 decays are "visible" about $\frac{2}{3}$ of the time because of the $\pi^+ + \pi^-$ mode.) Recently, in their study of $p\bar{p}$ systems annihilating into $K\bar{K}$ pairs with no extra pions, CERN, Paris bubble chamber groups (Armenteros *et al.*, 1962b) have observed 53 examples of "single V" K pair events; on the other hand, they have failed to observe "double V" K pair events that fit the kinematics for $p + \bar{p} \to K^0 + \bar{K}^0$. Thus the $K_1^0 + K_1^0$ final state seems to be forbidden in agreement with the idea of Day, Snow, and Sucher.

Let us now consider another capture reaction

$$\bar{p} + p \to K^{*0} + \bar{K}^0, \quad \bar{K}^{*0} + K^0$$

where K^* stands for the $T = \frac{1}{2}$ $K\pi$ resonance with mass 880 Mev. (first observed by Alston *et al.*, 1961a). We focus our attention on those events in which K^* ($\bar{K}^*$) subsequently decays into $K^0 + \pi^0(\bar{K}^0 + \pi^0)$. Let us assume that the K^* spin is zero. Then $^3S_1$ state captures would be forbidden by parity conservation. (Note that the relative K^*K parity is odd for a spinless K^* that decays into $K + \pi$.) For a ^1S state capture which is allowed by parity conservation, charge conjugation invariance requires that the final system be in the $C = +1$ state. We must then have

$$\bar{p} + p \to \begin{Bmatrix} K^{*0} + \bar{K}^0 \\ \bar{K}^{*0} + K^0 \end{Bmatrix} \longrightarrow \begin{Bmatrix} K_1^0 + \pi^0 + K_1^0 \\ K_2^0 + \pi^0 + K_2^0 \end{Bmatrix}$$

where we have used the fact that π^0 is even under C, and a $C = +1$, $K^0\bar{K}^0$ system must be of the form

$$K_a^0\bar{K}_b^0 + \bar{K}_a^0K_b^0 = K_{1a}^0K_{1b}^0 - K_{2a}^0K_{2b}^0$$

Experimentally the CERN-Paris collaboration group (Armenteros *et al.*, 1962b) has demonstrated that the measured ratio of "double V" to "single V" events in the above reactions are inconsistent with the ratio expected on the basis of the $K_1^0 + \pi^0 + K_1^0$ hypothesis; in this way the spin zero assignment for K^* has been ruled out. (More recently the spin-1 assignment for K^* has been confirmed by Chinowsky *et al.* (1962) who observed a large $\cos^2\theta$ term in the decays of K^* produced in the reaction $K^+ + p \to K^{0*} + \pi^+ + p$.)

We may generalize the considerations we have discussed above. Since $C = (-1)^\ell$ for a $K^0\bar{K}^0$ system, $\ell =$ even $K^0\bar{K}^0$ pairs are forbidden to decay into $K_1^0 + K_2^0$, and $\ell =$ odd $K^0\bar{K}^0$ pairs are forbidden to decay into $K_1^0 + K_1^0$ and $K_2^0 + K_2^0$. This is a very powerful selection rule. (Goldhaber, Lee, and Yang, 1958.)

10.5. Decay Interactions of Strange Particles

We have already remarked that for weak decays $|\Delta S| = 1$ has been suggested. We may naturally ask whether some weak processes also obey $|\Delta S| = 2$. Consider

$$\Xi^- \to n + \pi^-$$

Without the $|\Delta S| = 1$ rule this decay mode should be more favorable, because of the available phase space, than

$$\Xi^- \rightarrow \Lambda^0 + \pi^-$$

To date, no $n\pi^-$ decay of Ξ^- has been observed. But since there have been not many Ξ^-'s observed anyway, this point requires further investigation.

Another argument against $|\Delta S| > 1$ is that if $|\Delta S| = 2$ transitions were allowed, the K_1^0-K_2^0 mass difference would be about 10^7 times larger than the observed mass difference. The argument goes as follows. If we represent K^0 and $\bar{K}^0$ by a column matrix, the Hamiltonian matrix for K^0 and $\bar{K}^0$ would be given by

$$H \quad \begin{array}{cc|c} K^0 & \bar{K}^0 & \\ \hline \mu & v & K^0 \\ v & \mu & \bar{K}^0 \end{array}$$

The equality of the diagonal elements is a consequence of the *CPT* theorem which guarantees the equality of the K^0 and the $\bar{K}^0$ mass. The off-diagonal matrix elements v arise because of intermediate states that connect K^0 and $\bar{K}^0$, e.g.

$$K^0 \xrightarrow{\text{weak}} \begin{Bmatrix} 2\pi \\ N + \bar{N} \text{ etc.} \end{Bmatrix} \xrightarrow{\text{weak}} \bar{K}^0$$

The transformation matrix that carries the (K^0, $\bar{K}^0$) representation into (K_1^0, K_2^0) representation given by

$$\frac{1}{\sqrt{2}}\begin{pmatrix} 1 & 1 \\ 1 & -1 \end{pmatrix}$$

diagonalizes H as follows,

$$H' = \begin{array}{cc|c} K_1^0 & K_2^0 & \\ \hline \mu + v & 0 & K_1^0 \\ 0 & \mu - v & K_2^0 \end{array}$$

Note that the $K_1 - K_2$ mass difference is given by $2v$, which is twice the matrix element that connects K^0 and $\bar{K}^0$. Now in a theory in which the $|\Delta S| = 1$ rule holds, v is of second order in the weak interaction constant G since the strangeness of K^0 differs from that of $\bar{K}^0$ by two units. Meanwhile the decay width of K_1, $\Gamma = \hbar/\tau_1$, is also of the order G^2; so we expect that the K_1-K_2 mass difference is of the order of the K_1 decay width, which is indeed the case experimentally. On the other hand, Okun and Pontecorvo (1957) point out that if $|\Delta S| = 2$ transitions were allowed by a first-order weak interaction, then the matrix element v would now be of the order G, hence about 10^7 times larger than $\hbar/\tau_1$. In other words, the

K_1-K_2 mass difference would be about 10^7 times larger than the observed value, and the oscillation time between K^0 and $\bar{K}^0$ would be of the order of 10^{-17} sec. rather than 10^{-10} sec. (i.e., a K^0 would lose its strangeness identity immediately after it is created). So we must conclude that $|\Delta S| = 2$ transitions are forbidden in lowest order in weak interactions. (See, however, Glashow (1961) for an example of weak interaction Lagrangian that allows $\Xi \rightarrow N + \pi$ but still makes the K_1-K_2 mass difference small.)

Since both electric charge and baryon number are rigorously conserved, the $|\Delta S| = 1$ rule is completely equivalent to the rule $|\Delta T_3| = \frac{1}{2}$. The simplest way to guarantee this is to assume $|\Delta \mathbf{T}| = \frac{1}{2}$ which, of course, is a more stringent rule, and which leads to more specific predictions that can be tested by experiments. The $|\Delta \mathbf{T}| = \frac{1}{2}$ rule was first suggested by Gell-Mann and Pais (1954).

What are some of the consequences of this $|\Delta \mathbf{T}| = \frac{1}{2}$ rule? Consider

$$K \rightarrow 2\pi$$

The rule says that the final two-pion system must be either in $T = 0$ or $T = 1$ since the K particle has a half unit of isospin. From Bose-Einstein statistics and from the fact that the spatial part of the two-pion wave function must be symmetric for $L = 0$, the isospin wave function for the two-pion system must also be symmetric. But the $T = 1$ wave function is necessarily anti-symmetric (cf. Appendix A), e.g.

$$|T = 1, T_3 = 0\rangle = \frac{1}{\sqrt{2}}(-|1, -1\rangle + |-1, 1\rangle)$$

where $|1, -1\rangle$ means that the first pion has $T_3 = 1$, the second $T_3 = -1$. (A more elegant way to prove this is to note that the only bilinear combination of two unit vectors $\hat{n}_1$ and $\hat{n}_2$ we can form that transforms like one unit of angular momentum, namely $\hat{n}_1 \times \hat{n}_2$, is necessarily odd under $\hat{n}_1 \rightleftarrows \hat{n}_2$.) Thus if the $|\Delta \mathbf{T}| = \frac{1}{2}$ rule holds, the 2π must be in $T = 0$. But as far as

$$K^+ \rightarrow \pi^+ + \pi^0$$

is concerned, $T = 0$ in the final state is impossible since $T_3 = 1$. In contrast the decay products of K_1^0 $(\rightarrow \pi^+ + \pi^-, 2\pi^0)$ can be in $T = 0$. Hence

$$R = \frac{K^+ \rightarrow \pi^+ + \pi^0}{K_1^0 \rightarrow 2\pi} = 0$$

Experimentally the $\pi^+\pi^0$ rate is 2×10^7 sec.$^{-1}$ whereas the K_1^0 decay rate is 10^{10} sec.$^{-1}$. The allowedness of the $\pi^+\pi^0$ mode might be accounted for by electromagnetic corrections to the $|\Delta \mathbf{T}| = \frac{1}{2}$ rule. However, simple considerations lead to $R \sim (1/137)^2$ rather than 1/500 as experimentally observed. (If we analyze $K^+ \rightarrow \pi^+ + \pi^0$ using $|\Delta \mathbf{T}| = \frac{3}{2}$, the $|\Delta \mathbf{T}| = \frac{3}{2}$ amplitude necessary to account for the $\pi^+\pi^0$ rate turns out to be about

4% of that leading to the $K_1^0 \to 2\pi$ via the $|\Delta \mathbf{T}| = \frac{1}{2}$ channel.) As for the $\pi^+\pi^-$ to $2\pi^0$ ratio, the strict $|\Delta \mathbf{T}| = \frac{1}{2}$ rule predicts (cf. Appendix A)

$$\frac{K_1^0 \to 2\pi^0}{K_1^0 \to \pi^+ + \pi^-, 2\pi^0} = \frac{1}{3}$$

whereas if we allow the small admixture of $|\Delta \mathbf{T}| = \frac{3}{2}$ necessary to account for $K^+ \to \pi^+ + \pi^0$, this ratio can be any number between 0.28 and 0.38. Experimental results of the Berkeley and Michigan groups essentially give 1/3 for this ratio (Crawford *et al.*, 1959a; Brown *et al.*, 1959).

The $|\Delta \mathbf{T}| = \frac{1}{2}$ rule can also be tested in $K_{\pi 3}$ decay

$$K^+ \to \pi^+ + \pi^+ + \pi^- \quad (\tau \text{ mode})$$
$$K^+ \to \pi^+ + \pi^0 + \pi^0 \quad (\tau' \text{ mode})$$
$$K_2^0 \to \pi^+ + \pi^- + \pi^0, 3\pi^0$$

especially if the final state wave function of the three pions has a simple symmetry property (Wentzel, 1960a). The predictions for the decay rate ratios are unique if the spatial wave function is totally symmetric, which condition seems approximately satisfied for the τ^+ mode. We shall not discuss $K_{\pi 3}$ decay any further. The reader may refer to Gell-Mann and Rosenfeld (1957), Dalitz (1959), and Weinberg (1960a).

We now turn our attention to hyperon decay. For Λ^0 decay the $|\Delta \mathbf{T}| = \frac{1}{2}$ rule says that the πN system must be in $T = \frac{1}{2}$, from which follows

$$\frac{\Lambda^0 \to n + \pi^0}{\Lambda^0 \to p + \pi^-, n + \pi^0} = \frac{1}{3}$$

The Berkeley and Michigan groups have shown that this expected ratio is in excellent agreement with experiments to an accuracy of about 10%. In addition the $|\Delta \mathbf{T}| = \frac{1}{2}$ rule predicts that the asymmetry parameters must be the same for the $p\pi^-$ mode and the $n\pi^0$ mode, which has also been confirmed experimentally:

$$\frac{\alpha(\Lambda^0 \to n + \pi^0)}{\alpha(\Lambda^0 \to p + \pi^-)} = 1.10 \pm 0.27$$

according to Cork *et al.* (1960).

More interesting and striking are the predictions of the $|\Delta \mathbf{T}| = \frac{1}{2}$ rule for the various asymmetry parameters in Σ decay:

$$\Sigma^+ \to p + \pi^0$$
$$\Sigma^+ \to n + \pi^+$$
$$\Sigma^- \to n + \pi^-$$

With each decay channel we associate a pair of amplitudes corresponding to the s state and the p state emission of π. Assume CP invariance; since the final state interactions are not too important, the s and p wave amplitudes are roughly real. It is convenient to consider a two-dimensional vector $\vec{M}$ in an s-p plane such that the components of $\vec{M}$ along the orthogonal s and p axes correspond respectively to the s and p wave amplitudes in Σ decay (Gell-Mann and Rosenfeld, 1957). Now the change of isospin by a half unit in Σ decay is completely equivalent to the conservation of isospin in the process

$$\text{“spurion”} + \Sigma \rightarrow \pi + N \tag{10.18}$$

where the "spurion" stands for a hypothetical particle that bears no energy, no momentum, no angular momentum, no electric charge, but just a half unit of isospin (Wentzel, 1956b). The spurion is like K^0 as far as its isospin properties are concerned so that the isospin properties of (10.18) are identical with those of associated production $\pi + N \rightarrow K + \Sigma$ discussed earlier. Hence we can derive a triangular relation analogous to (10.5), which takes the form

$$\sqrt{2}\,\vec{M}_0 + \vec{M}_+ - \vec{M}_- = 0$$

where the subscripts 0, +, and − refer to $\pi^0 p$, $\pi^+ n$, and $\pi^- n$ respectively. The magnitude of $\vec{M}$ is the square root of the decay rate in question since

$$|\vec{M}|^2 = |a_s|^2 + |a_p|^2$$

Experimentally the decay rates of the three modes are roughly equal. This means that $\sqrt{2}\,\vec{M}_0$, $\vec{M}_+$, and $\vec{M}_-$ must form the three sides of a right isosceles triangle in the s-p plane such that $\vec{M}_+$ and $\vec{M}_-$ subtend the 90° corner.

The orientation of the triangle is *not* known *a priori* by the $|\Delta \mathbf{T}| = \frac{1}{2}$ rule. But suppose $\vec{M}_0$ makes 45° with the s axis, which physically means that $\alpha(p\pi^0) = [a_s a_p/(|a_s|^2 + |a_p|^2)]_{p\pi^0}$ is very large in magnitude. Then for $n\pi^-$ and $n\pi^+$ the asymmetry parameters must be vanishingly small since $n\pi^-$ with $\vec{M}_-$ along the $p(s)$ axis is in a pure p state (s state), and $n\pi^+$, in a pure s state (p state).

Experimentally the $p\pi^0$ up-down asymmetry for Σ^+ produced in π^+-p collisions is indeed large,

$$|\alpha(p\pi^0)P_{\Sigma^+}| \sim 0.75$$

whereas the $n\pi^+$ asymmetry parameter is practically zero, $|\alpha(n\pi^+)P_{\Sigma^+}| < 0.15$, as is also the $n\pi^-$ up-down asymmetry. The absence of the up-down asymmetry in the $n\pi^-$ mode may be due to the lack of Σ^- polarization in the strong production reaction $\pi^- + p \rightarrow K^+ + \Sigma^-$; this point can be checked by studying whether $\Sigma^+ \rightarrow p + \pi^0$ from the charge-symmetric reaction $\pi^+ + \text{“}n\text{”} \rightarrow K^0 + \Sigma^+$ (where "n" stands for the neutron in the deuterium nucleus) exhibits a large up-down asymmetry, since we already know that the $p\pi^0$ mode is an excellent analyzer of the Σ^+ polarization.

In any case the *ratio* of the $p\pi^0$ up-down asymmetry to the $n\pi^+$ up-down asymmetry is real since the parent Σ^+'s must be in the same condition in both cases. These experimental observations (Cool *et al.*, 1959; Cork *et al.*, 1960) fit beautifully to the $|\Delta \mathbf{T}| = \frac{1}{2}$ rule, but, of course, there may be other ways of explaining this situation. If we assume the $|\Delta \mathbf{T}| = \frac{1}{2}$ rule, the experimental situation is summarized in Fig. 10.5. The fact that the

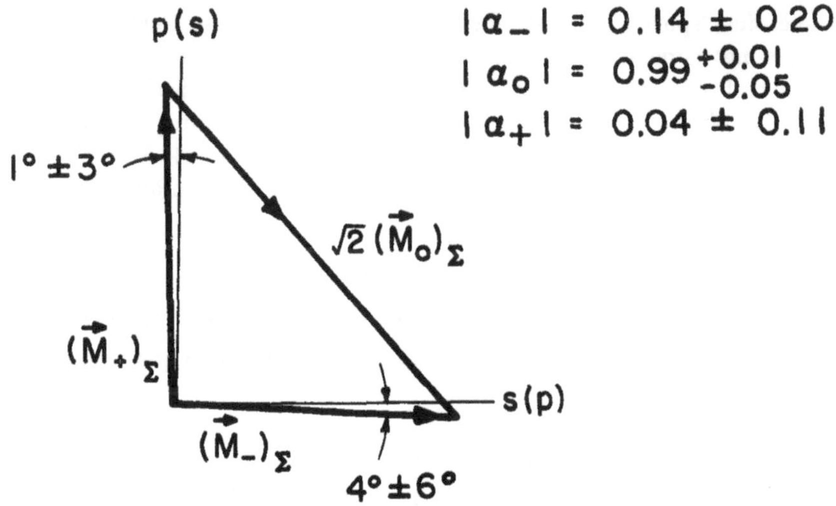

Figure 10.5. $|\Delta \mathbf{T}| = \frac{1}{2}$ rule and Σ decay

two sides of the triangle coincide almost exactly with the s and the p direction may be an indication of a symmetry principle even more stringent than the $|\Delta \mathbf{T}| = \frac{1}{2}$ rule. (On the other hand, the possibility that this remarkable feature is accidental is not completely excluded either.)

The empirical fact that the $|\Delta \mathbf{T}| = \frac{1}{2}$ rule works well in many instances is somewhat puzzling for the following reason. We have already emphasized that beta decay, muon decay and muon capture fit nicely to the view that the basic weak interaction Hamiltonian can be written as an interaction of a charged *VA* current with itself. If we extend this view to strange particles, we have $(\bar{\Lambda}p)(\bar{p}n)$ etc. We may ask whether the total four-fermion interaction $(\bar{\Lambda}p)(\bar{p}n)$ gives rise to nonleptonic decays of strange particles satisfying $|\Delta \mathbf{T}| = \frac{1}{2}$. Now $(\bar{p}n)$ leads to an isospin change of one unit since n with $T_z = -\frac{1}{2}$ is converted into p with $T_z = \frac{1}{2}$. Similarly $(\bar{\Lambda}p)$ changes isospin by $\frac{1}{2}$ unit. If we add $T = 1$ and $T = \frac{1}{2}$, in general we get both $T = \frac{1}{2}$ and $T = \frac{3}{2}$. This means that the four-fermion interaction $(\bar{\Lambda}p)(\bar{p}n)$ gives rise to both $|\Delta \mathbf{T}| = \frac{1}{2}$ and $|\Delta \mathbf{T}| = \frac{3}{2}$. It may be that the strong couplings are so arranged that the $|\Delta \mathbf{T}| = \frac{1}{2}$ channel gets enhanced in some mysterious manner (see, e.g., Oneda, Pati, and Sakita, 1960).

It is possible to write down current-current interactions that result in the $|\Delta \mathbf{T}| = \frac{1}{2}$, $|\Delta S| = 1$ rule in a rigorous fashion independently of the details of strong interactions. However, such attempts necessarily involve the use of neutral baryonic currents, e.g., $(\overline{\Lambda}p)(\bar{p}n) + (\overline{\Lambda}n)(\bar{n}n)$. If we include neutral currents for strongly interacting particles, then the absence of neutral leptonic currents (e.g., $(\bar{e}e)$, $(\bar{e}\mu)$) becomes even more mysterious.

To sum up, there are essentially two schools of thought concerning the $|\Delta \mathbf{T}|$ rule. The exponents of one school argue that although the basic weak coupling Lagrangian admits both $|\Delta \mathbf{T}| = \frac{1}{2}$ and $|\Delta \mathbf{T}| = \frac{3}{2}$, the strong interactions are of such a nature as to enhance the $|\Delta \mathbf{T}| = \frac{1}{2}$ channel (or suppress the $|\Delta \mathbf{T}| = \frac{3}{2}$ channel). The exponents of the second school prefer to accept the $|\Delta \mathbf{T}| = \frac{1}{2}$ rule as a basic rule inherent in the weak coupling Lagrangian. The first approach relies heavily on the properties of the strong couplings about which we know practically nothing. The second approach seems somewhat *ad hoc* in the sense that neutral baryonic currents are allowed to participate in the basic weak couplings whereas neutral leptonic currents are excluded. (For yet other approaches, see Salam and Ward, 1960; McCliment and Nishijima, 1962).

In the following, we restrict ourselves to discussions of strangeness changing currents which also change electric charge. In writing down $|\Delta Q| = 1$ strangeness changing current like $(\overline{\Lambda}p)$ it is customary to assume that the direction of electric charge change coincides with the direction of strangeness change, i.e., $\Delta S/\Delta Q = +1$ not -1. To see how this rule works let us consider $(\overline{\Lambda}p)$:

$$\frac{S(\Lambda) - S(p)}{Q(\Lambda) - Q(p)} = +1$$

In contrast, for $(\overline{\Sigma^+}n)$ we have

$$\frac{S(\Sigma^+) - S(n)}{Q(\Sigma^+) - Q(n)} = -1$$

Thus a current like $(\overline{\Sigma^+}n)$ violates this rule. Suppose $(\overline{\Sigma^+}n)$ were present. Then we would get $\Delta S = 2$ for weak processes since $(\bar{n}\Sigma^+)(\overline{\Sigma^0}\Xi^-)$ would give

$$\Xi^- \rightarrow \Sigma^0 + \overline{\Sigma^+} + n \rightarrow \pi^- + n$$

For this reason the $\Delta S/\Delta Q$ rule is sometimes called the "extended" $|\Delta S| = 1$ rule. This rule forbids

$$\Sigma^+ \rightarrow n + e^+ + \nu,\ n + \mu^+ + \nu$$

but allows

$$\Sigma^- \rightarrow n + e^- + \bar{\nu},\ n + \mu^- + \bar{\nu}$$

Recently the existence of the beta decay of Σ^- has been confirmed experimentally whereas no example of Σ^+ beta decay has been reported (Franzini and Steinberger, 1961; Humphrey *et al.*, 1961). Further the rule allows

$$K^0 \rightarrow \pi^- + \begin{pmatrix} \mu^+ \\ e^+ \end{pmatrix} + \nu$$

$$\bar{K}^0 \rightarrow \pi^+ + \begin{pmatrix} \mu^- \\ e^- \end{pmatrix} + \bar{\nu}$$

but forbids

$$K^0 \rightarrow \pi^+ + \begin{pmatrix} \mu^- \\ e^- \end{pmatrix} + \bar{\nu}$$

$$\bar{K}^0 \rightarrow \pi^- + \begin{pmatrix} \mu^+ \\ e^+ \end{pmatrix} + \nu$$

If these selection rules are experimentally confirmed, we may use the time variation of the $(\pi^+e^-)/(\pi^-e^+)$ ratio to determine the K_1^0-K_2^0 mass difference along the lines discussed in Section 4 (Treiman and Sachs, 1956).

Whatever the final word on the $|\Delta \mathbf{T}| = \frac{1}{2}$ rule for nonleptonic processes may be, the hypothesis that the strangeness changing current (not the total four-fermion interaction) has the property of changing $\mathbf{T}$ by a half unit leads to interesting predictions for the leptonic modes of K particles. For instance,

$$\frac{K_2^0 \rightarrow \pi^\pm + \mu^\pm + \nu}{K^+ \rightarrow \pi^0 + \mu^+ + \nu} = \frac{K^0 \rightarrow \pi^- + \mu^+ + \nu}{K^+ \rightarrow \pi^0 + \mu^+ + \nu} = 2$$

which is in rough accord with experiments. (See, e.g., Okubo *et al.*, 1958; Kobzarev and Okun, 1958.) In addition, the spectra shapes of the various $(\pi\mu\nu)$ modes must be identical.

Note added in proof. Recent experiments seem to show that the $\Delta S/\Delta Q = 1$ rule might be violated. To date there are three experimental indications against this rule.

(i) Barbaro-Galtieri *et al.* (1962) at Berkeley report a hyperon decay event which is most likely an example of the $\Delta S/\Delta Q \neq 1$ process

$$\Sigma^+ \rightarrow n + \mu^+ + \nu$$

(ii) In a neutral K beam produced by charge exchange scattering of $K^+(K^+ + n \rightarrow K^0 + p)$, we may study the branching ratio $(\pi^+e^-\bar{\nu})/(\pi^-e^+\nu)$ as a function of time of traversal. If the $\Delta S/\Delta Q = 1$ rule were valid, this ratio would start with zero and remain nearly zero for a time interval much smaller than $\tau_1 \approx 1/|\mu_1 - \mu_2|$, as can be seen from (10.16) and (10.17). In a Berkeley-Wisconsin-Padua collaboration experiment Ely *et al.* (1962a) have found a few $\pi^+e^-\bar{\nu}$ decays at $t \ll \tau_1$. Moreover, the distribution of $\pi^-e^+\nu$ events does not seem to follow at all the distribution of the $\bar{K}^0$ component (measured by the rate of hyperon production).

(iii) On the basis of the $\Delta S/\Delta Q = 1$ rule and CP invariance, we expect that $\bar{K}^0 \to \pi^+ + e^- + \bar{\nu}$ and $K^0 \to \pi^- + e^+ + \nu$ are allowed with a *common* rate whereas $\bar{K}^0 \to \pi^- + e^+ + \nu$ and $K^0 \to \pi^+ + e^- + \bar{\nu}$ are forbidden. It then follows from (10.16) and (10.17) that the total number of electronic (both e^- and e^+) decays of neutral K particles originated by K^+ charge exchange scattering (or by the process $\pi^- + p \to \Lambda + K^0$) must be of the form

$$N^{(+)}(t) + N^{(-)}(t) \propto \exp(-\lambda_1 t) + \exp(-\lambda_2 t)$$

Experimentally the ratio of the coefficient of $\exp(-\lambda_1 t)$ to that of $\exp(-\lambda_2 t)$ (commonly known as Γ_1/Γ_2) is $11.9^{+7.5}_{-5.6}$ according to Ely *et al.* (1962a) and $6.6^{+6.0}_{-4.1}$ according to Alexander, Almeida, and Crawford (1962) at Berkeley. Since these numbers are significantly different from unity, the $\Delta S/\Delta Q = 1$ rule does not seem very likely.

Note that a violation of the $\Delta S/\Delta Q = 1$ rule also implies a violation of the $T = \frac{1}{2}$ current rule since T_3 must change by $\frac{3}{2}$ in $\Delta S/\Delta Q = -1$ transitions (e.g. $\bar{K}^0$ with $T_3 = \frac{1}{2}$ changes into π^- with $T_3 = -1$). Theoretical implications of $T = \frac{3}{2}$ currents in strangeness changing weak interactions can be found in Behrends and Sirlin (1962).

CHAPTER 11

Unsolved Problems

We have discussed various invariance principles and their applications to elementary particle physics. There are still several important problems unexplored and numerous questions unanswered.

First of all, both from the experimental and theoretical point of view, it is of considerable importance to examine the limits on the validity of various symmetry laws especially at higher energies or, equivalently, at shorter distances. For instance, parity conservation and isospin conservation in strange particle reactions in Bev. regions should be tested with greater accuracies. The reason why we place so much emphasis on *higher* energies is that some physicists have speculated on the possibility that local field theory might not be applicable to energies of the order of a few Bev. or more, or to distances of the Compton wave-length of the proton or shorter. Such a possibility would be very serious. After all, local field theory, which has had some success in quantum electrodynamics and low energy pion physics is the only language theoreticians know. Should we encounter an unfamiliar world in which the properties of space-time as we know them today must be modified in a fundamental manner, we would not even have an adequate language to express the strange situations.

It is customary to ask two questions about the present-day field theory: (1) Is it consistent? (2) Is it correct? We can conceive of the following possibility which would be extremely challenging. Field theory is neither consistent nor correct, yet the symmetry laws are still valid, e.g. parity conservation and strangeness conservation in the strong interactions, *CPT* invariance in all interactions etc. hold exactly. It may well be that the requirements imposed by the *CPT* theorem turn out to be of a far greater generality than our present field theory itself on the basis of which the theorem has been "proved." In such cases only arguments based on symmetries are reliable and are on a firm and permanent basis.

Turning now to more specific problems, we may ask whether there is any other symmetry we have missed. In discussing strong interactions of strange particles we have seen that if we restrict ourselves to Yukawa type couplings, there are eight coupling constants to be determined. The eight constants are entirely independent if charge independence turns out to be the only internal symmetry realized in the strong interactions. One may argue on aesthetic grounds that nature could not possibly have put as many as eight unrelated constants. There have been proposals to impose further restrictions among these constants. Such restrictions imply that the strong interactions exhibit symmetries higher than the symmetry implied by charge independence. Or conversely, just as charge independence imposes restrictions such as $G(\pi^+\bar{p}n) = \sqrt{2}G(\pi^0\bar{p}p)$, a higher symmetry would impose restrictions among the eight constants which are otherwise unrelated.

It is important to note that too high a symmetry is incompatible with experiments. Pais (1958a,b) has shown that the following coupling constant combination with even $\Lambda\Sigma$ parity leads to contradictions with experiments:

$$G(\pi\bar{\Sigma}\Sigma) = G(\pi\bar{\Lambda}\Sigma), \quad G(K\bar{N}\Lambda) = G(K\bar{N}\Sigma), \quad G(K^G\bar{\Xi}\Lambda) = G(K^G\Xi\Sigma) \tag{11.1}$$

This relation implies that as far as the interaction Lagrangian is concerned, Λ and Σ can be regarded as two doublets

$$Y = \begin{pmatrix} \Sigma^+ \\ \frac{1}{\sqrt{2}}(\Lambda^0 - \Sigma^0) \end{pmatrix}, \qquad Z = \begin{pmatrix} \frac{1}{\sqrt{2}}(\Lambda^0 + \Sigma^0) \\ \Sigma^- \end{pmatrix}$$

as well as one singlet and one triplet, and the K particle can be regarded as two singlets, K^+ and K^0, as well as one doublet. For instance,

$$i\boldsymbol{\pi}\cdot(\bar{\boldsymbol{\Sigma}}\times\boldsymbol{\Sigma}) + \boldsymbol{\pi}\cdot\bar{\Lambda}\boldsymbol{\Sigma} + \boldsymbol{\pi}\cdot\bar{\boldsymbol{\Sigma}}\Lambda = (\bar{Y}\boldsymbol{\tau}Y)\cdot\boldsymbol{\pi} + (\bar{Z}\boldsymbol{\tau}Z)\cdot\boldsymbol{\pi}$$

$$\bar{N}K\Lambda + \bar{N}\boldsymbol{\tau}\cdot\boldsymbol{\Sigma}K = K^0\bar{N}Y + K^+\bar{N}Z$$

In this manner we can construct a new isospin scheme such that each strongly interacting particle can be characterized by two kinds of isospins T_A and T_B with the *usual* isospin given by $\mathbf{T} = \mathbf{T}_A + \mathbf{T}_B$ as in Table 11.1.

TABLE 11.1

Pais' Doublet Scheme

	π	K	N	Λ	Σ	Ξ
T_A	1	0	$\frac{1}{2}$	$\frac{1}{2}$	$\frac{1}{2}$	$\frac{1}{2}$
T_B	0	$\frac{1}{2}$	0	$\frac{1}{2}$	$\frac{1}{2}$	0
T	1	$\frac{1}{2}$	$\frac{1}{2}$	0	1	$\frac{1}{2}$

It is assumed that T_A and T_B commute. In this scheme $\mathbf{T}_A$ and $\mathbf{T}_B$ are separately conserved. Now consider

$$K^- + p \rightarrow \pi^- + \Sigma^+$$

Since $T_{A3}(K^-) = 0$, $T_{A3}(p) = \frac{1}{2}$, $T_{A3}(\pi^-) = -1$, $T_{A3}(\Sigma^+) = \frac{1}{2}$, this reaction should be forbidden to the extent that Pais' doublet symmetry is exact. Roughly, we expect that, if (11.1) were correct, the reaction would have zero cross section to the extent that the $\Lambda\Sigma$ mass difference could be ignored, or the reaction cross section would be of the order of

$$\left(\frac{M_\Sigma - M_\Lambda}{2M_\Lambda}\right)^2 = 0.005$$

compared to the fully allowed reaction $K^- + p \rightarrow \pi^+ + \Sigma^-$. Experimentally the two reaction cross sections are of the same order of magnitude. It is easy to show that this doublet symmetry would "forbid" many other reactions (such as $\pi^+ + p \rightarrow K^+ + \Sigma^+$, $K^+ + n \rightarrow K^0 + p$) that have been frequently observed in our laboratory.

Pais (1958b) further shows that if there is any relation stronger than the usual triangular relation among associated production processes

$$\pi + N \rightarrow \Lambda + K, \quad \Sigma + K$$

then it must necessarily be the relation that follows from the doublet symmetry implied by (11.1), which forbids $\pi^+ + p \rightarrow \Sigma^+ + K^+$ and makes the angular distributions of $\Sigma^- K^+$ production identical to that of $\Lambda^0 K^0$ production, in disagreement with experiments.

In spite of this many symmetry models have been invented in which relations among various coupling constants are postulated. One symmetry model that has received more attention than others is the so-called "global symmetry" model, in which the pion-baryon couplings are assumed to be universal, $G(\pi\bar{N}N) = G(\pi\bar{\Lambda}\Sigma) = G(\pi\bar{\Sigma}\Sigma) = G(\pi\bar{\Xi}\Xi)$ (Wigner, 1949, 1952; Gell-Mann, 1957; Schwinger, 1957). This symmetry model necessarily requires that the relative Λ-Σ parity be even; otherwise the equality between the $\pi\bar{\Lambda}\Sigma$ coupling and the $\pi\bar{N}N$ coupling would be meaningless. In the global symmetry model the K couplings must be of such a nature as to destroy the high symmetry of the pion couplings; otherwise the symmetry would be too high. It is conjectured that in the absence of the K couplings all baryons would be degenerate just as in the absence of the electromagnetic couplings the proton and the neutron would be degenerate. If the K couplings were weak, global symmetry might be useful in the sense that certain definite predictions among strange particle reactions may be made. The various "tests" of the global symmetry model proposed for $K^- + p \rightarrow \Sigma + \pi$ are in disagreement with experiments. (See, e.g., Salam, 1959.) Crudely speaking, the disagreement stems from the fact that the global symmetry model, *if it is to be useful*, predicts that the phase difference between the $T = 0$ and $T = 1$ S wave $\pi\Sigma$ scattering is necessarily small ($\sim 10°$) at the K^-p threshold while the phase shift difference deduced from experiments by Dalitz and Tuan (1960) is as large as 65° in magnitude. (We have assumed that the relative $K\Sigma$ parity is odd. Essentially the same argument holds for the even parity case.) Dalitz and Tuan further find that any perturbative treatment of the K couplings in low energy K^-p reactions is completely unjustified. To sum up then, either the fundamental assumptions of the global symmetry model are wrong, or the postulated symmetry of the pion couplings is completely masked by the K couplings which are unfortunately also "strong."

In any critical evaluation of a symmetry model of strong interactions the following questions arise. Are the pion and the K particle "truly elementary," or are they bound states of baryon-antibaryon pairs? Are all baryons elementary, or are there some baryons which are more elementary than others? If the pion is a compound state of some baryon-antibaryon pair, the Yukawa-type couplings of the pion to baryons are not fundamental; rather they are to be regarded as phenomenological manifestations of some basic couplings.

In strong interaction physics we are accustomed to regard states that are stable against decay via strong interactions as "elementary particles." It is instructive to recall that if the Σ mass were higher by 60 Mev., the Σ hyperon would look like a resonant state of the $T = 1$ $\pi\Lambda$ system. Each strongly interacting elementary particle *appears to be* elementary simply because it happens to be the ground state of a system with given baryon number, strangeness and isospin. Is there any distinction between a compound particle and an elementary particle? Is there any distinction between a resonance due to the decay of an unstable elementary particle and a "dynamical resonance" due to an attractive force? Whatever the final answers to these deep questions may be, there is no doubt that resonant states are of the same importance as stable particle states.

Note added in proof. This chapter was written immediately before the rash of new highly unstable particles and resonances in strong interaction physics. These newly discovered states serve to remind us once again that short-lived strongly interacting isobars are at least as important as stable (within the framework of strong interactions) particles in our dynamical understanding of strong interaction physics. Stimulated by these new discoveries, a number of models of higher symmetry have been discussed. Among the various models proposed to date, the octet version of unitary symmetry, commonly called the "eightfold way" (Gell-Mann, 1962; Ne'eman, 1961), seems most promising at the present moment. (See also review papers of Glashow (1963) and Sakurai (1963).)

Let us now consider the weak interactions. With the discovery of parity nonconservation, the re-examination of recoil experiments in beta decay, and the confirmation of the existence of the $\pi^+ \rightarrow e^+ + \nu$ mode at the theoretically predicted rate, it is fair to say that we understand the weak decays of non-strange particles in a much more unified manner than we used to before 1957. However, there still remain a few important questions concerning the weak decays of "old" particles. First of all, there is the question of the validity of the conserved vector hypothesis discussed in

Chapter 9. As yet, there is no conclusive experimental evidence for or against this interesting hypothesis. If the hypothesis does not hold, two very serious questions are raised:

(1) Why are $0 \rightarrow 0$ (no parity change) beta transitions characterized by the same ft values?

(2) Why does g_{VA} in muon decay agree rather well with C_V in beta decay?

Another important question concerns the ratio of C_A to C_V. The slight $\sim 20\%$) departure of this ratio from -1 is usually attributed to renormalization effects due to strong interactions. However, we have no reasons to expect that the renormalization effects are so small. In this connection note that we can formally construct an axial vector current that satisfies the continuity equation

$$M_\mu^{(A)} = i\bar{u}_N(p')\gamma_5\gamma_\mu u_p(p) - 2M_N\bar{u}_n(p')\gamma_5 u_p(p)q_\mu/q^2$$
$$q \equiv p' - p \qquad q_\mu M_\mu^{(A)} = 0 \tag{11.2}$$

Let us postulate that this $M_\mu^{(A)}$ is coupled to the leptonic currents. The first term in (11.2) is what we usually associate with the axial vector current for the nucleon, and gives rise to the A part of beta decay in the conventional theory. The second term contains the propagator for a pseudoscalar meson of zero rest mass. Physically speaking the coupling of the second term to the leptonic current would produce the $e\nu$ or $\mu\nu$ decay of the hypothetical "cloud pion" of zero rest mass. It has been conjectured by Nambu (1960) and by Gell-Mann and Lévy (1960) that the smallness of the renormalization effects for the C_A coupling originates in the approximate nature of axial-vector conservation and that in the limit of zero pion mass the conserved axial-vector current hypothesis would hold exactly.

Perhaps the most attractive feature of the almost-conserved-axial-vector-current hypothesis is that the pion decay rate calculated in the limit where the axial-vector conservation hypothesis holds exactly turns out to agree remarkably well with the observed pion decay rate. Let $G_{\pi N}$ and $G_{\pi\mu}$ be the *ps-ps* πN Yukawa (strong coupling) constant and the *ps-pv* pion decay constant respectively. If we interpret the $C_A 2M_N$ term in

$$C_A M_{5\mu}[\bar{u}_\nu\gamma_\mu(1 + \gamma_5)u_e] = C_A\left[i\bar{u}_n\gamma_5\gamma_\mu u_p - \frac{2M_N\bar{u}_n\gamma_5 u_p q_\mu}{q^2}\right][\bar{u}_\nu\gamma_\mu(1 + \gamma_5)u_e] \tag{11.3}$$

as giving rise to the decay of the "cloud pion," then comparison between this term and the diagram shown in Fig. 11.1 immediately gives a relation among $G_{\pi N}$, $G_{\pi\mu}$, C_A, and M_N,

$$\sqrt{2}\, G_{\pi N}G_{\pi\mu} = 2M_N C_A \tag{11.4}$$

The calculated rate for the pion lifetime on the basis of (11.4) turns out to be 2.7×10^{-8} sec. with $G_{\pi N}^2/4\pi = 13.5$ (corresponding to $f^2 = 0.077$), which is to be compared with the observed lifetime 2.56×10^{-8} sec. This is an extremely interesting result. It essentially states that our "real" world in which the pion mass is finite and in which the axial vector current is only approximately conserved does not differ appreciably from a fictitious world in which the pion mass is zero and axial-vector conservation holds exactly, or, to put it differently, the limit $\mu_\pi^2 \to 0$ is a "gentle" one. (See Bernstein *et al.*, 1960.) In view of this interesting result it seems worth examining a class of theories of strong and weak interactions in which axial-vector conservation is approximately satisfied. (It is to be

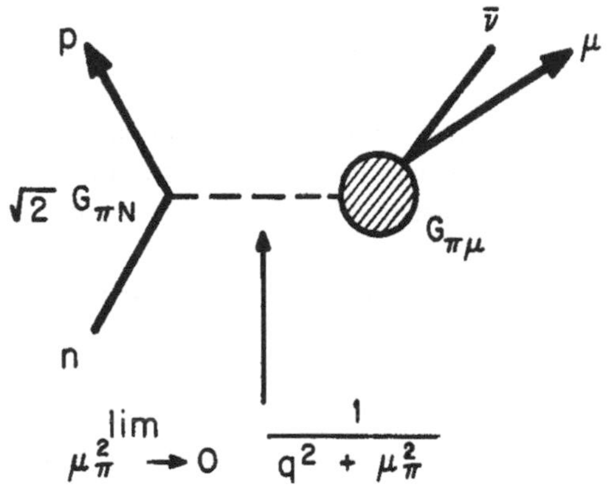

Figure 11.1. Axial-vector conservation and pion decay

pointed out that the formula (11.4) was first obtained by Goldberger and Treiman (1958b) using an entirely different dispersion-theoretic approach with approximations that cannot be rigorously justified.)

The (partial?) success of the almost-conserved-axial-vector-current hypothesis naturally leads to speculations concerning the role played by approximate conservation laws in elementary particle physics. The conservation of axial-vector current is not exact in so far as the pion mass is finite; also approximate is the conservation law of isospin which is not rigorous in so far as there exists the n-p mass difference of 1.3 Mev. and the π^+-π^0 mass difference of 4.6 Mev. In other words, these conservation laws are violated by the mass terms in the Lagrangian. It may be, as Gell-Mann and Zachariasen (1961) speculated, that at extreme high energies where all energies and momenta are considerably greater than the masses of the particles involved, these approximate conservation laws become exact, and symmetries that are obeyed by the kinetic part of the Lagrangian (e.g., $\bar{N}\gamma_\mu \partial_\mu N$), but violated by the mass terms, eventually get recovered.

In the weak interactions of strange particles we still have a number of questions to be settled. First of all, there is no explanation for the empirical fact that the leptonic decay modes of hyperons (e.g., $\Lambda \rightarrow p + e^- + \bar{\nu}$) are rare compared to the rates predicted on the basis of the seemingly attractive hypothesis that $(\overline{\Lambda} p)$ is coupled to $(\bar{e}\nu)$ with the same strength as $(\bar{n}p)$ is coupled to $(\bar{e}\nu)$. Secondly, the $|\Delta \mathbf{T}| = \frac{1}{2}$ rule appears to work very well in many cases. This point is attractive in a certain sense but somewhat surprising as we have emphasized in Chapter 10. It is generally true that if the $|\Delta \mathbf{T}| = \frac{1}{2}$, $|\Delta S| = 1$ rule is to be exact independently of the details of strong interactions, we are forced to include neutral currents $(\bar{n}n)$, $(\overline{\Lambda} n)$, etc. But if we start adding neutral baryonic currents to the tetrahedron, we understand even less the puzzling fact that neutral leptonic currents, e.g., $(\bar{\nu}\nu)$, $(\bar{e}e)$, $(\bar{\mu}e)$ (which would lead to unobserved processes such as $\mu^- + p \rightarrow e^- + p$, $K^+ \rightarrow \mu^+ + e + \pi^+$) are peculiarly absent.

In discussing Σ decay we have noted that the observed relation among the asymmetry parameters may be an indication of a symmetry principle more stringent than the usual $\Delta \mathbf{T}$ rule. Experimentally $\Sigma^+ \rightarrow n + \pi^+$ goes via either the pure p wave channel *or* the pure s wave channel, which is another way of saying that the $n\pi^+$ mode is parity conserving. In contrast, $\Sigma^+ \rightarrow p + \pi^0$ strongly violates parity conservation (in the sense that the asymmetry parameter appears to be almost maximal in magnitude). Meanwhile, because of virtual strong interactions, Σ^+ is part of the time in $\Lambda^0 + \pi^+$, and we know experimentally that the decays of Λ particles into $n + \pi^0$ and into $p + \pi^-$ strongly violate parity conservation. Hence we expect *on general grounds* that both the $n\pi^+$ mode and the $p\pi^0$ mode violate parity conservation

$$\Sigma^+ \rightarrow \Lambda^0 + \pi^+ \rightarrow n + \pi^0 + \pi^+ \rightarrow n + \pi^+$$

$$\Sigma^+ \rightarrow \Lambda^0 + \pi^+ \rightarrow p + \pi^- + \pi^+ \rightarrow p + \pi^0$$

It appears then that if we are to explain in a *rational* way the near parity conservation in the $n\pi^+$ mode and the maximal parity violation in the $p\pi^0$ mode, we must first understand the nature of *both* strong and weak interactions. Quite generally speaking, we cannot construct a theory of non-leptonic decays of strange particles until we have a reliable theory of strong interactions. A variety of interesting proposals have been made along these lines. (See, e.g., Wolfenstein, 1961; Pais, 1961.)

Other interesting unsolved problems in weak interaction physics center around the question of whether the four-fermion interaction proceeds via a two-step process in which a virtual intermediate heavy vector boson, W, is created and annihilated, e.g., $n \rightarrow p + W^- \rightarrow p + e^- + \bar{\nu}$. Historically this idea actually formed part of the original Yukawa theory of nuclear forces even though it later became evident that the boson W cannot be identified with the particle responsible for nuclear forces. To transmit the VA type interaction W must have spin unity. The W field is coupled linearly to the V and A currents discussed in Chapter 7 with a dimensionless coupling constant $G\mu_W^2/4\pi$ (where G is the four-fermion coupling constant in beta decay) which turns out to be $\sim 10^{-6}$ for $\mu_W = M_N$. The W particle must be more massive than the K particle (otherwise the decay process $K \rightarrow \gamma + W$ (or $\pi + W$) would be "medium fast" with lifetime $\sim 10^{-15}$ sec.), and is expected to be produced in reactions like $\pi^+ + p \rightarrow W^+ + p$ with a cross section $\sim 10^{-6}$ times the typical strong interaction cross section. But note that W^+ decays instantly ($\sim 10^{-17}$ sec.) into $e^+ + \nu$, $\mu^+ + \nu$, $\pi^+ + \pi^0$, and, if energetically possible, into $K^+ + \pi^0$. The existence of a W boson may result in a much higher cross section for neutrino-nucleus collisions. This is because the reaction

$$\nu + Z \rightarrow W^+ + e^-(\mu^-) + Z$$

is expected to occur through the pair creation of W^+ and $e^-(\mu^-)$ by ν in the Coloumb field of a target nucleus (Pontecorvo and Ryndin, 1959; Lee and Yang, 1960a; Lee, Markstein, and Yang, 1961).

The intermediate boson theory of weak interactions implies that the four-fermion interaction is non-local in the sense that $(\bar{e}\nu)$ and the $(\bar{n}p)$ current do not interact at the same space-time point. In analogy with the Yukawa theory of nuclear forces, the radius of non-locality is seen to be of the order of the Compton wavelength of the W particle. A very accurate measurement of the Michel parameter in muon decay may set a lower limit on the W mass since the expected deviation of its value from 0.75 can be calculated to be

$$\rho - 0.75 \approx \tfrac{1}{3}(m_\mu/\mu_W)^2$$

It has been pointed out that in order to reconcile with the experimental fact that the reaction $\mu^+ \rightarrow e^+ + \gamma$ seems forbidden (Davis, Roberts, and Zipf, 1959; Berley, Lee, and Bardon, 1959)

$$\frac{\mu^+ \rightarrow e^+ + \gamma}{\mu^+ \rightarrow e^+ + \nu + \bar{\nu}} < 10^{-6}$$

the "neutrino" that appears together with μ must be different from the "neutrino" that appears together with e if the idea of the intermediate boson is correct. Otherwise the reaction $\mu^+ \rightarrow e^+ + \gamma$ occurs via diagrams shown in Fig. 11.2 with branching ratio 10^{-3}–10^{-4} for a reasonable value of cutoff (Feinberg, 1958b).

Note added in proof. We would like to emphasize that even without the W boson hypothesis the process $\mu \rightarrow e + \gamma$ is expected to take place in any theory in which the four-fermion interaction is non-local, and the "muon neutrino" is identical to the "electron neutrino." On the other hand, a strictly local four-fermion interaction would lead to a difficulty with the unitarity requirement for the process $\nu + \mu^- \rightarrow \nu + e^-$ etc. at high energies. Lee and Yang (unpublished) point out that a reasonable radius of non-locality that does not contradict unitarity gives a branching ratio for $\mu \to e + \gamma$ not too different from Feinberg's value $\sim 10^{-4}$, which is to be compared with the most recently obtained upper limit $\sim 10^{-8}$ (Bartlett, Devons, and Sachs, 1962; Frankel *et al.*, 1962). Fortunately the recent high-energy neutrino experiment of Danby *et al.* (1962) indicates that the "muon neutrino" is different from the "electron neutrino." Hence the absence of $\mu \to e + \gamma$ is no longer puzzling.

Another interesting speculation is concerned with the charge and isospin properties of the W particle. We have already remarked that to make the $|\Delta \mathbf{T}| = \frac{1}{2}$ rule rigorous independently of details of strong interactions,

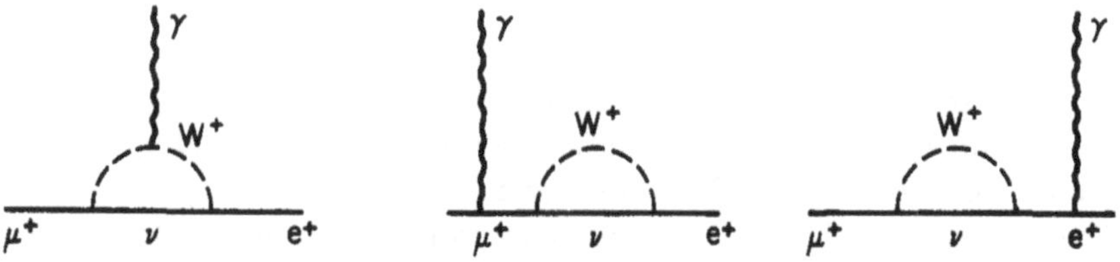

Figure 11.2. $\mu^+ \to e^+ + \gamma$ in the intermediate boson theory with the one neutrino hypothesis

we must include neutral baryonic currents. Lee and Yang (1960b) have shown that, if the $|\Delta \mathbf{T}| = \frac{1}{2}$ rule holds, and if the strict $|\Delta S| = 1$ rule is valid even in the presence of the electromagnetic couplings, it is essential that the intermediate boson W appear as a charge triplet when it is coupled to the strangeness conserving current and as a charge doublet when it is coupled to the strangeness changing current. This implies that there are *two neutral* intermediate bosons W^0 and $\bar{W}^0$ (also $W_a^0 = -\frac{1}{\sqrt{2}}(W^0 + \bar{W}^0)$ and $W_b^0 = \frac{i}{\sqrt{2}}(W^0 - \bar{W}^0)$ in analogy with K_1^0 and K_2^0) in addition to W^+ and W^-. The triplet coupled to the strangeness conserving current is formed by W^+, W_a^0, and W^-, and the doublet coupled to the strangeness nonconserving current is formed by W^+ and W^0 (or by W^- and $\bar{W}^0$). When the strangeness nonconserving current interacts with the strangeness conserving current via the W particles, the $|\Delta \mathbf{T}| = \frac{1}{2}$ rule emerges because of the difference between the two isospin transformation properties of the W particles. Because of this dual character of the isospin properties, the W bosons that can accommodate the $|\Delta \mathbf{T}| = \frac{1}{2}$ rule is called a "schizon."

Note added in proof. We have assumed the $\Delta S/\Delta Q = 1$ rule and the $T = \frac{1}{2}$ current rule as well as the $|\Delta \mathbf{T}| = \frac{1}{2}$ rule for nonleptonic processes. If we try to construct a theory of W particles which leads to the $|\Delta \mathbf{T}| = \frac{1}{2}$ rule for nonleptonic decays and also allows $\Delta S \neq \Delta Q$ transitions in leptonic decays, we need at least six distinct W particles (three particles and their charge conjugates). See Lee (1962) and Takeda (1962).

If the idea of the intermediate boson theory turns out to be correct, it is tempting to speculate on a possible analogy between electric charge conservation and the universal electromagnetic couplings of the photon on the one hand and fermion conservation and the postulated VA coupling of the W particle to the fermion current on the other (Bludman, 1958). Our conjecture might go even further; for every internal conservation law there exists a vector field such that it is coupled linearly to the conserved current in question. What are the properties of the vector mesons connected with baryon conservation, hypercharge conservation, and isospin conservation? Simple invariance considerations show that the vector mesons associated with baryon conservation and hypercharge conservation must have $T = 0$, $S = 0$, odd G, and the vector meson associated with isospin conservation must have $T = 1$, $S = 0$, even G. (Yang and Mills, 1954; Fujii, 1959; Sakurai, 1960a) The postulated vector mesons must be at least as massive as the pion and most likely more massive than three pions at rest, otherwise they would have been found as nearly stable particles. If the couplings of these mesons are strong, they "decay strongly" and appear as resonances of the two-pion and three-pion systems. The need for such resonances in our understanding of the nucleon structure and nuclear forces has already been emphasized in Chapter 9. Perhaps the Yukawa couplings of pions and K particles to baryons are phenomenological manifestation of the couplings of the postulated vector mesons to the appropriate conserved currents of strong interactions.

Note added in proof. The recently discovered ρ and ω mesons (whose properties are discussed in Chapter 9, Section 7) may correspond to two of the three vector mesons conjectured here. Further, in the unitary symmetry model of Gell-Mann (1962) and Ne'eman (1961) there is room for an $S = \pm 1$, $T = \frac{1}{2}$ vector meson coupled to a quasi-conserved strangeness-changing current (strangeness-changing current which would be conserved in the limit where the baryon mass differences etc. go to zero); this may correspond to the experimentally observed 880 Mev. K^* meson. (See also Salam and Ward, 1961.)

In electromagnetism the dynamics of the photon field interacting with the electrically charged fields is intimately interlocked with the conservation law of electric charge. This connection is already apparent in classical electrodynamics in which the continuity equation for the electric charge-current density is a consequence of the Maxwell equations, or conversely, Maxwell introduced the notion of "displacement current" in order that electric charge be conserved even in non-steady state cases. In contrast, conventional theories of strong interactions, whether we consider theories based on Lagrangians of the γ_5 type or ones based on dispersion-theoretic techniques, are essentially phenomenological from a fundamental point of view because they do not establish any deep-lying connection between symmetries and dynamics, nor do they touch the basic question of the origin of strong interactions. One may start with a "globally symmetric" Lagrangian in which all the pion-baryon constants are set to be equal, but there does not seem to be a profound principle that justifies such an approach. Recent attempts to formulate a dynamical theory of strong interactions on the basis of double dispersion relations and unitarity do not seem satisfactory either because if we ask, "Why are there strong interactions?" the answer we get is, "Because there are strong interactions in two other channels." It appears that no genuine progress in the theory of elementary particles will be made until we understand how the laws governing the dynamical behavior of elementary particles are related to the various laws of invariance.

Then there are more deep mysteries. Why does the muon, which has the same electromagnetic and weak couplings as the electron, exist at all, with a mass 207 times the electron mass? Are the masses of elementary particles due solely to the interactions? Why are the strong interactions 10^{14} times stronger than the weak interactions? Why are electric charge and "baryonic charge" quantized? Why is the proton charge equal to the positron charge? Why are some conservation laws approximate? Why does the electromagnetic coupling destroy the isotropy of isospin space in such a way that Q is sometimes displaced with respect to T_3? These and perhaps more perplexing questions which we cannot even conceive of today will fascinate and occupy us in the future. After all, who would have dreamed in 1953 that studies of the decay properties of K particles would lead to a new revolution in our understanding of invariance principles and weak interactions? As we further investigate higher and higher energies, there will probably be many interesting and challenging surprises in store for us.

Appendices

APPENDIX A

Useful Tables of Clebsch-Gordan Coefficients

Given $\vec{J} = \vec{j}_1 + \vec{j}_2$, $M = m_1 + m_2$, where M, m_1, m_2 stand for the z-components of $\vec{J}$, $\vec{j}_1$, $\vec{j}_2$ respectively, the Clebsch-Gordan coefficients $C_{j_1 j_2}(JM; m_1 m_2)$ are defined in such a way that

$$|JM\rangle = \sum_{m_1}\sum_{m_2}^{m_1+m_2=M} C_{j_1 j_2}(JM; m_1 m_2)|j_1 m_1\rangle|j_2 m_2\rangle \tag{A.1}$$

We can regard the Clebsch-Gordan coefficients as elements of the unitary $N \times N$ transformation matrix that carries the (m_1, m_2) representation into the (J, M) representation where

$$N = (2j_1 + 1)(2j_2 + 1) = \sum_{J=|j_1-j_2|}^{J=j_1+j_2} (2J + 1)$$

If we use the usual phase convention for the matrix elements of the ladder operators given by

$$\langle j, m \pm 1|j \pm|j, m\rangle = \sqrt{(j \mp m)(j \pm m + 1)}$$

then the Clebsch-Gordan coefficients are all real. The unitarity and the reality (hence the orthogonality) of the $N \times N$ matrix imply that the inverse transformation from the (J, M) representation to the (m_1, m_2) representation is given simply by

$$|j_1 m_1\rangle|j_2 m_2\rangle = \sum_{J=|j_1-j_2|}^{j_1+j_2} \sum_{M=-J}^{J} C_{j_1 j_2}(J, M; m_1 m_2)|JM\rangle \tag{A.2}$$

where the M summation is to be carried over only once ($M = m_1 + m_2$).

Sometimes $C_{j_1 j_2}(JM; m_1 m_2)$ is denoted by $(j_1 j_2 m_1 m_2|j_1 j_2 JM)$ as in Condon and Shortley (1951) or by $s^{j_1 j_2}_{Jm_1 m_2}$ as in Wigner (1959) or by $C(j_1 j_2; m_1 m_2 m)$ as in Rose (1957). The 3-j symbols which have a number of interesting symmetry properties (Wigner, 1959) are related to the Clebsch-Gordan coefficients by

$$\begin{pmatrix} j_1 j_2 j_3 \\ m_1 m_2 m_3 \end{pmatrix} = \frac{(-1)^{j_1 - j_2 - m_3}}{\sqrt{2j_3 + 1}} C_{j_1 j_2}(j_3, -m_3; m_1 m_2)$$

Although it is possible to obtain a closed expression for the Clebsch-Gordan coefficient, the general formula is rather involved (see Wigner,

1959, p. 191). In elementary particle physics $j_2 = \frac{1}{2}$ and $j_2 = 1$ occur frequently, so we give here the explicit formulas for the two cases. For $j_2 = \frac{3}{2}$ and $j_2 = 2$ the reader may consult Condon and Shortley (1951) p. 76. For an extensive tabulation of the 3-j symbols with $j_1, j_2, j_3 \leq 8$, see Rotenberg *et al.* (1959). Note that if the 3-j symbol is used, (A.1) becomes

$$|JM\rangle = (-1)^{j_2 - j_1 - M} \sum_{m_1} \sum_{m_2} \sqrt{2J+1} \begin{pmatrix} j_1 & j_2 & J \\ m_1 & m_2 & -M \end{pmatrix} |j_1 m\rangle |j_2 m\rangle$$

CASE 1. $j_2 = \frac{1}{2}$

$$\begin{aligned} |J = j_1 + \tfrac{1}{2}, M\rangle &= \sqrt{\frac{j_1 + M + \frac{1}{2}}{2j_1 + 1}}\, |m_1 = M - \tfrac{1}{2}, m_2 = \tfrac{1}{2}\rangle \\ &\quad + \sqrt{\frac{j_1 - M + \frac{1}{2}}{2j_1 + 1}}\, |m_1 = M + \tfrac{1}{2}, m_2 = -\tfrac{1}{2}\rangle \\ |J = j_1 - \tfrac{1}{2}, M\rangle &= -\sqrt{\frac{j_1 - M + \frac{1}{2}}{2j_1 + 1}}\, |m_1 = M - \tfrac{1}{2}, m_2 = \tfrac{1}{2}\rangle \\ &\quad + \sqrt{\frac{j_1 + M + \frac{1}{2}}{2j_1 + 1}}\, |m_1 = M + \tfrac{1}{2}, m_2 = -\tfrac{1}{2}\rangle \end{aligned} \tag{A.3}$$

CASE 2. $j_2 = 1$

$$\begin{aligned} |J = j_1 + 1, M\rangle &= \sqrt{\frac{(j_1 + M)(j_1 + M + 1)}{(2j_1 + 1)(2j_2 + 2)}}\, |m_1 = M - 1, m_2 = 1\rangle \\ &\quad + \sqrt{\frac{(j_1 - M + 1)(j_1 + M + 1)}{(2j_1 + 1)(j_1 + 1)}}\, |m_1 = M, m_2 = 0\rangle \\ &\quad + \sqrt{\frac{(j_1 - M)(j_1 - M + 1)}{(2j_1 + 1)(2j_1 + 2)}}\, |m_1 = M + 1, m_2 = -1\rangle \\ |J = j_1, M\rangle &= -\sqrt{\frac{(j_1 + M)(j_1 - M + 1)}{2j_1(j_1 + 1)}}\, |m_1 = M - 1, m_2 = 1\rangle \\ &\quad + \frac{M}{\sqrt{j_1(j_1 + 1)}}\, |m_1 = M, m_2 = 0\rangle \\ &\quad + \sqrt{\frac{(j_1 - M)(j_1 + M + 1)}{2j_1(j_1 + 1)}}\, |m_1 = M + 1, m_2 = -1\rangle \\ |J = j_1 - 1, M\rangle &= \sqrt{\frac{(j_1 - M)(j_1 - M + 1)}{2j_1(2j_1 + 1)}}\, |m_1 = M - 1, m_2 = 1\rangle \\ &\quad - \sqrt{\frac{(j_1 + M)(j_1 + M)}{j_1(2j_1 + 1)}}\, |m_1 = M, m_2 = 0\rangle \\ &\quad + \sqrt{\frac{(j_1 + M + 1)(j_1 + M)}{2j_1(2j_1 + 1)}}\, |m_1 = M + 1, m_2 = -1\rangle \end{aligned} \tag{A.4}$$

These formulas are extremely useful in adding isospins. In the following we give three typical examples of practical importance.

(a) $T^{(1)} = \frac{1}{2}$, $T^{(2)} = \frac{1}{2}$; KN, NN, $K^G N$ (where $K^G = (\bar{K}^0, -K^-)$), etc.

$$
\begin{aligned}
|T = 1, T_3 = 1\rangle &= |K^+p\rangle \\
|T = 1, T_3 = 0\rangle &= \sqrt{\tfrac{1}{2}}|K^0p\rangle + \sqrt{\tfrac{1}{2}}|K^+n\rangle \\
|T = 1, T_3 = -1\rangle &= |K^0n\rangle \\
|T = 0, T_3 = 0\rangle &= -\sqrt{\tfrac{1}{2}}|K^0p\rangle + \sqrt{\tfrac{1}{2}}|K^+n\rangle
\end{aligned}
\tag{A.5}
$$

(b) $T^{(1)} = 1$, $T^{(2)} = \frac{1}{2}$; πN, πK, ΣK, ΣN, etc.

$$
\begin{aligned}
|T = \tfrac{3}{2}, T_3 = \tfrac{3}{2}\rangle &= |\pi^+p\rangle \\
|T = \tfrac{3}{2}, T_3 = \tfrac{1}{2}\rangle &= \sqrt{\tfrac{2}{3}}|\pi^0p\rangle + \sqrt{\tfrac{1}{3}}|\pi^+n\rangle \\
|T = \tfrac{3}{2}, T_3 = -\tfrac{1}{2}\rangle &= \sqrt{\tfrac{1}{3}}|\pi^-p\rangle + \sqrt{\tfrac{2}{3}}|\pi^0n\rangle \\
|T = \tfrac{3}{2}, T_3 = -\tfrac{3}{2}\rangle &= |\pi^-n\rangle \\
|T = \tfrac{1}{2}, T_3 = \tfrac{1}{2}\rangle &= -\sqrt{\tfrac{1}{3}}|\pi^0p\rangle + \sqrt{\tfrac{2}{3}}|\pi^+n\rangle \\
|T = \tfrac{1}{2}, T_3 = -\tfrac{1}{2}\rangle &= -\sqrt{\tfrac{2}{3}}|\pi^-p\rangle + \sqrt{\tfrac{1}{3}}|\pi^0n\rangle
\end{aligned}
\tag{A.6}
$$

(c) $T^{(1)} = 1$, $T^{(2)} = 1$; $\pi\Sigma$, $\pi\pi$, etc.

$$
\begin{aligned}
|T = 2, T_3 = 2\rangle &= |\pi^+\Sigma^+\rangle \\
|T = 2, T_3 = 1\rangle &= \sqrt{\tfrac{1}{2}}|\pi^0\Sigma^+\rangle + \sqrt{\tfrac{1}{2}}|\pi^+\Sigma^0\rangle \\
|T = 2, T_3 = 0\rangle &= \sqrt{\tfrac{1}{6}}|\pi^-\Sigma^+\rangle + \sqrt{\tfrac{2}{3}}|\pi^0\Sigma^0\rangle + \sqrt{\tfrac{1}{6}}|\pi^+\Sigma^-\rangle \\
|T = 2, T_3 = -1\rangle &= \sqrt{\tfrac{1}{2}}|\pi^-\Sigma^0\rangle + \sqrt{\tfrac{1}{2}}|\pi^0\Sigma^-\rangle \\
|T = 2, T_3 = -2\rangle &= |\pi^-\Sigma^-\rangle \\
|T = 1, T_3 = 1\rangle &= -\sqrt{\tfrac{1}{2}}|\pi^0\Sigma^+\rangle + \sqrt{\tfrac{1}{2}}|\pi^+\Sigma^0\rangle \\
|T = 1, T_3 = 0\rangle &= -\sqrt{\tfrac{1}{2}}|\pi^-\Sigma^+\rangle + \sqrt{\tfrac{1}{2}}|\pi^+\Sigma^-\rangle \\
|T = 1, T_3 = -1\rangle &= -\sqrt{\tfrac{1}{2}}|\pi^-\Sigma^0\rangle + \sqrt{\tfrac{1}{2}}|\pi^0\Sigma^-\rangle \\
|T = 0, T_3 = 0\rangle &= \sqrt{\tfrac{1}{3}}|\pi^-\Sigma^+\rangle - \sqrt{\tfrac{1}{3}}|\pi^0\Sigma^0\rangle + \sqrt{\tfrac{1}{3}}|\pi^+\Sigma^-\rangle
\end{aligned}
\tag{A.7}
$$

APPENDIX B

Dirac Matrices

(I) Summary of Gamma Matrices

(a) Our notation (originally introduced by Pauli (1933)):

$$\{\gamma_\mu, \gamma_\nu\} = 2\delta_{\mu\nu} \qquad \mu = 1, \ldots 4$$

$$\gamma_5 = \gamma_1\gamma_2\gamma_3\gamma_4; \qquad \gamma_\mu, \gamma_5 \text{ are all Hermitian}$$

$$a\cdot b = \vec{a}\cdot\vec{b} + a_4b_4 = \vec{a}\cdot\vec{b} - a_0b_0$$

$$\gamma_k = -i\beta\alpha_k \overset{\text{D.P.}}{=} \begin{pmatrix} 0 & -i\sigma_k \\ i\sigma_k & 0 \end{pmatrix}, \quad k = 1, 2, 3; \qquad \gamma_4 = \beta \overset{\text{D.P.}}{=} \begin{pmatrix} I & 0 \\ 0 & -I \end{pmatrix}$$

$$\gamma_5 = \gamma_1\gamma_2\gamma_3\gamma_4 \overset{\text{D.P.}}{=} -\begin{pmatrix} 0 & I \\ I & 0 \end{pmatrix}, \quad \gamma_5^2 = 1; \qquad \alpha_k \overset{\text{D.P.}}{=} \begin{pmatrix} 0 & \sigma_k \\ \sigma_k & 0 \end{pmatrix}$$

$$\Lambda_+ = \frac{-i\gamma\cdot p + m}{2m} = \frac{-\beta\vec{\alpha}\cdot\vec{p} + \beta p_0 + m}{2m}, \qquad \bar{\psi} = \psi^\dagger\gamma_4$$

(b) The notation of Schweber and Bethe and de Hoffman (1956) (originally introduced by Feynman (1949)):

$$\{\gamma_\mu, \gamma_\nu\} = 2g_{\mu\nu}, \qquad g_{00} = 1, \qquad g_{jk} = -\delta_{jk} \qquad \begin{matrix} \mu = 0, 1, 2, 3 \\ j, k = 1, 2, 3 \end{matrix}$$

$$\gamma_5 = \gamma_0\gamma_1\gamma_2\gamma_3 \overset{\text{D.P.}}{=} i\begin{pmatrix} 0 & I \\ I & 0 \end{pmatrix}, \qquad \gamma_5^2 = -1$$

$$\gamma_k^\dagger = -\gamma_k, \qquad \gamma_0^\dagger = \gamma_0, \qquad \gamma_5^\dagger = -\gamma_5$$

$$a\cdot b = a_0b_0 - \vec{a}\cdot\vec{b}$$

$$\gamma_k = \beta\alpha_k \overset{\text{D.P.}}{=} \begin{pmatrix} 0 & \sigma_k \\ -\sigma_k & 0 \end{pmatrix}, \qquad \gamma_0 = \beta \overset{\text{D.P.}}{=} \begin{pmatrix} I & 0 \\ 0 & -I \end{pmatrix}, \qquad \bar{\psi} = \psi^\dagger\gamma_0$$

$$\Lambda_+ = \frac{\not{p} + m}{2m} = \frac{\gamma\cdot p + m}{2m} = \frac{\beta p_0 - \beta\vec{\alpha}\cdot\vec{p} + m}{2m}$$

$\gamma_0\gamma_\mu^\dagger\gamma_0 = \gamma_\mu$ (to evaluate $\gamma_0 0^\dagger\gamma_0$ merely reverse the order of the γ matrices and let $c \to c^*$)

(II) Useful Relations:

$$\mathrm{Tr}\,\underbrace{\{\gamma_\kappa\gamma_\rho\cdots\gamma_\nu\}}_{\text{odd}} = 0$$

$$(\gamma\cdot a)(\gamma\cdot b) = 2a\cdot b - (\gamma\cdot b)(\gamma\cdot a)$$

$$(\vec{\gamma}\cdot\vec{a})(\vec{\gamma}\cdot\vec{b}) = \vec{a}\cdot\vec{b} + i\vec{\Sigma}\cdot(\vec{a}\times\vec{b}), \qquad \vec{\Sigma} = (\sigma_{23}, \sigma_{31}, \sigma_{12})$$

$$\mathrm{Tr}\,(\Gamma_a) = 0 \text{ where } \Gamma_a = \gamma_\mu, i\gamma_5\gamma_\mu, \gamma_5, \sigma_{\mu\nu} = \frac{1}{2i}(\gamma_\mu\gamma_\nu - \gamma_\nu\gamma_\mu)$$

$$\mathrm{Tr}\,(\underbrace{\gamma_5\gamma_\kappa\cdots}_{\text{less than 4 } \gamma\text{'s.}}) = 0$$

For 4 γ's, nonvanishing only if all four are different.

(III) Other Representations

(a) Majorana Representation:

$$\gamma_1 = \gamma_4^{(\mathrm{D.P.})} = \begin{pmatrix} I & 0 \\ 0 & -I \end{pmatrix}, \gamma_2 = \gamma_2^{(\mathrm{D.P.})} = \begin{pmatrix} 0 & -i\sigma_2 \\ i\sigma_2 & 0 \end{pmatrix},$$

$$\gamma_3 = \gamma_5^{(\mathrm{D.P.})} = \begin{pmatrix} 0 & -I \\ -I & 0 \end{pmatrix}$$

$$\gamma_4 = \gamma_1^{(\mathrm{D.P.})} = \begin{pmatrix} 0 & -i\sigma_1 \\ i\sigma_1 & 0 \end{pmatrix}, \qquad \gamma_5 = \gamma_3^{(\mathrm{D.P.})} = \begin{pmatrix} 0 & -i\sigma_3 \\ i\sigma_3 & 0 \end{pmatrix}$$

(b) Weyl Representation:

$$\gamma_k = \begin{pmatrix} 0 & i\sigma_k \\ -i\sigma_k & 0 \end{pmatrix}, \qquad \gamma_4 = \begin{pmatrix} 0 & I \\ I & 0 \end{pmatrix}, \qquad \gamma_5 = \begin{pmatrix} -I & 0 \\ 0 & I \end{pmatrix}$$

APPENDIX C

Physical Constants

Planck's constant

$$\hbar = 1.054 \times 10^{-27} \text{ erg-sec.} = 6.582 \times 10^{-22} \text{ Mev.-sec.}$$

$$\hbar c = 1.973 \times 10^{-11} \text{ Mev.-cm.}$$

$$\rightarrow 1.973 \times 10^{-14} \text{ cm.} = \lambda\!\!\!^{-} \text{ for } p = 1 \text{ Bev.}/c.$$

Pion-nucleon coupling constant

$$ps\text{-}pv;\ f^2 = 0.076$$

$$ps\text{-}ps;\ \frac{G^2}{4\pi\hbar c} = f^2\left(\frac{2M_N}{\mu_\pi}\right)^2 = 13.5$$

Fine structure constant

$$\alpha = \frac{1}{137.037}$$

Beta decay constant (vector)

$$\begin{aligned} C_V &= \sqrt{(|C_V|^2 + |C_V'|^2)/2} \\ &= 1.00 \times 10^{-49} \text{ erg-cm}^3 \\ &= 6.25 \times 10^{-44} \text{ Mev.-cm}^3 \\ &= \frac{10^{-5}}{\sqrt{2}} M_p c^2 \left(\frac{\hbar}{M_p c}\right)^3 \end{aligned}$$

Proton magnetic moment

$(1.000 + 1.793)(e\hbar/2M_p c)$ (in Gaussian units)

Neutron magnetic moment

$-1.913(e\hbar/2M_p c)$

Dirac radius of the proton

$$\langle r_p^2 \rangle^{1/2} = 0.8 \times 10^{-13} \text{ cm.}$$

Dirac radius of the neutron

$$\langle r_n^2 \rangle^{1/2} \approx 0.0 \times 10^{-13} \text{ cm.}$$

Compton wavelength of the charged pion

$$\frac{\hbar}{\mu_\pi c} = 1.413 \times 10^{-13} \text{ cm.} \approx \sqrt{2} \times 10^{-13} \text{ cm. (to an accuracy of 0.07\%)}$$

$$\left(\frac{\hbar}{\mu_\pi c}\right)^2 = 19.97 \text{ mb. } (1 \text{ mb.} = 10^{-27} \text{ cm}^2)$$

Mass spectrum in units of the charged pion mass

Ξ	9.5
Σ (average)	8.53
Λ	7.98
N (average)	6.71
K (average)	3.55
π (neutral)	0.96
μ	0.756
e	$0.0037 \approx \frac{1}{2 \times 137}$
ν	0
γ	0

REFERENCES

Abashian, A., and Hafner, E. M. *Phys. Rev. Letters 1*, 255 (1958).

Adair, R. K. *Phys. Rev. 100*, 1540 (1955).

Aizu, K. *Proceedings of the International Conference on Theoretical Physics* (Science Council of Japan, Tokyo, 1954).

Akimov, Yu. K., Savchenko, O. V., and Soroko, L. M. *Proceedings of the 1960 Annual International Conference on High Energy Physics at Rochester* (University of Rochester, 1960), p. 49.

Alexander, G., Almeida, S. P., and Crawford, F. S. *Phys. Rev. Letters 9*, 69 (1962).

Alikanov, A. T., Galaktionov, Tu. V., Gorodkov, Yu. V., Eliseyev, G. P., and Lubimov, V. A. *J. Exptl. Theoret. Phys.* (USSR) *38*, 1918 (1960) [translation: Soviet Phys. *JETP 11*, 1380 (1960)].

Alitti, J., Baton, J. P., Berthelot, A., Daudin, A., Deler, B., Goussu, O., Jabiol, M. A., Lewin, C., Neveu-René, M., Rogzinski, A., Shively, F., Laberrigue-Frolow, J., Ouannès, O., Sené, M., Vigneron, L., Albattista, M., Mongelli, S., Romano, A., Waloschek, P., Alles Borelli, V., Benedetti, E., Litvak, J., Puppi, G., and Whitehead, M. *Nuovo Cimento 25*, 365 (1962).

Allen, J. S., Burman, R. L., Herrmannsfeldt, W. B., Stähelin, P., and Braid, T. H. *Phys. Rev. 116*, 134 (1959).

Alston, M. H., Alvarez, L. W., Eberhard, P., Good, M. L., Graziano, W., Ticho, H. K., and Wojcicki, S. G. *Phys. Rev. Letters 5*, 520 (1960).

Alston, M. H., Alvarez, L. W., Eberhard, P., Good, M. L., Graziano, W., Ticho, H. K., and Wojcicki, S. G. *Phys. Rev. Letters 6*, 300 (1961a).

Alson, M. H., Alvarez, L. W., Eberhard, P., Good, M. L., Graziano, W., Ticho, H. K., and Wojcicki, S. G. *Phys. Rev. Letters 6*, 698 (1961b).

Alvarez, L. W., Crawford, F. S., Good, M. L., and Stevenson, M. L. *Phys. Rev. 101*, 503 (1956).

Alvarez, L. W. *Proceedings of the Ninth International Annual Conference on High Energy Physics* (Academy of Sciences, USSR, Moscow, 1960) Vol. 1, p. 47.

Alvarez, L. W., Berge, J. P., Kalbfleisch, G. R., Button-Shafer, J., Solmitz, F. T., Stevenson, M. L., and Ticho, H. K. *Proceedings of the 1962 International Conference on High Energy Physics at CERN* (CERN, Geneva, 1962) p. 433.

Amati, D., and Vitale, B. *Nuovo Cimento 2*, 719 (1955).

Ammar, R. G. *Nuovo Cimento 14*, 1226 (1959).

Anderson, C. D. *Phys. Rev. 43*, 491 (1933).

Anderson, E. W., Bleser, E., Meyer, S., Rosen, J., Rothberg, J., and Wang, I. T. *Proceedings of the 1962 International Conference on High Energy Physics at CERN* (CERN, Geneva, 1962) p. 417.

Anderson, H. L., Fermi, E., Martin, R., and Nagle, D. E. *Phys. Rev. 91*, 155 (1953).

Anderson, H. L., Davidon, W., and Kruse, U. *Phys. Rev. 100*, 339 (1955).

Anderson, H. L., Fujii, T., Miller, R. H., and Tau, L. *Phys. Rev. 119*, 2050 (1960).

Armenteros, R., Budde, R., Montanet, L., Morrison, D. R. O., Nilsson, S., Shapira, A., Vandenmeulen, J., d'Andlau, C., Astier, A., Chesquière, C., Gregory, B., Rahm, D., Rivet, P., and Solmitz, F. *Proceedings of the 1962 International Conference on High Energy Physics at CERN* (CERN, Geneva, 1962a) p. 90.

Armenteros, R., Montanet, L., Morrison, D. R. O., Nilsson, S., Shapira, A., Vandermeulen, J., d'Andlau, Ch., Astier, A., Ballam, J., Chesquière, C., Gregory, B. P., Rahm, D., Rivet, P., and Solmitz, F. *Proceedings of the 1962 International Conference on High Energy Physics at CERN* (CERN, 1962b), p. 351.

Ashkin, J., and Vosko, S. H. *Phys. Rev. 92*, 1248 (1953).

Bacastow, R., Elioff, T., Larsen, R., Wiegand, C., and Ypsillantis, T. *Phys. Rev. Letters 9*, 400 (1962).

Backenstoss, G. K., Frauenfelder, H., Hyams, B. D., Koester, L. J., Jr., and Marin, P. C. *Nuovo Cimento 16*, 1169 (1960).

Backenstoss, G. K., Hyams, B. D., Knop, G., Marin, P. C., and Stierlin, U. *Phys. Rev. Letters 6*, 415 (1961).

Baldo-Ceolin, M., Bonetti, A., Greening, D. B., Limentani, S., Merlin, M., and Vanderhaege, V. *Nuovo Cimento 6*, 84 (1957).

Baltay, C., Courant, H., Fickinger, W. J., Fowler, E. C., Kraybill, H. L., Sandweiss, J., Sanford, J. R., Stonehill, D. L., and Taft, H. D. *Revs. Mod. Phys. 33*, 374 (1961).

Barbaro-Galtieri, A., Barkas, W. H., Heckman, H. H., Patrick, J. W., and Smith, F. M. *Phys. Rev. Letters 9*, 26 (1962).

Bardin, R. K., Barnes, C. A., Fowler, W. A., and Seeger, P. A. *Phys. Rev. 127*, 583 (1962).

Bardon, M., Landé, K., and Lederman, L. M. *Ann. Phys. 5*, 156 (1958).

Bardon, M., Berley, D., and Lederman, L. M. *Phys. Rev. Letters 2*, 56 (1959).

Bardon, M., Franzini, P., and Lee, J. *Phys. Rev. Letters 7*, 23 (1961).

Bartlett, D., Devons, S., and Sachs, A. M. *Phys. Rev. Letters 8*, 120 (1962).

Bastien, P. L., Berge, J. P., Dahl, O. I., Ferro-Luzzi, M., Miller, D. H., Murray, J. J., Rosenfeld, A. H., and Watson, M. B. *Phys. Rev. Letters 8*, 114 (1962).

Beall, E. F., Cork, B., Keefe, D., Murray, P. G., and Wentzel, W. A. *Phys. Rev. Letters 7*, 285 (1961).

Bég, M. A. B. *Phys. Rev. Letters 9*, 67 (1962).

Behrends, R. E., and Sirlin, A. *Phys. Rev. Letters 8*, 221 (1962).

Benczer-Koller, N., Schwarzschild, A., Vise, J. B., and Wu, C. S. *Phys. Rev. 109*, 85 (1958).

Berley, D., Lee, J., and Bardon, M. *Phys. Rev. Letters 2*, 357 (1959).

Berley, D., and Gidal, G. *Phys. Rev. 118*, 1086 (1960).

Berman, S. M. *Phys. Rev. 112*, 267 (1958).

Bernstein, J., Lee, T. D., Yang, C. N., and Primakoff, H. *Phys. Rev. 111*, 313 (1958).

Bernstein, J., Gell-Mann, M., and Michel, L. *Nuovo Cimento 16*, 560 (1960).

Bernstein, J., Fubini, S., Gell-Mann, M., and Thirring, W. *Nuovo Cimento 17*, 757 (1960).

Bertanza, L., Brisson, V., Connolly, P. L., Hart, E. L., Mittra, I. S., Moneti, G. C., Rau, R. R., Samios, N. P., Skillicorn, I. O., Yamamoto, S. S., Goldberg, M., Gray, L., Leitner, J., Lichtman, S., and Westgard, J. *Phys. Rev. Letters 9*, 229 (1962).

Bertolini, E., Citron, A., Gialamella, G., Focardi, S., Mukhin, A., Rubbia, C., and Saporetti, S. *Proceedings of the 1962 International Conference on High Energy Physics at CERN* (CERN, Geneva, 1962) p. 421.

Bethe, H. A., and Morrison, P. *Elementary Nuclear Theory* (John Wiley and Sons, New York, 1956).

Biedenharn, L. C., and Rose, M. E., *Phys. Rev. 83*, 459 (1951).

Birge, R. W., Perkins, D. H., Peterson, J. R., Stork, D. H., and Whitehead, M. N. *Nuovo Cimento 4*, 834 (1956).

Birge, R. W., Ely, R. P., Powell, W. M., Huzita, H., Fry, W. F., Gaides, J. A., Natali, S. V., Willman, R. B., and Camerini, U. *Proceedings of the 1960 Annual International Conference on High Energy Physics at Rochester* (University of Rochester, 1960) p. 601.

Birge, R. W., and Fowler, W. B. *Phys. Rev. Letters 5*, 254 (1960).

Biswas, N. N., Ceccarelli-Fabbrichesi, L., Ceccarelli, M., Gottstein, K. *Nuovo Cimento 3*, 825 (1956).

Bjorklund, R., Crandall, W. E., Moyer, B. J., and York, H. F. *Phys. Rev. 77*, 213 (1950).

Blatt, J. M., and Weisskopf, V. F. *Theoretical Nuclear Physics* (John Wiley and Sons, New York, 1952).

Bleser, E., Lederman, L., Rosen, J., Rothberg, J., and Zarattini, E. *Phys. Rev. Letters 8*, 288 (1962).

Bleuler, K. *Helv. Phys. Acta 23*, 567 (1950).

Block, M. M., Brucker, E. B., Hughes, I. S., Kikuchi, T., Meltzer, C., Anderson, F., Pevsner, A., Harth, E. M., Leitner, J., and Cohn, H. O. *Phys. Rev. Letters 3*, 291 (1959).

Block, M. M., Brucker, E. B., Gessaroli, R., Kikuchi, T., Meltzer, C. M., Pevsner, A., Schlein, P., Strand, R., Cohn, H. O., Harth, E. M., Leitner, J., Minguzzi-Ranzi, A., Monari, L., and Puppi, G. *Phys. Rev. 120,* 570 (1960).

Block, M. M., Lendinara, L., and Monari, L. *Proceedings of the 1962 International Conference on High Energy Physics at CERN* (CERN, Geneva, 1962) p. 371.

Bludman, S. A. *Nuovo Cimento 9,* 433 (1958).

Bodansky, L., Eccks, S. F., Farwell, G. W., Rickney, M. E., and Robinson, P. C. *Phys. Rev. Letters 2,* 101 (1959).

Bogoliubov, N. N., and Shirkov, D. V. *Introduction to the Theory of Quantized Fields* (Interscience Pub., New York, 1959).

Burgoyne, N. *Nuovo Cimento 8,* 607 (1958).

Bowcock, J., Cottingham, W. N., and Lurié, D. *Phys. Rev. Letters 5,* 386 (1960).

Briet, G., Condon, E. U., and Present, R. D. *Phys. Rev. 50,* 825 (1936).

Breit, G. *Proc. Natl. Acad. Sci.* (U.S.) *46,* 746 (1960).

Breit, G., Hull, M. H., Jr., Lassila, K., and Ruppel, H. M. *Phys. Rev. Letters 5,* 274 (1960).

Brisson, J. C., Detoef, J., Falk-Varient, P., Van Rossum, L., Valladas, G., and Yuan, L. C. *Phys. Rev. Letters 3,* 561 (1960).

Brown, J. L., Bryant, H. C., Burnstein, R. A., Hartung, R. W., Glaser, D. A., Kadyk, J. A., Sinclair, D., Trilling, G. H., Vander Velde, J. C., and Van Putten, J. D. *Phys. Rev. Letters 3,* 563 (1959).

Brown, J. L., Kadyk, J. A., Trilling, G. H., Van de Walle, R. T., Roe, B. P., and Sinclair, D. *Phys. Rev. Letters 7,* 423 (1961).

Brown, J. L., Kadyk, J. A., Trilling, G. H., Van de Walle, R. T., Roe, B. P., and Sinclair, D. *Phys. Rev. Letters 8,* 450 (1962).

Brueckner, K. A., and Case, K. M. *Phys. Rev. 83,* 1141 (1951).

Brueckner, K. A., Serber, R., and Watson, K. M. *Phys. Rev. 81,* 575 (1951).

Brueckner, K. A., and Watson, K. M. *Phys. Rev. 86,* 923 (1952).

Burgy, M. T., Krohn, V. E., Novey, J. B., Ringo, G. R., and Telegdi, V. L. *Phys. Rev. 110,* 1214 (1958a).

Burgy, M. T., Krohn, V. E., Novey, J. B., Ringo, G. R., and Telegdi, V. L. *Phys. Rev. Letters 1,* 324 (1958b).

Burrowes, H. C., Caldwell, D. O., Frisch, D. H., Hill, D. A., Ritson, D. M., Schluter, R. A., and Wahlig, M. A. *Phys. Rev. Letters 2,* 119 (1959).

Butler, J. W., and Bondelid, R. O. *Phys. Rev. 121,* 1770 (1961).

Button-Shafer, J., Crawford, F. S., Hubbard, R., Stevenson, M. L., Block, M., Engler, A., Gessaroli, R., Kovacs, A., Meltzer, C., Onley, D., Kraemer, R., Nussbaum, M., Pevsner, A., and Schlein, P. *Proceedings of the 1962 International Conference on High Energy Physics at CERN* (CERN, Geneva, 1962) p. 272.

Byers, N., and Burkhardt, H. *Phys. Rev. 121,* 281 (1961).

Camerini, U., Fry, W. F., Gaidos, J. A., Huzita, H., Natali, S. V., Willmann, R. B., Birge, R. W., Ely, R. P., Powell, W. M., and White, H. S. *Phys. Rev. 128,* 362 (1962).

Capps, R. H. *Phys. Rev. 126,* 1574 (1962).

Carmony, D. D., Rosenfeld, A. H., and Van de Walle, R. T. *Phys. Rev. Letters 8,* 117 (1962).

Carmony, D. D., and Van de Walle, R. T. *Phys. Rev. Letters 8,* 73 (1962).

Cartwright, W. F., Richman, C., Whitehead, W., and Wilcox, H. *Phys. Rev. 81,* 652 (1951).

Case, K. M. *Phys. Rev. 103,* 1449 (1956).

Case, K. M. *Phys. Rev. 107,* 307 (1957).

Cassen, B., and Condon, E. U. *Phys. Rev. 50,* 846 (1936).

Chamberlain, O., Segrè, E., Tripp, R., Wiegand, C., and Ypsilantis, T. *Phys. Rev. 93,* 1430 (1954).

Chamberlain, O., Segrè, E., Wiegand, C., and Ypsilantis, T. J. *Phys. Rev. 100,* 947 (1955).

Chamberlain, O., Segrè, E., Tripp, R. D., Wiegand, C., and Ypsilantis, T. *Phys. Rev. 105,* 288 (1957).

Charpak, G., Farley, F. J. M., Garwin, R. L., Muller, T., Sens, J. C., and Zichichi, A. *Physics Letters 1,* 16 (1962).

Cheston, W. B. *Phys. Rev. 83,* 1118 (1951).

Chew, G. F., and Low, F. E. *Phys. Rev. 101,* 1570 (1956a).

Chew, G. F., and Low, F. E. *Phys. Rev. 101,* 1579 (1956b).

Chew, G. F. Goldberger, M. L., Low, F. E., and Nambu, Y. *Phys. Rev. 106,* 1337 (1957a).

Chew, G. F., Goldberger, M. L., Low, F. E., and Nambu, Y. *Phys. Rev. 106,* 1345 (1957b).

Chew, G. F., Karplus, R., Gasiorowicz, S., and Zachariasen, F. *Phys. Rev. 110,* 265 (1958).

Chew, G. F., and Low, F. E. *Phys. Rev. 113,* 1640 (1959).

Chew, G. F. *Phys. Rev. Letters 4,* 142 (1960).

Chinowsky, W., and Steinberger, J. *Phys. Rev. 95,* 1561 (1954).

Chinowsky, W., Goldhaber, G., Goldhaber, S., Lee, W., and O'Halloran, T. *Phys. Rev. Letters 9,* 330 (1962).

Chrétien, M., Bulos, F., Crouch, H. R., Jr., Lanou, R. E., Jr., Massimo, J. T., Shapiro, A. M., Averell, J. A., Bordner, C. A., Jr., Brenner, A. E., Firth, D. R., Law, M. E., Ronat, E. E., Strauch, K., Street, J. C., Szymanski, J. J., Weinberg, A., Nelson, B., Pless, I. A., Rosenson, L., Salandin, G. A., Yamamoto, R. K., Guerriero, L., and Wladner, F. *Phys. Rev. Letters 9,* 127 (1962).

Clark, D. L., Roberts, A., and Wilson, R. *Phys. Rev. 83,* 649 (1951).

Cocconi, G., and Salpeter, E. E. *Nuovo Cimento 10,* 646 (1958).

Cocconi, V. T., Fazzini, T., Fidecaro, G., Legros, M., Lipman, N. H., and Merrison, A. W. *Phys. Rev. Letters 5,* 19 (1960).

Coester, F. *Phys. Rev. 84,* 1259 (1951).

Condon, E. U., and Shortley, G. H. *The Theory of Atomic Spectra* (Cambridge Univ. Press, Cambridge, England, 1951).

Conversi, M., Pancini, E., and Piccioni, O. *Phys. Rev. 71,* 209 (1947).

Cool, R. L., Cork, B., Cronin, J. W., and Wenzel, W. A. *Phys. Rev. 114,* 912 (1959).

Coombes, C. A., Cork, B., Galbraith, W., Lambertson, G. R., and Wentzel, W. A. *Phys. Rev. 108,* 1348 (1957).

Cork, B., Keith, L., Wenzel, W. A., Cronin, J. W., and Cool, R. L. *Phys. Rev. 120,* 1000 (1960).

Cowan, C. L., Jr., Reines, F., Harrison, F. B., Kruse, H. W., and McGuire, A. D. *Science 124,* 103 (1956).

Crawford, F. S., Jr., Cresti, M., Good, M. L., Gottstein, K., Lyman, E. M., Solmitz, F. T., Stevenson, M. L., and Ticho, H. K. *Phys. Rev. 108,* 1102 (1957).

Crawford, F. S., Jr., Cresti, M., Good, M. L., Solmitz, F. T., and Stevenson, M. L. *Phys. Rev. Letters 1,* 209 (1958a).

Crawford, F. S., Jr., Cresti, M., Good, M. L., Kalbfleisch, G. R., Stevenson, M. L., and Ticho, H. K. *Phys. Rev. Letters 1,* 377 (1958b).

Crawford, F. S., Jr., Cresti, M., Douglass, R. L. Good, M. L., Halbfleisch, G. R., Stevenson, M. L., and Ticko, H. K. *Phys. Rev. Letters 2,* 266 (1959a).

Crawford, F. S., Jr., Cresti, M., Good, M. L., Stevenson, M. L., and Ticho, H. H. *Phys. Rev. Letters 2,* 114 (1959b).

Crewe, A. V., Garwin, E., Ledley, B., Lillethun, E., March, R., and Marcowitz, S. *Phys. Rev. Letters 2,* 269 (1959).

Cronin, J. W., and Overseth, O. E. *Proceedings of the 1962 International Conference on High Energy Physics at CERN* (CERN, Geneva, 1962) p. 453.

Crowe, K. M. *Nuovo Cimento 5,* 541 (1957).

Culligan, G., Frank, S. G. F., Kluyver, J. C., and Holt, J. R. *Nature 180,* 751 (1957).

Culligan, G., Lathrop, J. F., Telegdi, V. L., Winstin, R., and Lundy, R. A. *Phys. Rev. Letters 7,* 458 (1961).

Cziffra, P., MacGregor, M. H., Moravcsik, M. J., and Stapp, H. P. *Phys. Rev. 114,* 880 (1959).

Dalitz, R. H. *Proc. Phys. Soc. London (A) 65,* 175 (1952).

Dalitz, R. H. *Phil. Mag. 44,* 1068 (1953).

Dalitz, R. H. *Phys. Rev. 94,* 1046 (1954).

Dalitz, R. H. *Phys. Rev. 99,* 915 (1955).

Dalitz, R. H. *Reports on Progress in Physics 20,* 163 (1957).

Dalitz, R. H., and Liu, L. *Phys. Rev. 116,* 1312 (1959).

Dalitz., R. H., and Tuan, S. F. *Phys. Rev. Letters 2,* 425 (1959).

Dalitz, R. H. "Enrico Fermi," *Rend. scuola internaz. fis. 11,* 298 (1960).

Dalitz, R. H., and Tuan, S. F. *Ann. of Phys. 10,* 307 (1960).

Dalitz, R. H. *Proceedings of the Aix-en-Province Conference on Elementary Particles* (C.E.N. Saclay, France, 1961).

Dalitz, R. H. *Strange Particles and Strong Interactions* (Tata Institute of Fundamental Research, Bombay, 1962).

Dallaporta, N. *Nuovo Cimento 1,* 962 (1953).

Davidon, W. C., and Goldberger, M. L. *Phys. Rev. 104,* 1119 (1956).

Davis, H. F., Roberts, A., and Zipf, T. F. *Phys. Rev. Letters 2,* 211 (1959).

Day, T. B., Snow, G. A., and Sucher, J. *Phys. Rev. Letters 3,* 61 (1959).

Depommier, P., Heintze, J., Mukhin, A., Rubbia, C., Soergel, V., and Winter, K. *Proceedings of the 1962 International Conference on High Energy Physics at CERN* (CERN, Geneva, 1962) p. 411.

D'Espagnat, B. *Nuovo Cimento 20,* 1217 (1961).

Dirac, P. A. M. *Proc. Phys. Soc. London (A) 126,* 360 (1931).

Dobrohotov, E. I., Lazarenko, V. R., and Luk'ianov, S. Iu. *Doklady Acad. Nauk.* (USSR) *110,* 966 (1956) [translation: Soviet Phys. "Doklady" *1,* 600 (1957)]

Doede, J. H., Hildebrand, R. H., Israel, M. H., and Pyka, M. R. *Phys. Rev.* (to be published) (1963).

Drell, S. D. *Proceedings of the 1958 Annual International Conference on High Energy Physics at CERN* (CERN, Geneva, 1958) p. 27.

Duerr, H. P., and Heisenberg, W. *Nuovo Cimento 23,* 807 (1962).

Dunaitsev, A. F., Petrukhin, V. I., Prokoshkin, Yu., D., and Ryaklin, V. I. *Physics Letters 1,* 138 (1963).

Durbin, R., Loar, H., and Steinberger, J. *Phys. Rev. 83,* 646 (1951).

Edwards, S. F., and Matthews, P. T. *Phil. Mag. 2,* 166 (1956).

Eisler, F., Plano, R., Samios, N., and Steinberger, J. *Nuovo Cimento 5,* 1700 (1957a).

Eisler, F., Plano, R., Prodell, A., Samios, N., Schwartz, M., Steinberger, J., Bassi, P., Borelli, V., Puppi, G., Tanaka, G., Woloschek, P., Zoboli, V., Conversi, M., Franzini, P., Mannelli, I., Santangelo, R., Silvestrini, V., Glaser, D. A., Greaves, C., and Perl, M. L. *Phys. Rev. 108,* 1353 (1957).

Eisler, F., Plano, R., Prodell, A., Samios, N., Schwartz, M., Steinberger, J., Bassi, P., Borelli, V., Puppi, G., Tanaka, H., Waloschek, P., Zoboli, V., Conversi, M., Franzini, P., Manelli, I., Santangelo, R., Silvestrini, V., Brown, G. L., Glaser, D. A., and Graves, C. *Nuovo Cimento 7,* 222, (1958).

Ely, R. P., Fung, S. Y., Gidal, G., Pan, Y.-L., Powell, W. M., and White, H. S. *Phys. Rev. Letters 7,* 461 (1961).

Ely, R. P., Powell, W. M., White, H., Baldo-Ceolin, M., Calimani, E.,

Ciampolillo, S., Fabri, O., Farini, F., Filippi, C., Huzita, H., Miari, G., Camerini, U., Fry, W. F., and Natali, S. *Phys. Rev. Letters 8*, 132 (1962a).

Ely, R. P., Gidal, G., Oswald, L., Singleton, W., Powell, W. M., Bullock, F. W., Kalmus, G. E., Henderson, C., and Stannard, F. R. *Proceedings of the 1962 International Conference on High Energy Physics at CERN* (CERN, Geneva, 1962b) p. 445.

Eötvös, R. V., Pekár, D., and Fekete, E. *Ann. d. Physik 68*, 11 (1922).

Erwin, A. R., March, R., Walker, W. D., and West, E. *Phys. Rev. Letters 6*, 628 (1961).

Fabri, E. *Nuovo Cimento 11*, 479, (1954).

Fazzini, T., Fidecaro, G., Merrison, A. W., Paul, H., and Tollestrup, A. V. *Phys. Rev. Letters 1*, 247 (1958).

Federbusch, P., Goldberger, M. L., and Treiman, S. B. *Phys. Rev. 112*, 642 (1958).

Feinberg, G. *Phys. Rev. 108*, 878 (1957).

Feinberg, G. *Phys. Rev. 109*, 1019 (1958a).

Feinberg, G. *Phys. Rev. 110*, 1482 (1958b).

Feinberg, G., and Goldhaber, M. *Proc. Natl. Acad. Sci.* (U.S.) *45*, 1301 (1959).

Feld, B. T. *Phys. Rev. 89*, 330 (1953).

Feldman, D. *Phys. Rev. 103*, 254 (1955).

Feldman, G., and Fulton, T. *Nucl. Phys. 8*, 106 (1958).

Fermi, E. *Zeit. Physik 88*, 161 (1934).

Fermi, E., and Amaldi, E. *Ricerca scientifica 1*, 1 (1936).

Fermi, E., and Yang, C. N. *Phys. Rev. 76*, 1739 (1949).

Fermi, E. *Elementary Particles* (Yale Univ. Press, New Haven, 1951).

Fermi, E. *Suppl. Nuovo Cimento 2*, 17 (1955).

Ferretti, B. *Report on the International Conference on Low Temperature and Fundamental Particles* (Cambridge, England, 1947) p. 75.

Feynman, R. P. *Phys. Rev. 76*, 749 (1949).

Feynman, R. P., and Gell-Mann, M. *Phys. Rev. 109*, 193 (1958).

Fields, T. H., Yodh, G. B., Derrick, M., and Fektovich, J. G. *Phys. Rev. Letters 5*, 69 (1960).

Fierz, M., *Zeit. Physik 104*, 553 (1937).

Fitch, V., and Motley, R. *Phys. Rev. 101*, 496 (1956).

Fowler, W. B., Shutt, R. P., Thorndike, A. M., and Whittemore, W. L. *Phys. Rev. 93*, 861 (1954).

Fowler, W. B., Birge, R. W., Eberhard, P., Ely, R., Good, M. L., Powell, W. M., and Ticho, H. K. *Phys. Rev. Letters 6*, 134 (1961).

Frankel, S., Halpern, J., Holloway, L., Wales, W., Yearian, M., Chamberlain, O., Lemonick, A., and Pipkin, F. M. *Phys. Rev. Letters 8*, 123 (1962).

Franzinetti, C., and Morpurgo, G. *Suppl. Nuovo Cimento 6*, 469 (1957).
Franzini, P., and Steinberger, J. *Phys. Rev. Letters 6*, 281 (1961).
Frauenfelder, H., Bobone, R., von Goeler, E., Levine, N., Lewis, H. R., Peacock, R. N., Rossi, A., and De Pasquali, G. *Phys. Rev. 106*, 386 (1957).
Frazer, W. R., and Fulco, J. R. *Phys. Rev. Letters 2*, 365 (1959).
Friedman, J. I., and Telegdi, V. L. *Phys. Rev. 105*, 1681 (1957).
Fröhlich, H., Heitler, W., and Hemmer, N. *Proc. Roy. Soc. (A) 166*, 154 (1938).
Fry, W. F., and Sachs, R. G. *Phys. Rev. 109*, 2212 (1958).
Fry, W. F., Schneps, J., and Swami, M. S. *Phys. Rev. 103*, 1904 (1956).
Fujii, Y. *Progr. Theoret. Phys. 21*, 232 (1959).
Fujimoto, Y., and Miyazawa, H. *Progr. Theoret. Phys. 5*, 1052 (1950).
Furry, W. F. *Phys. Rev. 51*, 125 (1937).
Furuichi, S., Kodama, T., Ogawa, S., Sugakara, Y., Wakase, A., and Yonezawa, M. *Progr. Theoret. Phys. 17*, 89 (1957).

Gammel, J. L., and Thaler, R. M. *Phys. Rev. 107*, 291 (1957).
Gartenhaus, S. *Phys. Rev. 100*, 900 (1955).
Garwin, R. L., Lederman, L. M., and Weinrich, M. *Phys. Rev. 105*, 1415 (1957).
Gatto, R. *Phys. Rev. 108*, 1103 (1957).
Gell-Mann, M. *Phys. Rev. 92*, 833 (1953).
Gell-Mann, M., and Goldberger, M. L. *Phys. Rev. 91*, 398 (1953).
Gell-Mann, M., and Pais, A. *Proceedings of the Glasgow Conference* (Pergamon Press, London, 1954).
Gell-Mann, M., and Watson, K. M. *Ann. Rev. Nucl. Sci. 4*, 219 (1954).
Gell-Mann, M., and Pais, A. *Phys. Rev. 97*, 1387 (1955).
Gell-Mann, M. *Suppl. Nuovo Cimento 4*, 848 (1956a).
Gell-Mann, M. *Proceedings of the Sixth Annual Rochester Conference on High Energy Physics* (University of Rochester, 1956b).
Gell-Mann, M. *Phys. Rev. 106*, 1296 (1957).
Gell-Mann, M., and Rosenfeld, A. H. *Ann. Rev. Nucl. Sci. 7*, 407 (1957).
Gell-Mann, M. *Phys. Rev. 111*, 372 [*112*, 2139 (E)] (1958).
Gell-Mann, M., and Lévy, M. *Nuovo Cimento 16*, 705 (1960).
Gell-Mann, M., and Zachariasen, F. *Phys. Rev. 123*, 1065 (1961).
Gell-Mann, M. *Phys. Rev. 125*, 1067 (1962).
Gershtein, S. S., and Zeldovich, J. B. *J. Expt. Theoret. Phys.* (USSR) *29*, 698 (1955) [translation: Soviet Phys. *JETP 2*, 576 (1957)].
Giamati, C. C., and Reines, F. *Phys. Rev. 126*, 2178 (1962).
Glashow, S. L. *Phys. Rev. Letters 6*, 196 (1961).
Glashow, S. L. *Proceedings of the International Summer School in Theoretical Physics, Istanbul 1962* (Gordon and Breach, London, 1963).
Glauber, R. J. *Progr. Theoret. Phys. 9*, 295 (1953).

Goebel, C. *Phys. Rev. 103*, 258 (1956).
Goebel, C. *Phys. Rev. Letters 1*, 337 (1958).
Goldberger, M. L., Miyazawa, H., and Oehme, R. *Phys. Rev. 99*, 986 (1955).
Goldberger, M. L., and Treiman, S. B. *Phys. Rev. 110*, 1178 (1958a).
Goldberger, M. L., and Treiman, S. B. *Phys. Rev. 111*, 354 (1958b).
Goldberger, M. L. *Proceedings of the Midwest Conference on Theoretical Physics* (Purdue University, Lafayette, Ind., 1960) p. 50.
Goldhaber, M., Grodzins, L., and Sunyar, A. W. *Phys. Rev. 109*, 1015, (1958).
Goldstein, H. *Classical Mechanics* (Addison-Wesley, Cambridge, Mass., 1953).
Good, M. L. *Phys. Rev., 106*, 591 (1957).
Good, M. L. *Phys. Rev. 110*, 550 (1958).
Good, M. L. *Phys. Rev. 121*, 311 (1961).
Greider, K. R. *Phys. Rev. 122*, 1919 (1961).
Grodzins, L. *Progr. Nucl. Phys. 7*, 163 (1959).
Gupta, S. N. *Proc. Phys. Soc. London (A) 63*, 681 (1950).
Gupta, S. N. *Can. J. Phys. 35*, 1309 (1957).

Haas, R., Leipuner, L. B., and Adair, R. K. *Phys. Rev. 116*, 1221 (1959).
Hall, D., and Wightman, A. S. *Kgl. Danske Vidensk. Selsk. Mat.-Fys. Medd. 31* #5 (1957).
Harting, D., Kluyver, J. C., Kusumegi, A., Rigopoulos, R., Sachs, A. M., Tibell, G., Vanderhaeghe, G., and Weber, G. *Phys. Rev. 119*, 1716 (1960).
Heisenberg, W. *Zeit. Physik 77*, 1 (1932).
Heisenberg, W. *Zeit. Physik. 90*, 209 (1934).
Heitler, W. *Proc. Roy. Irish Acad., 51*, 33 (1946).
Henley, E. M., and Jacobsohn, B. A. *Phys. Rev. 113*, 225 (1959).
Henley, E. M., and Thirring, W. *Elementary Quantum Field Theory* (McGraw-Hill, New York, 1962).
Herrmannsfeldt, W. B., Maxson, D. R., Stähelin, P., and Allen, J. S. *Phys. Rev. 107*, 641 (1957).
Hildebrand, R. H. *Phys. Rev. 89*, 1090 (1953).
Hildebrand, R. H. *Phys. Rev. Letters 8*, 34 (1962).
Hofstadter, R. *Ann. Rev. Nucl. Sci. 7*, 231 (1957).
Hofstadter, R., Bumiller, F., and Croissiaux, M. *Phys. Rev. Letters 5*, 263 (1960).
Huang, K., and Low, F. E. *Phys. Rev. 109*, 1400 (1958).
Humphrey, W. E., Kirz, J., Rosenfeld, A. H., Leitner, J., and Rhee, Y. I. *Phys. Rev. Letters 6*, 478 (1961).

Ioffe, B. *J. Exptl. Theoret. Phys.* (USSR) *38*, 1608 (1960) [translation: Soviet Phys. JETP *11*, 1158 (1960)].

Jackson, J. D., Treiman, S. B., and Wyld, H. W., Jr. *Phys. Rev. 106*, 517 (1957).

Jackson, J. D., Ravenhall, D. G., and Wyld, H. W., Jr. *Nuovo Cimento 9*, 834 (1958).

Jacob, M., and Wick, G. C. *Ann. of Phys. 7*, 404 (1959).

Jauch, J. M., and Rohrlich, F. *The Theory of Photons and Electrons* (Addison-Wesley, Cambridge, Mass., 1955).

Jordan, P., and Wigner, E. P. *Zeit. Physik 47*, 631 (1928).

Jost, R. *Helv. Phys. Acta 30*, 409 (1957).

Kawakami, I. *Progr. Theoret. Phys. 19*, 459 (1957).

Kemmer, N. *Proc. Camb. Phil. Soc. 34*, 354 (1938).

Kinoshita, T., and Sirlin, A. *Phys. Rev. 113*, 1652 (1959).

Klein, O. *Zeit. Physik 53*, 157 (1929).

Klein, O. *Nature 161*, 897 (1948).

Kobzarev, I. Iu., and Okun', L. B. *J. Exptl. Theoret. Phys.* (USSR) *34*, 736 (1958) [translation: Soviet Phys. *JETP 7*, 524 (1958)].

Koester, L. J., Jr., and Mills, F. E. *Phys. Rev. 105*, 1900, (1957).

Köhler, H. S. *Phys. Rev. 118*, 1345 (1960).

Konopinski, E. J., and Mahmoud, H. M. *Phys. Rev. 92*, 1045 (1953).

Kramers, H. A. *Proc. Acad. Amst. 33*, 959 (1930).

Kramers, H. A. *Proc. Acad. Amst. 40*, 814 (1937).

Kroll, N. M., and Ruderman, M. A. *Phys. Rev. 93*, 233 (1954).

Kroll, N. M., and Wada, W. W. *Phys. Rev. 98*, 1355 (1955).

Landau, L. D. *Doklady Akad. Nauk* (USSR) *60*, 207 (1948).

Landau, L. D. *Nucl. Phys. 3*, 127 (1957).

Landé, K., Booth, E. T., Impeduglia, J., Lederman, L., M., and Chinowsky, W. *Phys. Rev. 103*, 1901 (1956).

Laporte, O. *Zeit. Physik 23*, 135 (1924).

Lattes, C. M. G., Occhialini, P. S., and Powell, C. F. *Nature 160*, 453 (1947).

Lee, T. D., Rosenbluth, M., and Yang, C. N. *Phys. Rev. 75*, 905 (1949).

Lee, T. D., and Yang, C. N. *Phys. Rev. 98*, 1501 (1955).

Lee, T. D., and Yang, C. N. *Nuovo Cimento 3*, 749 (1956a).

Lee, T. D., and Yang, C. N. *Phys. Rev. 104*, 254 (1956b).

Lee, T. D., Oehme, R., and Yang, C. N. *Phys. Rev. 106*, 340 (1957).

Lee, T. D., Steinberger, J., Feinberg, G., Kabir, P. K., and Yang, C. N. *Phys. Rev. 106*, 1367 (1957).

Lee, T. D., and Yang, C. N. *Phys. Rev. 105*, 1671 (1957a).

Lee, T. D., and Yang, C. N. "Elementary Particles and Weak Interactions," BNL 443 (T91), Brookhaven National Laboratory Report (unpublished) (1957b).

Lee, T. D., and Yang, C. N. *Phys. Rev. 108*, 1645 (1957c).

Lee, T. D., and Yang, C. N. *Phys. Rev. 109*, 1755 (1958).
Lee, T. D., and Yang, C. N. *Phys. Rev. Letters 4*, 307 (1960a).
Lee, T. D., and Yang, C. N. *Phys. Rev. 119*, 1410 (1960b).
Lee, T. D., Markstein, P., and Yang, C. N. *Phys. Rev. Letters 7*, 429 (1961).
Lee, T. D. *Phys. Rev. Letters 9*, 319 (1962).
Lehmann, H. *Nuovo Cimento 11*, 342 (1954).
Leipuner, L. B., and Adair, R. K. *Phys. Rev. 109*, 1358 (1958).
Leitner, J. *Nuovo Cimento 8*, 68 (1958).
Leitner, J., Nordin, P., Jr., Rosenfeld, A. H. Solmitz, F. T., and Tripp, R. D. *Phys. Rev. Letters 3*, 238 (1959).
Leitner, J., Gray, L., Harth, E., Lichtman, S., Westgard, J., Block, M., Brucker, B., Engler, A., Gessaroli, R., Kovacs, A., Kikuchi, T., Meltzer, C., Cohn, H. O., Bugg, W., Pevsner, A., Schlein, P., Meer, M., Grinellini, N. T., Lendinara, L., Monari, L., and Puppi, G. *Phys. Rev. Letters 7*, 264 (1961).
Lindenbaum, S. J., Love, W. A., Niederer, J. A., Ozaki, S., Russell, J. J., and Yuan, L. C. L. *Phys. Rev. Letters 7*, 352 (1961).
Lippmann, B. A., and Schwinger, J. *Phys. Rev. 79*, 469 (1950).
London, F. *Zeit. Physik 42*, 375 (1927).
Lüders, G. *Kgl. Dansk. Vidensk. Selsk. Mat.-Fys. Medd. 28*, #5 (1954).
Lüders, G. *Ann. of Phys. 2*, 1 (1957).
Lüders, G., and Zumino, B. *Phys. Rev. 110*, 1450 (1958).
Luers, D., Mittra, I. S., Willis, W. J., and Yamamoto, S. S. *Phys. Rev. Letters 7*, 255 (1961).
Lundy, R. A. *Phys. Rev. 125*, 1686 (1962).
Lyttleton, R. A., and Bondi, H. *Proc. Roy. Soc. London (A) 252*, 313 (1959).

MacDowell, S. W. *Nuovo Cimento 6*, 1445 (1957).
MacGregor, M. H., Moravcsik, M. J., and Stapp, H. P. *Phys. Rev. 116*, 1248 (1959).
Macq, P. C., Crowe, K. M., and Haddock, R. P. *Phys. Rev. 112*, 2061 (1958).
McCliment, E. R., and Nishijima, K. *Phys. Rev. 128*, 1970 (1962).
McLennan, J. A. *Phys. Rev. 106*, 821 (1957).
Maglić, B. C., Kalbfleisch, G. R., and Stevenson, M. L. *Phys. Rev. Letters 7*, 137 (1961).
Maglić, B. C., Alvarez, L. W., Rosenfeld, A. H., and Stevenson, M. L. *Phys. Rev. Letters 7*, 178 (1961).
Majorana, E. *Nuovo Cimento 14*, 171 (1937).
Maloy, J. O., Salandin, G. A., Manfredini, A., Peterson, V. Z., Friedman, J. I., and Kendall, H. K. *Phys. Rev. 122*, 1338 (1961).
Marshak, R. E., and Bethe, H. A. *Phys. Rev. 72*, 506 (1947).

Marshak, R. E. *Phys. Rev. 82*, 313 (1951).
Mayer, M. G., and Jensen, J. H. D. *Elementary Theory of Nuclear Structure* (John Wiley and Sons, New York, 1955).
Mayer-Kuckuk, T., and Michel, F. C. *Phys. Rev. Letters 7*, 167 (1961).
Messiah, A. M. *Phys. Rev., 86*, 430 (1952).
Michel, L. *Proc. Phys. Soc. London (A) 63*, 514 (1950).
Michel, L. *Nuovo Cimento 10*, 319 (1953).
Michel, L., and Wightman, A. S. *Phys. Rev. 98*, 1190 (1955).
Michel, L., and Rouhaninejad, H. *Phys. Rev. 122*, 242 (1961).
Minami, S. *Progr. Theoret. Phys. 11*, 213 (1954).
Morpurgo, G. *Nuovo Cimento 4*, 1222 (1956).
Muller, F., Birge, R. W., Fowler, W. B., Good, R. H., Hirsch, W., Matsen, R. P., Oswald, L., Powell, W. M., White, H. S., and Piccioni, O. *Phys. Rev. Letters 4*, 418 (1960).

Nakano, T., and Nishijima, K. *Progr. Theoret. Phys. 10*, 581 (1953).
Nambu, Y. *Phys. Rev. 106*, 1366 (1957a).
Nambu, Y. *Nuovo Cimento 6*, 1064 (1957b).
Nambu, Y. *Phys. Rev. Letters 4*, 380 (1960).
Nambu, Y., and Jona-Lasinio, G. *Phys. Rev. 122*, 345 (1961).
Neddermeyer, S. H., and Anderson, C. D. *Phys. Rev. 51*, 884 (1937).
Ne'eman, Y. *Nucl. Phys. 26*, 222 (1961).
Nelson, D. F., Schupp, A. A., Pidd, R. W., and Crane, H. R. *Phys. Rev. Letters 2*, 492 (1959).
Niskijima, K. *Progr. Theoret. Phys. 13*, 285 (1955).
Nishijima, K. *Phys. Rev. 108*, 907 (1957).
Nordin, P., Orear, J., Reed, L., Rosenfeld, A. H., Solmitz, F. T., Taft, H. D., and Tripp, R. D. *Phys. Rev. Letters 1*, 380 (1958).
Noyes, H. P. *Phys. Rev. 119*, 1736 (1960a).
Noyes, H. P. *Proceedings of the 1960 Annual International Conference on High Energy Physics at Rochester* (*University of Rochester*, 1960b) p. 117.

Okubo, S. *Phys. Rev. 109*, 984 (1958).
Okubo, S., Marshak, R. E., Sudarshan, E. C. G., Teutsch, W. B., and Weinberg, S. *Phys. Rev. 112*, 655 (1958).
Okubo, S. *Nuovo Cimento 13*, 292 (1959).
Okun', L., and Pontecorvo, B. *J. Expt. Theoret. Phys.* (USSR) *32*, 1587 (1957) [translation: Soviet Phys. *JETP 5*, 1297 (1957)].
Oneda, S., Patti, J. C., and Sakita, B. *Phys. Rev. 119*, 482 (1960).
Oppenheimer, J. R. *Phys. Rev. 35*, 562 (1930).
Orear, J. *Phys. Rev. 96*, 1417 (1954).
Orear, J., Harris, G., and Taylor, S. *Phys. Rev. 102*, 1676 (1956a).
Orear, J., Harris, G., and Taylor, S. *Phys. Rev. 104*, 1463 (1956b).

Pais, A. *Phys. Rev. 86*, 663 (1952).

Pais, A., and Jost, R. *Phys. Rev. 87*, 871 (1952).

Pais, A., and Piccioni, O. *Phys. Rev. 100*, 1487 (1955).

Paid, A., and Treiman, S. B. *Phys. Rev. 105*, 1616 (1957a).

Pais, A. *Phys. Rev. 110*, 574 (1958a).

Pais, A. *Phys. Rev. 110*, 1480 (1958b).

Pais, A. *Phys. Rev. Letters 3*, 242 (1959).

Pais, A. *Revs. Mod. Phys. 33*, 493 (1961).

Panofsky, W. K. H., Aamodt, R. L., and Hadley, J. *Phys. Rev. 81*, 565 (1951).

Pauli, W. *Handbuch der Physik* (Verlag. J. Springer, Berlin, 1933) Vol. 24.

Pauli, W. *Ann. de l'Inst. H. Poincaré 6*, 137 (1936).

Pauli, W. *Phys. Rev. 58*, 716 (1940).

Pauli, W. *Progr. Theoret. Phys. 5*, 526 (1950).

Pauli, W. *Niels Bohr and the Development of Physics* (McGraw-Hill, New York, and Pergamon, London, 1955).

Pauli, W. *Nuovo Cimento 6*, 204 (1957).

Peierls, R. F. *Phys. Rev. Letters 1*, 174 (1958).

Petermann, A. *Helv. Phys. Acta 30*, 407 (1957).

Pevsner, A., Kraemer, R., Nussbaum, M., Richardson, C., Schlein, P., Strand, R., Toohig, T., Block, M., Engler, A., Gessaroli, R., and Meltzer, C. *Phys. Rev. Letters 7*, 421 (1961).

Pickup, E., Robinson, P. K., and Salant, E. O. *Phys. Rev. Letters 7*, 192 (1961).

Pjerrou, G. M., Prowse, D. J., Schlein, P., Slater, W. E., Stork, D. H., and Ticho, H. K. *Proceedings of the 1962 International Conference on High Energy Physics at CERN* (CERN, Geneva, 1962) p. 289.

Plano, R., Prodell, A., Samios, N., Schwartz, M., and Steinberger, J. *Phys. Rev. Letters 3*, 525 (1959).

Plano, R. J. *Phys. Rev. 119*, 1400 (1960).

Poirier, J. A., Pripstein, M., Carroll, J. B., and Bowman, W. C. *Bull. Am. Phys. Soc. 7*, 488 (1962).

Pontecorvo, B. *J. Expt. Theoret. Phys.* (USSR) *37*, 1751 (1959) [translation: Soviet Phys. *JETP 10*, 1236 (1960)].

Pontecorvo, B., and Ryndin, R. H. *Proceedings of the Ninth International Annual Conference on High Energy Physics* (Academy of Sciences, USSR, Moscow, 1960) Vol. 2, p. 233.

Powell, C. F., Fowler, P. H., and Perkins, D. H. *The Study of Elementary Particles by the Photographic Method* (Pergamon, London, 1959).

Primakoff, H. *Revs. Mod. Phys. 31*, 802 (1959).

Primakoff, H., and Rosen, S. P. *Reports on Progress in Physics 22*, 121 (1959).

Puppi, G. *Nuovo Cimento 5*, 587 (1948).

Purcell, E. M., and Ramsey, N. F. *Phys. Rev. 78*, 807 (1950).
Pursey, D. L. *Nuovo Cimento 6*, 266 (1957).

Querzoli, R., Salvini, G., and Silverman, A. *Nuovo Cimento 19*, 53 (1961).

Reines, F., and Cowan, C. L., Jr. *Phys. Rev. 113*, 273 (1959).
Reiter, R. A., Romanowski, T. A., Sutton, R. B., and Chidley, B. G. *Phys. Rev. Letters 5*, 22 (1960).
Ridgeway, S. S. T., Berley, D., and Collins, G. B. *Phys. Rev. 104*, 513 (1956).
Rochester, G. D., and Butler, C. C. *Nature 160*, 855 (1947).
Rose, M. E. *Elementary Theory of Angular Momentum* (John Wiley and Sons, New York, 1957).
Rosen, L., and Brolley, J. E., Jr. *Phys. Rev. Letters 2*, 98 (1959).
Rosenbluth, M. N. *Phys. Rev. 29*, 615 (1950).
Rotenberg, M., Bivins, R., Metropolis, N., and Wooten, J. K., Jr. *The 3-j and 6-j symbols* (Technology Press, Cambridge, Mass., 1959).
Ruderman, M., and Finkelstein, R. *Phys. Rev. 76*, 1458 (1949).
Ruderman, M. *Phys. Rev., 87*, 383 (1952).

Sachs, R. G. *Nuclear Theory* (Addison-Wesley, Cambridge, Mass., 1953).
Sakata, S., and Inoue, T. *Progr. Theoret. Phys. 1*, 143 (1946).
Sakata, S. *Progr. Theoret. Phys. 16*, 686 (1956).
Sakurai, J. J. *Nuovo Cimento 7*, 649 (1958a).
Sakurai, J. J. *Phys. Rev. Letters 1*, 258 (1958b).
Sakurai, J. J. *Ann. of Phys. 11*, 1 (1960a).
Sakurai, J. J. *Phys. Rev. 119*, 1784 (1960b).
Sakurai, J. J. *Proceedings of the 1962 International Summer School of Physics*, "Enrico Fermi," (1963).
Salam, A. *Nuovo Cimento 5*, 299 (1957).
Salam, A. *Proceedings of the Ninth International Annual Conference on High Energy Physics* (Academy of Sciences, USSR, Moscow, 1960) Vol. 1, p. 540.
Salam, A., and Ward, J. C. *Phys. Rev. Letters 5*, 390 (1960).
Salam, A., and Ward, J. C. *Nuovo Cimento 20*, 418 (1961).
Samios, N. P. *Phys. Rev. Letters 4*, 470 (1960).
Savedoff, M. P. *Nature 178*, 688 (1956).
Schiff, L. I. *Quantum Mechanics* (McGraw-Hill, New York) (1949).
Schupp, A. A., Pidd, R. W., and Crane, H. R. *Phys. Rev. 121*, 1 (1961).
Schwartz, M. *Phys. Rev. Letters 4*, 306 (1960).
Schwartz, M. *Phys. Rev. Letters 6*, 556 (1961).
Schweber, S. S., Bethe, H. A., and de Hoffman, F. *Mesons and Fields*, Vol. I (Row-Peterson, Evanston, Ill., 1955).
Schweber, S. S. *An Introduction to the Relativistic Quantum Field Theory* (Row-Peterson, Evanston, Ill., 1961).

Schwinger, J. *Phys. Rev. 82*, 914 (1951).
Schwinger, J. *Phys. Rev. 91*, 713 (1953).
Schwinger, J. *Ann. of Phys. 2*, 407 (1957).
Schwinger, J. *Phys. Rev. 125*, 397 (1962a).
Schwinger, J. *Phys. Rev. 128*, 2425 (1962b).
Sens, J. C. *Phys. Rev. 113*, 679 (1959).
Serpe, J. *Physica 18*, 295 (1952).
Signell, P. S., and Marshak, R. E. *Phys. Rev. 106*, 832 (1957).
Smith, J. H., Purcell, E. M., and Ramsey, N. F. *Phys. Rev. 108*, 120 (1957).
Snow, G. A., and Sucher, J. *Nuovo Cimento 18*, 195 (1960).
Solov'ev, V. G. *J. Expt. Theoret. Phys.* (USSR) *33*, 796 (1957) [translation: Soviet Phys. *JEPT 6*, 613 (1958)].
Sommerfield, C. M. *Phys. Rev. 107*, 328 (1957).
Stapp, H. P., Ypsilantis, T. J., and Metropolis, N. *Phys. Rev. 105*, 302 (1957).
Stech, B., and Jensen, J. H. D. *Zeit. Physik 141*, 403 (1955).
Stein, P. C. *Phys. Rev. Letters 2*, 473 (1959).
Steinberger, J., Panofsky, W. K. H., and Steller, J. *Phys. Rev. 78*, 802 (1950).
Sternheimer, R. M. *Phys. Rev. 113*, 828 (1959).
Stonehill, D., Baltay, C., Courant, H., Fickinger, W., Fowler, E. C., Kraybill, H., Sandweiss, J., Sanford, J., and Taft, H. *Phys. Rev. Letters 6*, 624 (1961).
Street, J. C., and Stevenson, E. C. *Phys. Rev. 52*, 1003 (1937).
Stueckelberg, E. C. G. *Helv. Phys. Acta 11*, 225 (1938a).
Stueckelberg, E. C. G. *Helv. Phys. Acta 11*, 299 (1938b).
Sudarshan, E. C. G., and Marshak, R. E. *Phys. Rev. 109*, 1860 (1958).

Takeda, G. *Ann. of Phys. 18*, 310 (1962).
Taketani, M. *Progr. Theoret. Phys.*, Suppl. #3 (1956).
Tanner, N. *Phys. Rev. 107*, 1203 (1957).
Taylor, J. G., Moravcsik, M. J., and Uretzky, J. L. *Phys. Rev. 113*, 689 (1959).
Telegdi, V. L. *Phys. Rev. Letters 3*, 59 (1959).
Tiomno, J., and Wheeler, J. A. *Revs. Mod. Phys. 21*, 153 (1949).
Tiomno, J. *Nuovo Cimento 1*, 226 (1955).
Tolhoek, H. A. *Revs. Mod. Phys. 28*, 277 (1956).
Touschek, B. F. *Nuovo Cimento 5*, 754 (1957).
Treiman, S. B. *Phys. Rev. 101*, 1217 (1956).
Treiman, S. B., and Sachs, R. G. *Phys. Rev. 103*, 1545 (1956).
Treiman, S. B. *Phys. Rev. 113*, 355 (1959).
Tripp, R. D., Watson, M. B., and Ferro-Luzzi, M. *Phys. Rev. Letters 8*, 175 (1962).

Tuve, M. A., Heydenburg, N. P., and Hafstad, L. R. *Phys. Rev. 50*, 806 (1936).

Valuev, B., and Geshkenbein, B. *J. Exptl. Theoret. Phys.* (USSR) *39*, 1046 (1960) [translation: Soviet Phys. *JETP 12*, 728 (1961)].
von Dardel, G., Dekkers, D., Mermod, R., Vivargent, M., Weber, G., and Winter, K. *Phys. Rev. Letters 8*, 173 (1962).

Wali, K. C. *Phys. Rev. Letters 9*, 120 (1962).
Watanabe, S. *Phys. Rev. 106*, 1306 (1957).
Watson, K. M. *Phys. Rev. 95*, 228 (1954).
Ward, J. C. *Phys. Rev. 84*, 897 (1951).
Weinberg, S. *Phys. Rev. 110*, 782 (1958a).
Weinberg, S. *Phys. Rev. 112*, 1375 (1958b).
Weinberg, S. *Phys. Rev. Letters 4*, 87; 585(E) (1960a).
Weinberg, S. *Phys. Rev. Letters 4*, 575 (1960b).
Weisskopf, V. F. *Phys. Rev. 116*, 1615 (1959).
Wentzel, G. *Phys. Rev. 101*, 1214 (1956a).
Wentzel, G. *Proceedings of the Sixth Annual Rochester Conference on High Energy Physics* (University of Rochester, 1956b).
Weyl, H. *Zeit. Physik 56*, 330 (1929).
Weyl, H. *Theory of Groups and Quantum Mechanics* (Dover, New York, 1931).
Weyl, H. *Symmetry* (Princeton Univ. Press, Princeton, N.J., 1952).
Wheeler, J. A. *Ann. New York Acad. Sci. 48*, 219 (1946).
Wheeler, J. A. *Revs. Mod Phys. 21*, 133 (1949).
Wick, G. C. *Acad. Lincei, Alti 21*, 170 (1935).
Wick, G. C., Wightman, A. S., and Wigner, E. P. *Phys. Rev. 88*, 101 (1952).
Wick, G. C. *Revs. Mod. Phys. 27*, 339 (1955).
Wick, G. C. *Ann. Revs. Nucl. Sci. 9*, 1 (1958).
Wightman, A. S. *Phys. Rev. 101*, 860 (1956).
Wigner, E. P. *Zeit. Physik 43*, 624 (1927).
Wigner, E. P. *Nachrich v. der Gesselschaft der Wissenschaften zu Göttingen 32*, 35 (1932).
Wigner, E. P. *Phys. Rev. 51*, 106 (1937).
Wigner, E. P. *Ann. Math. 40*, 149 (1939).
Wigner, E. P. *Proc. Am. Phil. Soc. 93*, 529 (1949).
Wigner, E. P. *Proc. Natl. Acad. Sci.* (U.S.) *38*, 449 (1952).
Wigner, E. P. *Group Theory and its Application to the Quantum Mechanics of Atomic Spectra* (Academic Press, New York, 1959).
Wilkinson, D. H. *Phys. Rev. 109*, 1603, 1610, 1614 (1958).
Wilson, R. R. *Phys. Rev. 110*, 1212 (1958).
Wolfenstein, L., and Ashkin, J. *Phys. Rev. 85*, 947 (1952).
Wolfenstein, L., and Ravenhall, D. G. *Phys. Rev. 88*, 279 (1952).

Wolfenstein, L. *Phys. Rev. 96*, 1654 (1954).
Wolfenstein, L. *Bull. Am. Phys. Soc. 1*, 36 (1955).
Wolfenstein, L. *Ann. Revs. Nucl. Sci. 6*, 43 (1956).
Wolfenstein, L. *Nuovo Cimento 8*, 882 (1958).
Wolfenstein, L. *Phys. Rev. 121*, 1245 (1961).
Wood, C. D., Devlin, T. J., Helland, J. A., Longo, M. J., Moyer, B. J., and Perez-Mendez, V. *Phys. Rev. Letters 6*, 481 (1961).
Wu, C. S., and Shaknov, I. *Phys. Rev. 77*, 136 (1950).
Wu, C. S. Ambler, E., Hayward, R. W., Hoppes, D. D., and Hudson, R. P. *Phys. Rev. 105*, 1413 (1957).
Wu, C. S. *Theoretical Physics in the Twentieth Century: a Memorial Volume to Wolfgang Pauli*, ed. M. Fierz and V. F. Weisskopf (Interscience, New York, 1960) p. 249.

Xuong, N. H., Lynch, G. R., and Hinrichs, C. K. *Phys. Rev. 124*, 575 (1961).
Xuong, N. H., and Lynch, G. R. *Nuovo Cimento 25*, 923 (1962).

Yang, C. N. *Phys. Rev. 77*, 242 (1950).
Yang, C. N., and Mills, R. L. *Phys. Rev. 96*, 191 (1954).
Yennie, D. R., Ravenhall, D. G., and Wilson, R. N. *Phys. Rev. 95*, 500 (1954).
Yukawa, H. *Proc. Phys.-Math. Soc. Japan 17*, 48 (1935).
Yukawa, H., and Sakata, S. *Proc. Phys.-Math. Soc. Japan 19*, 1084 (1937).

Zorn, J. C., Chamberlain, G. E., and Hughes, V. W. *Proceedings of the 1960 Annual International Conference on High Energy Physics at Rochester* (University of Rochester, 1960) p. 790.

Index

abnormal statistics, 124–127
Adair method, 17–18
angular momentum, 11–12
  conservation, 15–17
  selection rule, 15–16
antigravity hypothesis, 275
antisymmetrized product, 123, 140
antiunitary operator, 81–84
associated production, 255, 257–258
asymmetry parameter
  in beta decay, 164, 167
  in Λ decay, 58–62, 280
  in muon decay, 157–159
  in Σ decay, 62, 65, 280–281
axial-vector conservation, 290

B matrix, 102–105, 118
baryon conservation, 185–187
beta decay, 74–77, 106–108, 161–167, 170, 190–193, 235–241
bilinear covariants, 26–27
  behavior under *CPT*, 141–142
  behavior under charge conjugation, 129–131
  behavior under G conjugation, 225
  behavior under parity, 71–72
  behavior under strong reflection, 140–142
  behavior under time reversal, 104–106
  in beta decay, 74–75

C invariance, *see* charge conjugation invariance
C matrix, 104, 117–119, 129
C parity, *see* charge conjugation parity
CP invariance, 150–151, 270, 275–276
CPT invariance, 133, 135, 147–150
CPT theorem, 133, 135, 136–151
charge, *see* electric charge
charge conjugation
  in the Dirac theory, 117–124
  in the Klein-Gordon theory, 111
  operator, 111–113
charge conjugation invariance
  and beta decay, 132–136
  and muon decay, 131–132
  validity of, 112–113
  violation of, 112–113, 149–150
charge conjugation parity, 113–117
  of the e^+e^- system, 115–116
  of the η meson, 257–258
  of the K^0K^0 system, 276–277
  of the ω meson, 250
  of the photon, 113
  of the $\pi^+\pi^-$ system, 116
  of the π^0 meson, 115–117
charge independence
  and meson theory, 196–197
  and the strangeness scheme, 257
  of nuclear forces, 195–196
  validity of, 198–202, 260–263
charge radius of the nucleon, 234–235
Chew-Low Hamiltonian, 208, 212–213
Chew-Low theory, 212–222
chiral spinor, 152–153, 159–160
chirality, 31, 104, 128–129
Clebsch-Gordan coefficients, 297–299
conserved vector current hypothesis, 237–239, 289–290
crossing symmetry, 210–212

Dalitz plot, 245–247
  of η meson, 252–253
  of ω meson, 247–249
$\Delta S/\Delta Q$ rule, 283–285
$|\Delta \mathrm{T}| = \frac{1}{2}$ rule, 279–283, 292, 294
  in *K* decay, 279–280
  in hyperon decay, 280–282
density matrix formalism, 15, 38
detailed balance, 89–91
Dirac equation, 23
  Lorentz invariance of, 23–24
  plane wave solution of, 24–26
Dirac matrices, 22, 300–301
  Dirac-Pauli representation of, 22, 102, 118–119
  equivalent representation of, 103
  Majorana representation of, 103, 119
  Weyl representation of, 103–104, 152–153
Dirac-Pauli representation, *see* Dirac matrices
Dirac spinors, 24–25
  normalization of, 28–29
Diracology, 101–102
dispersion relations, 212
displacement operator, 10

double beta decay, 190, 192–193
doublet symmetry, 237–238

effective range formula
  in πN scattering, 215
  in K^-p interaction, 264–265
  in the zero-range approximation, 264–265
eightfold way, 289
electric charge, 188–190
  conservation of, 177
  equality of, 188, 236
electric dipole moment
  and time reversal invariance, 105
  interaction, 106, 130
  of the electron, 106
  of the muon, 106
  of the neutron, 66, 106
  of the proton, 106
electromagnetic form factor, 231–235
electron-proton scattering, 231–233
equivalence principle, 275
exclusion principle, 120–124

ft value, 170
Fermi function, 170
Fermi transition, 163–164
Fermi-Yang ambiguity, 41
fine structure constant, 3, 302
  constancy of, 9

G conjugation, 223–235, 241
G parity, 223–226
  of the nucleon-antinucleon system, 229
  of the η meson, 252–253
  of the ω meson, 250
  of the pion, 226–227
  of the ρ meson, 242
gamma matrices, *see* Dirac matrices
Gamow-Teller transition, 163–164
gauge invariance, 178–183
gauge transformation, 177–187, 204–205
Gell-Mann-Nishijima scheme, 257–260
global symmetry, 189, 288–289
Goldberger-Treiman relation, 290–291

Heisenberg rule, 123, 140
helicity, 29–30, 128
  of beta particles, 77–78, 134, 165
  of the electron in muon decay, 68, 112–113, 158–159
  of the muon in $K_{\mu 2}$ decay, 176
  of the muon in pion decay, 156–157
  of the neutrino, 76, 152–154, 156, 166–167, 191–194
hole theory, 72–73, 117–118, 120, 131, 154
hypercharge, 256–257
hypernucleus, 265–267
hyperon beta decay, 176, 283–284
hyperon decay (nonleptonic), 58–65, 110, 150, 280–283, 292

induced pseudoscalar, 240–241
inertia
  anisotropy of, 10
infinitesimal rotation, 11–12
infinitesimal transformation, 9–11
interaction picture, 35–36
interactions
  classification of, 7
intermediate boson, 292–295
intrinsic parity
  and the super selection rule, 187–188
  of the charged pion, 46
  of the electromagnetic field, 44
  of the η meson, 252–253
  of the neutral pion, 45–46
  of the ω meson, 247–248
  of the photon, 42–44
  of the ρ meson, 243
isospin
  of the deuteron, 198
  of the η meson, 250–252
  of the K meson, 255–257, 261–262
  of the nucleon, 195
  of the ω meson, 244
  of the pion, 166–199
  of the ρ meson, 242
  of the strange particles, 255–257, 261–262
  rotations, theory of, 204–209
isotopic spin, *see* isospin

K matrix, 95–99
K_1^0 decay, 16, 127, 280
K_1-K_2 mass difference, 272–275, 278–279
$K^\pm$ decay, 47–51, 175–176, 280
Klein's paradox, 120
Kramer's degeneracy, 85–86

Λ decay, 58–62, 280
${}_\Lambda H^4$ (${}_\Lambda He^4$) decay, 63–65
Laporte's rule, 53
lepton conservation, 155–158, 190–194
lepton number, 155–156, 190
lifetime
  equality, 148–149
  of elementary particles, 7
  of the electron, 177
  of the proton, 185
longitudinal polarization, *see* helicity
Lorentz transformation, 19–23
  complex, 137–138
  generators of, 20–21
  inhomogeneous, 21
  orthochronous, 21

M matrix, 37
Mach principle, 151
magnetic moment
  equality, 148
  of the electron, 183
  of the meson, 183
  of the nucleon, 234–235
Majorana representation, *see* Dirac matrices
mass
  equality, 147–149, 275
  of elementary particles, 7, 303
Michel parameter, 158
microcausality, 146–147
micoscopic reversibility, 79
Minami ambiguity, 41
minimal electromagnetic coupling, 182–183, 259
Möller wave operator, 87–88
momentum conservation, 11
multipole expansion, 52–57
muon capture, 171–173
muon decay, 156–159, 170–171

neutral *K* particles, 269–279
nuclear beta decay, *see* beta decay
nuclear forces, 195–197, 220–223
nucleon-nucleon scattering, 93–95, 222–223

OPE (one-pion-exchange) potential, 220–222

Panofsky ratio, 46, 219
parity, 32–33
  in the Dirac theory, 70–73, 127–128
  in the Klein-Gordon theory, 69–70
  in the Schrödinger theory, 34–36
    see also intrinsic parity, parity conservation, parity nonconservation and relative parity
parity conservation, 65—69
  and the *M* matrix, 37–38
  and the *S* matrix, 34–36
  in the π^+-p scattering, 37–38
  validity of, 39, 51, 53, 65–68
    see also parity nonconservation
parity nonconservation,
  and the τ–θ puzzle, 50–51
  in beta decay, 67–68, 75–78, 135
  in hyperon decay, 58–61
  in the $\pi\mu e$ sequence, 68, 155–159
Pauli moment interaction, 105, 130
Pauli transformation, 191–192
Pauli's fundamental theorem, 101
Peripheral model, 242–243
phase shift analysis, 39–42, 200–203
photoproduction of the pion, 53–57, 87–98, 201–202, 217–220
physical constants, 302–303
$\pi^\pm$ decay, 68, 156, 173–175, 290–291
π^+-p scattering, *see* pion-nucleon scattering
π^0 decay, 15–16, 45, 115
$\pi\mu e$ sequence, 68, 155–159, 197
pion-hyperon resonance, 265
pion-nucleon coupling constant, 215, 217–219, 222, 302
pion-nucleon interaction, 208–209, 212–213
pion-nucleon scattering, 37–42, 200–204, 209–217
polarization
  in nucleon-nucleon scattering, 93–95
  in pion-nucleon scattering, 37–38, 41
  of a spin $\frac{1}{2}$ particle, 15
  of beta particles, 68, 77–78, 165
  of the hyperon in associated production, 58–63
  of the nucleon in hyperon decay, 61–62
    see also helicity
polarization-asymmetry equality, 94
polarization vector, 13
positronium, 74, 115–116
projection operator
  for angular momentum states, 38–40
  for isospin states, 209–210
  for spin states, 29–31
pseudoscalar (-pseudoscalar) coupling, 28, 130, 208
pseudovector (gradient) coupling, 28, 208, 130 (*see also* Chew-Low Hamiltonian)

quantization
  of a spin $\frac{1}{2}$ field, 120–124
  of a spin zero field, 125–126

radiative transition, 16
reciprocity relation, 89, 100
recoil experiments, 166–167
regeneration of K_1^0, 272 274–275
  of the e^+e^- system, 72–74
  of the the $K\Lambda$ system, 51–52, 65, 268
  of the $\Lambda\Sigma$ system, 268–269
  of the $N\Xi$ system, 268–269
Rosenbluth formula, 232
rotation operator, 11

S matrix, 35–36, 88, 98–100
scalar coupling, 106, 130
scalar field, 13
scattering theory, 86–89
schizon, 294
Schur's lemma, 101
Schwinger time reversal, 86, 142–143
second quantization, *see* quantization
$\Sigma^\pm$ decay, 62, 65, 280, 282

Σ^0 decay, 57, 259–260
space inversion, 32–34, *see also* parity
spectral function, 144–145
spectral representation, 144–145
spin
  of the η meson, 252–253
  of the K meson, 16
  of the K^*, 277
  of the Λ hyperon, 17–18, 63
  of the ω meson, 247–248
  of the π meson, 16, 90–91
  of the ρ meson, 242
  of the Σ hyperon, 18
  zero, 13
  one, 13
  $\frac{1}{2}$, 13–15, 29–31
spin-orbit force, 222
spin-statistics connection, 120–127, 140–141
spurion, 281
static Hamiltonian, *see* Chew-Low Hamiltonian
static model, *see* Chew-Low theory
Stern-Gerlach type experiment, 14
strangeness, 257–260, 269–275, 277–279
strong reflection, 136–139, 146–147
superselection rule, 187–188

$T = \frac{1}{2}$ current rule, 284–285
T invariance, *see* time reversal invariance
$T^{(+)}$ matrix, 86–90, 98–100
τ-θ puzzle, 47–51
3-j symbol, 297
3-3 resonance, 42, 57, 98, 201–202, 215–216
time reversal
  in classical physics, 79–81
  in quantum mechanics, 81–86
  in scattering theory, 87–100
  in the Dirac theory, 101–106
  in the Klein-Gordon theory, 100–101, 105
  in the Maxwell theory, 84–85
  see also Schwinger time reversal, time reversal invariance and Wigner time reversal
time reversal invariance
  in beta decay, 106–108, 135
  tests of in decay processes, 108–110
  validity of, 91, 94, 108
translation operator, 10
transversality condition, 13, 184
two component spinor, 13, 23, 25–27, 152–154
two-component theory of the neutrino, 152–155
two-meson hypothesis, 197
two-neutrino hypothesis, 194, 293–294
two-nucleon problem, 220–223, *see also* nuclear forces

unitary symmetry, 289
unitary unimodular group, 207
universal Fermi interaction, 167–176, 236

V-A theory, 159–161
vacuum expectation value, 123, 143–147
vector field, 13, 186–187, 208, 292–295

W particle, 292–295
weak local commutativity, 146–147
weak magnetism, 238–240
weak reflection, 143
Weyl representation, *see* Dirac matrices
Wightman product, 143–147
Wigner time reversal, 81–86

Y^*, *see* pion-hyperon resonance
Yukawa theory, 196, 208–209, 212–213

Milton Keynes UK
Ingram Content Group UK Ltd.
UKHW021955090624
443913UK00008B/341

9 780691 624808